£5·95 amazon.

8999

GW00630548

RESEARCH DEPT.

2 1 OCT 1987

ACK. | ANS.

LIBRARY

M

FRICTION AND WEAR OF POLYMER COMPOSITES

COMPOSITE MATERIALS SERIES

Series Editor: R. Byron Pipes, Center for Composite Materials, University of Delaware, Newark, Delaware, U.S.A.

Vol. 1. **Friction and Wear of Polymer Composites** (K. Friedrich, Editor)

The background photo on the book cover is taken from a fracture surface of a carbon fiber/polyetheretherketone (PEEK) matrix composite (material APC2, from ICI, Wilton, U.K.).

Fracture was performed with DCB specimens at room temperature, under mode I conditions and a crack opening velocity of $v = 10^{-7}$ m s^{-1}. The corresponding fracture energy measured was $G_{Ic} = 1.5$ kJ m^{-2} (courtesy T. Smiley, R.B. Pipes, Center for Composite Materials, University of Delaware, Newark, Del. 19716, U.S.A., September 1985).

Composite Materials Series, 1

FRICTION AND WEAR OF POLYMER COMPOSITES

edited by

Klaus Friedrich

Technical University Hamburg–Harburg, Hamburg, F.R.G.

ELSEVIER
Amsterdam – Oxford – New York – Tokyo   1986

ELSEVIER SCIENCE PUBLISHERS B.V.
Sara Burgerhartstraat 25
P.O. Box 211, 1000 AE Amsterdam, The Netherlands

Distributors for the United States and Canada:

ELSEVIER SCIENCE PUBLISHING COMPANY INC.
52, Vanderbilt Avenue
New York, NY 10017, U.S.A.

Library of Congress Cataloging-in-Publication Data

Friction and wear of polymer composites.

   (Composite materials series ; 1)
   Bibliography: p.
   Includes indexes.
   1. Polymeric composites. 2. Friction.
I. Friedrich, Klaus, 1945-   . II. Series.
TA418.9.C6F75 1986   620.1'92   86-2059
ISBN 0-444-42524-1

ISBN 0-444-42524-1 (Vol. 1)
ISBN 0-444-42525-X (Series)

© Elsevier Science Publishers B.V., 1986

All rights reserved. No part of this publication may be reproduced, stored in a retrieval system or transmitted in any form or by any means, electronic, mechanical photocopying, recording or otherwise, without the prior written permission of the publishers, Elsevier Science Publishers B.V./Science & Technology Division, P.O. Box 330, 1000 AH Amsterdam, The Netherlands.

Special regulation for readers in the USA — This publication has been registered with the Copyright Clearance Center Inc. (CCC), Salem, Massachusetts. Information can be obtained from the CCC about conditions under which photocopies of parts of this publication may be made in the USA. All other copyright questions, including photocopying outside of the USA, should be referred to the publishers.

Printed in The Netherlands

PREFACE

Over recent decades, the field of tribology, i.e. the study of friction, lubrication and wear of materials, has received increasing attention from the scientific, technical and practical points of view. It has become evident that the operation of many mechanical systems depends on friction and wear values. In addition, the annual dissipation of energy and waste of valuable raw materials resulting from high friction and wear are of great economical significance. Correspondingly, potential savings can be expected from improved tribological knowledge. In fact, in recent years, a variety of books have appeared which cover fundamental principles as well as application-related aspects of friction, lubrication and wear of materials [1–5].

Polymers and polymer composites are being used increasingly often as engineering materials for technical applications in which tribological properties are of considerable importance. In addition to the traditional fields (i.e. the use of reinforced rubbers for tires and transmission belts or of filled thermosetting resins as brake materials) polymers and polymer composites are well established in the field of bearings and slider materials. They are encountered in many different industries, e.g. automotive, aerospace, mining or biomedical. In all of these examples, a fundamental understanding of the interactions between the polymer-based components and the possible counterfaces can further enhance their potential application.

In an attempt to meet this objective, several books have already dealt with the subject of polymer tribology [6–9]. However, more work needs to be done to reveal the physical and chemical nature of their tribological characteristics and to generate reliable data for design. This is particularly true for high-performance, polymer composites, a very young group of materials specifically formulated for advanced applications in the future. Although several hundred publications have appeared, mostly concerned with the phenomenology of friction and wear of composites, and although some progress has been made in developing composites systematically for special friction and wear applications, there are many important questions still unanswered today.

In such a situation, it seems appropriate to try to establish a frame of reference. I believe that this can best be done not only by presenting the crucial experimental results and the conclusions drawn from them, but also by displaying the new analytical and experimental tools being used. Obviously, such a wide-ranging task cannot be accomplished by one or two authors.

However, I have been privileged to collaborate with a distinguished group of researchers who are playing a prominent part in the development and application of new techniques and fields of activity. Thus, it is hoped that this volume will not only provide a useful summary of current knowledge on the friction and wear properties of polymer composites, but also promote further studies in this important field.

Special consideration is given, first, to the fundamental aspects of tribology in general and of polymer composites in particular; second, to the effects of the micro-structure of composites on their friction and wear behavior under different external loading conditions; and third, to the problem of the control of friction and wear behavior in practical situations. Although polymer composites associated with bearing-type applications dominate the contributions, part of the volume is also devoted to the friction and wear of metal-based composites and rubber compounds.

In an introductory chapter, H. Czichos presents definitions and terminology used in friction and wear research and deals with general aspects of friction and wear mechanisms. In addition, he explains why friction and wear must be considered as system properties. The fundamental principles of interfacial friction of polymers and polymer composites are discussed by B.J. Briscoe. Special attention is focussed on the individual components of the frictional coefficient under abrasive and non-abrasive dry-sliding conditions, on other wear models, and on friction in lubricated systems. A concept to describe friction and wear properties of materials with heterogeneous microstructures in terms of tribological tensors, introduced by E. Hornbogen, can be considered as a new approach to describing the tribological behavior of anisotropic materials more systematically. It will help towards a better understanding of the effects of individual components and their geometrical arrangement in a composite on its tribological properties, thus leading to basic rules for systematic material development.

The second part of this volume is reserved for a discussion of tribological properties of polymeric and composite materials tested under different conditions. J.M. Thorp analyzes the tribological properties of selected polymer matrices with a special tri-pin-on-disc tribometer and against abrasive counterparts. His wear tests largely endorse manufacturers' recommendations regarding applications for the test materials. The friction and wear properties of PTFE, well known as a potential candidate for polymer bearing application, are surveyed by K. Tanaka. In particular, the author outlines the wear-reducing action of fillers in PTFE-based composites and the effects of water lubrication. A special chapter on friction and wear of metal matrix/graphite fiber composites is presented by Z. Eliezer. Typical wear mechanisms as a function of sliding direction are discussed, along with some potential tribological applications of this type of material.

T. Tsukizoe and N. Ohmae give an overview of the performance of unidirectionally oriented glass, carbon, aramid, and steel-fiber-reinforced plastics sliding against smooth steel counterparts. The authors propose a law-of-mixture approach for calculating the composites' coefficient of friction and they have developed a model wear equation for fiber-reinforced plastics. In a contribution by K. Friedrich, differences in wear behavior of short fiber-reinforced thermoplastics and continuous-fiber woven-fabric-reinforced thermosets against various abrasive counterparts (steel, severe abrasive paper) are discussed as a function of fiber volume fraction. Methods are suggested for correlating the wear rates of materials obtained under different conditions with the properties of the sliding partners and the conditions in the contact area. The mild wear processes of rubber-based compounds are elucidated by J.Å. Schweitz and L. Åhman. The authors concentrate on micro-mechanical detachment events with particular reference to modified surface layers, ageing, fatigue, crack formation and growth.

In a third section, specific friction and wear problems encountered in various practical applications of polymer composites are pointed out. In discussing the subject of commercially available polymer bearing composites, J.C. Anderson subdivides candidate materials into several categories. The classification considers the structural composition and appearance of the materials, their performance attributes, and other characteristics, and contributes towards the optimum selection of polymer composites for particular bearing applications. J.K. Lancaster, a pioneer in the study of friction and wear of polymer composites, reviews in detail the influences of operating parameters on the friction and wear properties and the load-carrying capacity of polymer composites for aerospace dry bearing applications. He also outlines some principles of materials development and life prediction for this particular application and discusses future trends in materials for airframe bearings. In a final highlight, M.N. Gardos reviews typical compositions of self-lubricating composites used for extreme environmental conditions. It was shown that the most important and decisive factor in composite versus counterface performance is the effectiveness of a preferentially accumulated, low shear strength surface layer on the worn composite.

For whom is this volume intended? Undoubtedly, it will be of interest to three groups of specialists: (1) those who are active or intend to become active in research on some aspects of polymer composite tribology (material scientists, physical chemists, mechanical engineers); (2) those who have encountered a practical friction or wear problem and wish to learn more methods of solving such problems (designers, engineers and technologists in industries dealing with selection, reprocessing and application of polymer engineering materials); (3) teachers and students at universities, in view of the fact that, today, the average mechanical engineer receives less than two hours of instruction on wear during his university studies.

Finally, many thanks are due to Professor R.B. Pipes, editor-in-chief of the new Elsevier composite materials series, for the invitation to organize this particular volume. In addition, my sincere thanks go to the chapter authors, who so kindly agreed to collaborate on this book, for their hard and efficient work.

January 1986   K. Friedrich

References

1 F.P. Bowden and D. Tabor, Friction and Lubrication of Solids, Claredon Press, Oxford, 1964.
2 D.S. Moore, Principles and Applications of Tribology, Pergamon Press, Oxford, 1975.
3 H. Czichos, Tribology, Elsevier, Amsterdam, 1978.
4 K.H. Habig, Verschleiss und Härte von Werkstoffen, Hanser Verlag, München, 1980.

5 D.A. Glaeser, Fundamentals of Friction and Wear of Materials, American Society for Metals, Metals Park, OH, 1981.

6 L.H. Lee, Advances in Polymer Friction and Wear, Plenum Press, New York, 1974.

7 D. Dowson, M. Godet and C.M. Taylor, The Wear of Non-Metallic Materials, Mechanical Engineering Publications, London, 1978.

8 G.M. Bartenev and V.V. Lavrentev, Friction and Wear of Polymers, Elsevier, Amsterdam, 1981.

9 H. Uetz and J. Wiedemeyer, Tribologie der Polymere, Carl Hanser Verlag, Munich, Vienna, 1985.

CONTENTS

A more detailed contents list is given at the beginning of each chapter.

Chapter 1

Introduction to Friction and Wear

HORST CZICHOS

Bundesanstalt für Materialprüfung (BAM), Berlin-Dahlem (F.R.G.)

Contents

Abstract

In the introductory section, a brief general description of the fundamentals of friction and wear is given.

The chapter opens with the definitions of basic terms concerning friction, wear, and lubrication. The section on the mechanisms of friction includes the following topics: real area of contact, adhesion component of friction, ploughing component of friction, and deformation component of friction. The section on the mechanisms of friction is concluded with a consideration on the complexity of friction processes and the mechanisms of energy dissipation.

The section on the mechanisms of wear includes the following topics: surface fatigue and delamination wear mechanisms, abrasive wear mechanisms, tribochemical wear mechanisms, and adhesive wear mechanisms. The section on the mechanisms of wear is concluded with a classification of wear phenomena.

Finally, it is pointed out that friction and wear are not intrinsic material properties but depend on so many influencing factors that, in any given situation, the "whole tribological system" must be considered, including the following basic groups of parameters: (i) the "structure" of the tribological system, i.e. the material components of the system and the tribologically relevant properties of the system's components; (ii) the operating variables, including load (or stress), kinematics, temperature, operating duration, etc.; and (iii) the tribological interactions between the systems components. To assist the systematic analysis of relevant tribological parameters in a given friction and wear situation, a methodology and data sheet are outlined.

1. Friction and wear: sub-areas of tribology

Friction and wear are nowadays considered as sub-areas of tribology. The term *tribology* was coined by a British committee in 1966 from the word *tribos*, which means rubbing in classic Greek [1]. Tribology is defined as the science and technology of interacting surfaces in relative motion and of related subjects and practices. Since its definition, tribology has been widely recognized as a general concept embracing all aspects of the transmission and dissipation of energy and materials in mechanical equipment including the various aspects of friction, wear, lubrication and related fields of science and technology.

In this chapter, a brief general introduction to the basic principles of friction and wear is given. The discussion concentrates on fundamental tribological mechanisms; the friction and wear behaviour of specific materials is considered in the following chapters.

2. Terminology

Owing to the multidisciplinary nature of tribology, the terms used in this field have many origins. Thus, some introductory remarks on terminology should be made in order to avoid misleading or ambiguous terms. In 1969, the International Research Group on Wear of Engineering Materials, working under the auspices of OECD, compiled a glossary of terms and definitions for the field of tribology. This glossary has been recognized as a key word index and is included in the new ASME Wear Control Handbook published in 1980 [2]. It contains definitions of 500 general tribological terms, 93 terms on bearing types, and 58 terms on oils and greases. In addition, it contains an eight-language index consisting of translations of the tribological terms into English, French, German, Italian, Spanish, Japanese (Romaji and Kanji), Arabic, and Portuguese.

With this glossary on terms and definitions, the basis for a multilingual terminology for the field of tribology has been laid. Because it is not possible to repeat details of the glossary within the frame of this chapter, only the definitions of the three most important tribological terms will be cited.

Friction: The resisting force tangential to the common boundary between two bodies when, under the action of an external force, one body moves, or tends to move, relative to the surface of the other.

Wear: The progressive loss of substance from the operating surface of a body occurring as a result of relative motion at the surface.

Lubrication: The reduction of frictional resistance and wear or other forms of surface deterioration between two load-bearing surfaces by the application of a lubricant.

Because of the great variety of types, modes, or processes of friction, lubrication and wear which determine the behaviour of any tribological entity, it is necessary to have a sub-classification of these basic terms. For these sub-classifications, different criteria such as

(a) kinematics or type of motion (sliding, rolling, impact, etc.),

(b) type of material (solid, fluid, metal, polymer, etc.), and

(c) type of interfacial tribological process (hydrodynamics, adhesion, abrasion, etc.)

are used.

In this chapter, the discussion is restricted to friction and wear of unlubricated solids in sliding motion.

3. Mechanisms of friction

3.1. General considerations

If two bodies are placed in contact under a normal load, F_N, a finite force is required to initiate or maintain sliding; this is the force of friction, F_F. In this situation, the following phenomenological rules have been observed experimentally.

(i) When relative motion between the contacting bodies occurs, the friction force, F_F, always acts in a direction opposite to that of the relative velocity of the surfaces.

(ii) The friction force, F_F, is proportional to the normal force, F_N

$$F_F = f \cdot F_N$$

Through this relationship, it is possible to define a coefficient of friction

$$f = \frac{F_F}{F_N}$$

(iii) The friction force is independent of the apparent geometric area of contact.

These relations, known as "Amontons–Coulomb laws" of dry sliding friction are used as simple guiding rules in tribological applications.

Despite the considerable amount of work that has been devoted to the study of friction since the early investigations of Leonardo da Vinci, Amontons, Coulomb and Euler [3], there is no "simple" model to predict or to calculate friction of a given pair of materials. From the results of various studies, it is obvious that friction originates from complicated molecular–mechanical interactions between contacting bodies [4]. In quoting a summarizing description by Suh and Sin [5], the "genesis of friction" may be characterized as "The coefficient of friction between sliding surfaces is due to the various combined effects of asperity deformation, ploughing by wear particles and hard surface asperities and adhesion between the flat surfaces. The relative contribution of these components depends on the condition of the sliding interface which is affected by the history of the sliding, the specific materials used, the surface topography and the environment".

Similarly, Tabor pointed out that, today, we recognize that three basic phenomena are involved in the friction of unlubricated solids [6]

(a) the area of real contact between the sliding surfaces;

(b) the type of strength of bond that is formed at the interface where contact occurs; and

(c) the way in which the material in and around the contacting regions is sheared and ruptured during sliding.

TABLE 1
Energy-based overview of friction phenomena

I. ┌──┐
   │ Introduction of mechanical energy into the contact zone │
   └──┘

   → formation of real area of contact

II. ┌──────────────────────────────────┐
    │ Transformation of mechanical energy │
    └──────────────────────────────────┘

    → elastic deformation and elastic hysteresis
    → plastic deformation
    → ploughing
    → adhesion

III. ┌──────────────────────────────┐
     │ Dissipation of mechanical energy │
     └──────────────────────────────┘

     (a) ┌──────────────────────┐
         │ Thermal transformation │
         └──────────────────────┘

     → generation of heat and entropy

     (b) ┌─────────┐
         │ Storage │
         └─────────┘

     → generation of point defects and dislocations
     → strain energy storage
     → phase transformations

     (c) ┌──────────┐
         │ Emission │
         └──────────┘

     → thermal radiation and conduction
     → phonons (acoustic waves, noise)
     → photons (triboluminescence)
     → electrons (exo-electrons)

Because friction is essentially an energy dissipation process, an energy consideration of friction may also be useful [4]. According to that, the whole course of the "loss" process of mechanical energy due to friction may be formally divided into different phases as compiled in Table 1. Firstly, mechanical energy is introduced into the contact zone by the formation of the real area of contact. Secondly, a transformation of mechanical energy takes place mainly by the effect of plastic deformation, ploughing and adhesion. Thirdly, the dissipation phenomena include the effects of thermal dissipation, storage or emission.

3.2. Real area of contact

Consider the contact of two nominally flat solid bodies. There has been considerable interest in recent years in surface topography and its role in tribocontact formation and performance. (See ref. 7 for a recent comprehensive book which contains numerous references.) These studies assume that the surfaces of tribological contacts are covered with asperities of a certain height distribution which deform elastically or plastically under the given load. The summation of individual contact spots gives the real area of contact which is generally much smaller than the apparent geometrical contact area. The basic properties and parameters relevant to the real area of a tribocontact are summarized in Fig. 1. There are two classes of properties, namely deformation properties and surface topography characteristics. Without going into details, it can be said that, for example, the behaviour of metals in contact is determined by a deformation criterion, namely the so-called plasticity

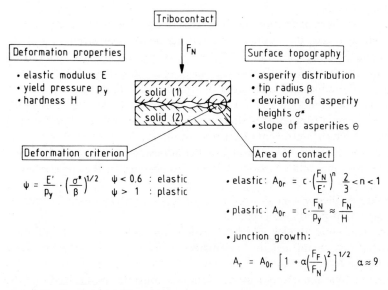

Fig. 1. Characteristics of tribocontact.

index. Depending on the value of the plasticity index, ψ, an elastic or plastic deformation mode results which can be characterized in a simplified manner by the formulas given in Fig. 1. If, in addition to the normal load, a tangential force is introduced, a junction growth of asperity contacts may occur leading to a considerably larger area of contact. For polymeric materials in contact, viscoelastic and viscoplastic effects and relaxation phenomena must be taken into consideration. These influences lead to a time-dependence of the contact area and to hysteresis losses in loading/unloading cycles.

3.3. Adhesion component of friction

The adhesion component of friction is due to the formation and rupture of interfacial adhesion bonds. There have been theoretical papers to explain this interaction, especially for the contact of clean metals, in terms of the electronic structures of the contacting partners (see, for example, refs. 8–11). Theoretically, the attractive interaction forces between two contacting solids include, at least in principle, all those types of interaction that contribute to the cohesion of solids, such as metallic, covalent and ionic, i.e. primary chemical bonds (short-range forces) as well as secondary van der Waals bonds (long-range forces). For example, two pieces of clean gold placed in contact will form metallic bonds over the regions of atomic contact and the interface will have the strength of bulk gold. With clean diamond, the surface forces will resemble valency forces. With rock salt, the surface forces will be partly ionic. All these forces are essentially short-range forces. Long-range van der Waals forces act in the adhesion between soft rubber-like materials and between polymeric solids. It is evident from these examples of metals, ceramics and polymers that interfacial adhesion is as natural as cohesion which determines the bulk strength of materials within a solid.

The adhesion component of friction has been described by Bowden and Tabor in a highly simplified model as the quotient of the interfacial shear strength and the yield pressure of the asperities (see Fig. 2). Because, for most materials, this ratio is of the order of 0.2, the friction coefficient may have this value. Because of junction growth (see Fig. 1), the adhesion component of friction may, however, increase for clean metals to values of about 10–100. If, on the other hand, the contacting surfaces are separated by a film with an effective shear strength of about half that of the parent metals, a friction coefficient of 0.1 may result.

The simplified Bowden and Tabor model of the adhesion component of friction has been extended by some other theories, taking into account further properties of solids (see Fig. 2), e.g.

(a) a surface energy theory which introduces the surface energy of the contacting partners as an important parameter [12] and

(b) a fracture mechanic model which considers the fracture of an adhesive junction and introduces as influencing parameters a critical crack-opening factor and a work-hardening factor [13].

In considering the adhesion component of friction, it must be emphasized that the relevant influencing properties, such as the interfacial shear strength or the surface

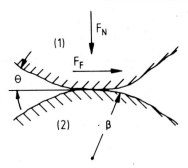

Adhesion component of friction

- simplest model:

$$f_a = \frac{F_F}{F_N} \approx \frac{\tau_{s_{12}}}{p_y}$$

$\tau_{s_{12}}$: interfacial shear strength
p_y : yield pressure

- surface energy theory: $f_a = \dfrac{\tau_{s_{12}}}{p_y} \left[1 - 2\,\dfrac{W_{12}\cdot\tan\Theta}{\beta\cdot p_y}\right]^{-1}$

where $W_{12} = \gamma_1 + \gamma_2 - \gamma_{12}$: surface energy

- fracture mechanic model: $f_a = c\,\dfrac{\sigma_{12}\cdot\delta_c}{n^2\cdot(F_N\cdot H)^{1/2}}$

σ_{12} : interfacial tensile strength
δ_c : critical crack opening factor
n : work hardening factor
H : hardness

Fig. 2. Characteristics of adhesion models of friction.

energy, are characteristics related to the given pair of materials rather than to the single components involved.

3.4. Ploughing component of friction

It is obvious that if one surface in a sliding tribocontact is considerably harder than the other, the harder asperities may penetrate into the softer surface. In tangentional motion, a certain force results because of the ploughing resistance. This may contribute considerably to the friction resistance, as pointed out by Gümbel in 1925 [14].

As compiled in a simplified manner in Fig. 3, the basic possibilities of ploughing are ploughing by asperities and ploughing by penetrated wear particles. In the simplest model, i.e. the case of a sliding conical asperity, the friction coefficient is related to the tangent of the slope of the ploughing asperity [12]. Because normal surface asperities seldom have an effective slope exceeding 5 or 6°, it follows that the coefficient of friction should have a value of $f \approx 0.04$. This value may be regarded, however, only as a lower friction coefficient limit for the ploughing component of friction, due to neglect of the experimentally observed fact that a pile-up of material ahead of the grooving path occurs in most cases of ploughing during sliding [15].

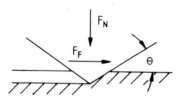

Ploughing component of friction

Ploughing by asperities

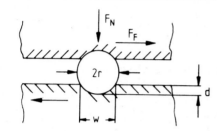

$$f_p = \frac{2}{\pi} \tan\theta$$

Ploughing and microcracking

$$f_p = \frac{F_F}{F_N} = c \frac{K_{Ic}^2}{E \cdot (H \cdot F_N)^{1/2}}$$

K_{Ic} : fracture toughness
E : elastic modulus
H : hardness

• simplest model:

Ploughing by penetrated wear particles

$$f_p = \frac{2}{\pi} \left\{ \left(\frac{2r}{w}\right)^2 \sin^{-1} \frac{w}{2r} - \left[\left(\frac{2r}{w}\right)^2 - 1\right]^{1/2} \right\}$$

Fig. 3. Characteristics of ploughing models of friction.

Because microcracking may occur during ploughing of a brittle-surface material, an extended model has been suggested by Zum Gahr [16]. In this fracture mechanic model of ploughing, properties of materials like the fracture toughness, the elastic modulus and the hardness are main influencing parameters.

The other possibility of ploughing, namely ploughing by penetrated wear particles, was investigated by Suh and co-workers [17]. Their analysis showed that the contribution of ploughing to friction is very sensitive to the ratio of the radius of curvature of the particle to the depth of penetration. This analysis indicates that, in addition to the properties of the materials, the geometric properties of the asperities or penetrated wear particles may considerably influence the friction behaviour of sliding surfaces.

3.5. Deformation component of friction

Because of the deformation during sliding contact, mechanical energy may be dissipated through plastic deformation effects. Green [18] analysed the deformation of the surface asperity contact using the slip-line field for a rigid–perfectly plastic material. In a similar way, in applying a two-dimensional stress analysis of Prandtl, Drescher has worked out a slip-line deformation model of friction as summarized in

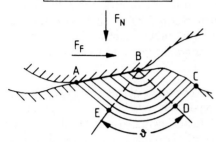

$$f_d = \frac{F_F}{F_N} = \lambda \cdot \tan \, arc \, \sin\left[\frac{\sqrt{2}}{4} \cdot \frac{(2+\vartheta)}{(1+\vartheta)}\right]$$

where $\lambda = \lambda \, (E',H)$ proportion of plastically supported load

E': elastic modulus
H : hardness

Energy-based plastic deformation model

$$f_d = \frac{A_r}{F_N} \tau_{max} \cdot F\left(\frac{\tau_s}{\tau_{max}}\right)$$

where $F\left(\frac{\tau_s}{\tau_{max}}\right) = 1 - 2 \, \dfrac{\ln\left(1+\frac{\tau_s}{\tau_{max}}\right) - \frac{\tau_s}{\tau_{max}}}{\ln\left[1-\left(\frac{\tau_s}{\tau_{max}}\right)^2\right]}$

A_r : real area of contact

τ_{max}: ultimate shear strength of material

τ_s : average interfacial shear strength

Fig. 4. Characteristics of deformation models of friction.

Fig. 4 [19]. In this model, it is assumed that, under an asperity contact (AB in Fig. 4), three regions of plastically deformed material may develop which are described in Fig. 4 by the regions ABE, BED, and BDC. The maximum shear stress in these areas is equal to the flow shear stress of the pertinent material. An important parameter in this model is the factor λ, the proportion of the plastically supported load which is related in a complicated manner to the ratio of the hardness to the elastic modulus. If the asperity contact is completely plastic and the asperity slope is 45°, a friction coefficient of $f = 1.0$ results. This value goes down to $f = 0.55$ if the asperity slope approaches zero. In discussing the deformation component of friction, Drescher pointed out that this model is a very simple one and that some other material properties such as the microstructure of the materials, work hardening effects, thermal softening, and the influence of interfacial layers should be considered.

A recent model of the slip-line theory of the deformation component of friction was advanced by Challen and Oxley [20].

Another model of the deformation component of friction, which relates friction mainly to plastic deformation, was suggested recently by Heilmann and Rigney [21]. The main assumption is that the frictional work performed is equal to the work of plastic deformation during steady-state sliding. As summarized in Fig. 4, there are three main parameters characterizing this model.

(a) The real area of contact (see Sect. 3.2).

(b) The ultimate shear strength of the material which can be achieved during shear.

(c) The average shear strength actually achieved at the interface during sliding. This quantity may depend on many experimental parameters such as operating conditions (load, sliding velocity, temperature) and other material characteristics such as crystal structure, microstructure, work-hardening rate and recovery rate.

3.6. The complexity of friction processes and the mechanisms of energy dissipation

The friction mechanisms so far discussed separately overlap in most practical tribological situations in a complicated manner. In reviewing, especially, the mechanisms of polymer friction, Briscoe [22] has recently compiled the various possible steps of friction-induced energy dissipation in a two-term non-interacting dissipation process model as shown in Fig. 5.

The frictional work is considered to be dissipated in two separate regions in the interfacial region. The interfacial zone is thought of as being very narrow and corresponds to high rates of energy dissipation. This class of process corresponds to

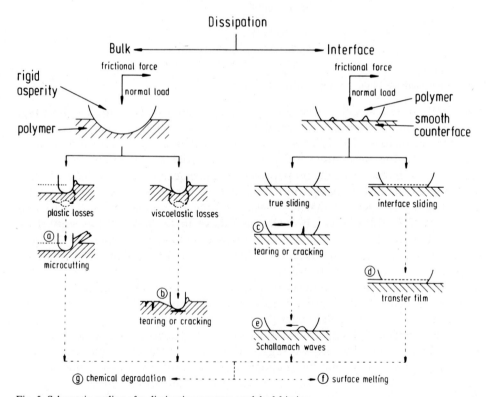

Fig. 5. Schematic outline of a dissipation process model of friction.

the dissipation mechanisms usually grouped under the heading of adhesion models of friction (see Sect. 3.3). The other process involves deformation within a larger volume of material and hence lower rates of energy dissipation. The ploughing and deformation mechanisms (see Sects. 3.4 and 3.5) are within this category. The distinction of two non-interacting processes is artificial and is no more than an expediency which greatly simplifies a complex range of processes. The partial processes involved in the model shown in Fig. 5 can be characterized as

(a) plastic grooving, leading to microcutting;

(b) viscoelastic grooving causing fatigue cracking and tearing with sub-surface heating and damage;

(c) true interfacial sliding: high, effective rates of surface strain and heating; potential for extensive chemical degradation;

(d) interface zone shear: rupture within the polymer and transfer wear; and

(e) a subgroup of true interfacial sliding, the propagation of Schallamach waves.

4. Mechanisms of wear

4.1. General considerations

Similar to friction, the wear behaviour of materials is also a very complicated phenomenon in which various mechanisms and influencing factors are involved. A great step forward in our understanding of wear was the classification of wear mechanisms given by Burwell in the 1950s [23], according to which wear mechanisms may be divided into four broad general classes under the headings of abrasion, adhesion, surface fatigue and tribochemical processes.

In recent years, an increasing number of studies have been devoted to wear (see, for example, refs. 24–27) which indicate that wear, i.e. "the removal of material from interacting surfaces in relative motion", results from various interaction processes. In quoting a summarizing description of Suh [28], it may be said that "wear of materials occurs by many different mechanisms depending on the materials, the environmental and operating conditions and the geometry of the wearing bodies. These wear mechanisms may be classified into two groups: those primarily dominated by the mechanical behaviour of solids and those primarily dominated by the chemical behaviour of materials. What determines the dominant wear behaviour are mechanical properties, chemical stability of materials, temperature and operating conditions".

Tabor, in his recent critical synoptic view of wear [29], for simplicity divided wear processes into three groups: "The first is that in which wear arises primarily from adhesion between the sliding surfaces, the second is that deriving primarily from non-adhesive processes and the third is that very broad class in which there is interaction between the adhesive and non-adhesive processes to produce a type of wear that seems to have characteristics of its own. The way in which these mechanisms interact with one another depends extremely sensitively on the specific operating conditions. In addition, the frictional process itself can produce profound

TABLE 2
Tribological interactions and wear mechanisms

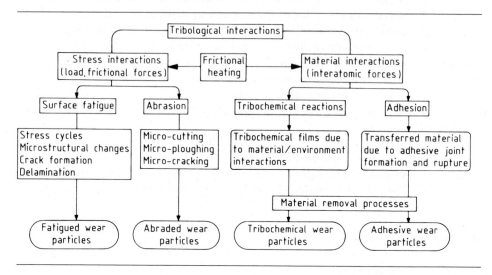

structural changes and modifications of the physical and chemical properties of the sliding surfaces. Consequently, unless a single wear process dominates, these surface changes and complex interactions must necessarily make wear predictions extremely difficult and elusive".

In order to discuss the processes of wear, the whole complex of the generation of loose wear particles must be considered. The chain of events that leads to the generation of wear particles and material removal from a given tribological system is initiated by two broad classes of tribological processes as summarized in Table 2.

(a) Stress interactions. These are due to the combined action of load forces and frictional forces and lead to wear processes described broadly as surface fatigue and abrasion.

(b) Material interactions. These are due to intermolecular forces either between the interacting solid bodies or between the interacting solid bodies and the environmental atmosphere (and/or the interfacial medium) and lead to wear processes, described broadly as tribochemical reactions and adhesion.

4.2. Surface fatigue and delamination wear mechanisms

As is known from the mechanical behaviour of bulk materials under repeated mechanical stressing, microstructural changes in the material may occur which result in gross mechanical failure. Similarly under repeated tribological loading, surface fatigue phenomena may occur leading finally to the generation of wear particles. These effects are mainly based on the action of stresses in or below the surfaces without needing a direct physical solid contact of the surfaces under consideration. This follows from the observation that surface fatigue effects are observed to occur

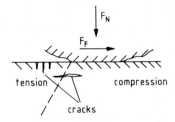

Surface fatigue and delamination wear mechanisms

F_N

F_F

tension   compression

cracks

• Fatigue wear model:

$$W_v = c \frac{\eta \cdot \gamma}{\overline{\varepsilon_1^2} \cdot H} \cdot F_N \cdot s$$

η : line distribution of asperities
γ : particle size constant
$\overline{\varepsilon}_1$: strain to failure in one loading cycle
H : hardness

• Delamination wear model:

(i) transmittance of stresses at contact points
(ii) incremental plastic deformation per cycle
(iii) subsurface void and crack nucleation
(iv) crack formation and propagation
(v) delamination of sheet-like wear particles

Fig. 6. Characteristics of surface fatigue and delamination wear models.

in journal bearings where the interacting surfaces are fully separated by a thick lubricant film [30]. As is known, the effect of fatigue wear is especially associated with repeated stress cycling in rolling contact. However, the asperities also undergo cyclic stressing in sliding, leading to stress concentration effects and the generation and propagation of cracks. On the basis of the dislocation theory, there are several possible mechanisms for crack initiation and propagation (see ref. 31). A contribution to the theory of surface fatigue wear mechanisms was put forward by Halling [32]. This model incorporates the concept of fatigue failure and also of simple plastic deformation failure, which is considered as fatigue failure in one cycle of loading. The main properties of materials relevant to surface fatigue in this model are listed in Fig. 6.

In studying the plastic–elastic stress fields in the sub-surface regions of sliding asperity contacts and the possible dislocation interaction, a "delamination theory of wear" has been put forward by Suh [33] in which the generation of sheet-like wear particles is explained on the basis of the chain of events compiled in Fig. 6. In refining this theory, some implications of crystal plasticity effects have been discussed including a specific dislocation model for hexagonal close packed metals under wear conditions and a correlation with stacking fault energy for face centred cubic and hexagonal close packed metals [34]. It appears that, by the use of dislocation models, it is now possible to explain the occurrence of sheet-like wear particles observed as early as in 1929 by Füchsel and interpreted through a process of stress-induced material separation (Abblättern in German) [35].

4.3. Abrasive wear mechanisms

The effect of abrasion occurs in contact situations, in which direct physical contact between two surfaces is given, where one of the surfaces is considerably

14

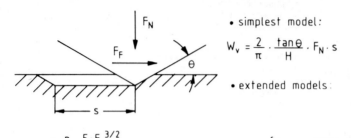

Abrasive wear mechanisms

F_N

F_F

θ

s

• simplest model:

$$W_v = \frac{2}{\pi} \cdot \frac{\tan\theta}{H} \cdot F_N \cdot s$$

• extended models:

$$W_v = n^2 \frac{p_y \cdot E \cdot F_N^{3/2}}{K_{Ic}^2 \cdot H^{3/2}} \cdot s$$

$$W_v = \frac{f_{abr}}{K_1 \cdot K_2 \tau_c} \cdot \frac{\cos\rho \cdot \sin\theta}{(\cos\frac{\theta}{2})^{1/2} \cdot \cos(\theta - \rho)} \cdot F_N \cdot s$$

E : elastic modulus
H : hardness
K_{Ic} : fracture toughness
n : workhardening factor
p_y : yield strength

f_{abr} : model factor (1 for micro-cutting)
K_1 : relation of normal and shear stress
K_2 : texture factor (1 for fcc-metals)
τ_c : shear stress for dislocation movement
ρ : friction angle at abrasive/material
     interface

Fig. 7. Characteristics of abrasive wear models.

harder than the other. The harder surface asperities press into the softer surface with plastic flow of the softer surface occurring around the asperities from the harder surface. When a tangential motion is imposed, the harder surface removes the softer material by combined effects of "micro-ploughing", "micro-cutting", and "micro-cracking".

As illustrated in Fig. 7, in the simplest model of abrasive wear processes the wear volume is related to the asperity slope of the penetrating abrasive particle and the hardness of the abraded material (see ref. 12). From the comprehensive experimental work of Khrushov [36], the following phenomenological observations on the influence of material properties on abrasive wear are drawn.

(a) Technically pure metals in an annealed state and annealed steels show a direct proportionality between the relative wear resistance and the pyramid hardness.

(b) For non-metallic hard materials and minerals, a linear relationship between wear resistance and hardness is similarly found.

(c) For metallic materials cold work-hardened by plastic deformation, the relative wear resistance does not depend on the hardness resulting from cold work-hardening.

(d) A heat treatment of structural steels (normal hardening and tempering) improves the abrasive wear resistance.

The simple abrasion model which includes only the hardness as material property has been recently extended. Hornbogen proposed a model to explain increasing relative wear rates with decreasing toughness of metallic materials [37]. This model is based on the comparison of the strain that occurs during asperity interactions with the critical strain at which crack growth is initiated. The basic properties of materials relevant to that model are listed in Fig. 7. Another model was proposed by Zum Gahr who considered the detailed processes of micro-cutting, micro-ploughing and micro-cracking in the abrasive wear of ductile metals [38]. This model includes other micro-structural properties of the materials worn beside the flow pressure or hardness. Influences of capacity of work-hardening, ductility, homogeneity of strain distribution, crystal anisotropy and mechanical instability have been identified to influence the abrasive wear mechanisms.

4.4. Tribochemical wear mechanisms

Whereas the mechanisms of surface fatigue and abrasion can be described mainly in terms of stress interactions and deformation properties, in tribochemical wear as third partner, the environment and the dynamic interactions between the material components and the environment determine the wear processes. These interactions may be expressed as cyclic stepwise processes.

(i) At the first stage, the material surfaces react with the environment. In this process, reaction products are formed on the surfaces.

(ii) The second step consists of the attrition of the reaction products as a result of crack formation and abrasion in the contact process interactions of the materials. When this occurs "fresh", i.e. reactive surface parts of the materials are formed and stage (i) continues.

As a consequence of thermal and mechanical activation, the asperities undergo the following changes.

(a) The reactivity is increased due to the increased asperity temperature. Therefore the formation of surface layers is accelerated.

(b) The mechanical properties of the surface asperity layers are changed: in general, they have a tendency to brittle fracture.

Starting with the assumption that tribochemically formed surface asperity layers are detached at a certain critical thickness, Quinn proposed a tribochemical wear hypothesis of the wear of steels [39]. In Fig. 8, the main characterizing parameters of tribochemical wear mechanisms according to Quinn are listed. Based on recent investigations on the oxidative wear of aluminium alloys, Razavizadeh and Eyre [40] concluded that tribochemical wear occurs by a process of oxidation, deformation, and fracture to produce layers which are compacted into grooves in the metal surface. Oxidative surfaces become smoother with time as the troughs become filled with oxide and oxidative wear then occurs by the fracture of plate-like debris.

4.5. Adhesive wear mechanisms

The adhesive wear processes are initiated by the interfacial adhesive junctions which form if solid materials are in contact on an atomic scale. As described in Sect. 3.3, depending on the nature of the solids in contact different adhesive junctions may result. The whole chain of events which leads to the generation of wear particles is summarized in Fig. 9. It is obvious that a number of properties of the contacting

16

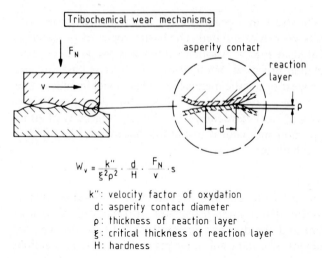

Fig. 8. Characteristics of tribochemical wear models.

solids influence the adhesive wear mechanisms. Obviously, the processes and parameters of adhesion (see Sect. 3.3) as well as those of fracture (see Sects. 4.2 and 4.3) must be taken into account. Since both adhesion and fracture are influenced by surface contaminants and the effect of the environment, it is quite difficult to relate adhesive wear processes with elementary bulk properties of materials. In vacuum,

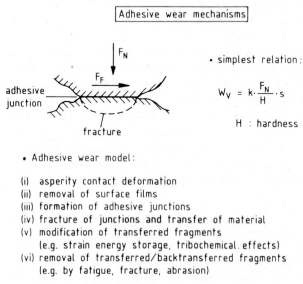

Fig. 9. Characteristics of adhesive wear models.

where these influences are eliminated, the following parameters have been observed to influence the adhesive wear processes of metal/metal pairs [41].

(a) Interfacial metallic adhesion bonding depends on the electronic structure of the contacting partners. It has been suggested that strong adhesion will occur if one metal can act as an electron donor, the other as an electron acceptor [9,42,43].

(b) Crystal structure exerts an influence on adhesive wear processes. Hexagonal metals, in general, exhibit lower adhesive wear characteristics than either body-centred cubic or face-centred cubic metals. This difference is assumed to be related to different plastic asperity contact deformation modes and the number of operable slip systems in the crystals [44,45].

(c) Crystal orientation influences the adhesive wear behaviour. In general, high atomic density low surface energy grain orientations exhibit lower adhesion and less adhesive wear than other orientations [41].

(d) When dissimilar metals are in contact, the adhesive wear process will generally result in the transfer of particles of the cohesively weaker of the two materials to the cohesively stronger [41].

The wear resulting from adhesive processes has been described phenomenologically by the well-known Archard equation (see Fig. 9) [46]. This equation contains the hardness of the material as the only material property. As is obvious from the above discussion, the Archard wear coefficient, k, depends on various properties of the materials in contact. However, it is not possible to describe the influence of the properties of the material on k in a simple manner.

4.6. Classification of wear phenomena

Because of the great variety of types, modes and processes of wear and its many influencing factors, there are considerable difficulties to classify or to denominate given wear situations in a logical, consistent, and comprehensive manner. In 1979, the German Institute for Standardization (DIN) published a standard on wear [47] in which a classification of the terms of wear is given which is based on an analogy to the (classical) field of the strength of materials.

Conventionally, in order to classify the strength, deformation, or failure of a material, the following specifications must be given:

(i) the type of material;

(ii) the type of external stress (e.g. compression, tension, bending, etc.); and

(iii) the type of internal deformation or fracture mechanism (e.g. ductile fracture, brittle fracture).

Analogously, in order to classify the type of wear occurring within a tribological entity, the following specifications must be given:

(i) the materials involved in wear (solid, fluid, etc.);

(ii) the type of kinematic tribological action (sliding, rolling, impact, etc.); and

(iii) the type of interfacial wear mechanism (adhesion, abrasion, surface fatigue, etc.).

The resulting classification of ref. 47 for the field of wear is shown in Table 3. In the classification scheme shown in Table 3, the types of wear are named according to

TABLE 3
Classification of wear phenomena

System structure	Tribological action (symbols)		Type of wear	Effective mechanisms (individually or combined)			
				Adhesion	Abrasion	Surface fatigue	Tribo-chemical reactions
Solid — interfacial medium (full fluid film separation) — solid	sliding rolling impact	*(symbol)*				X	X
Solid — solid (with solid friction, boundary, lubrication, mixed lubrication)	sliding	*(symbol)*	sliding wear	X	X	X	X
	rolling	*(symbol)*	rolling wear	X	X	X	X
	impact	*(symbol)*	impact wear	X	X	X	X
	oscillation	*(symbol)*	fretting wear	X	X	X	X
Solid — solid and particles	sliding	*(symbol)*	sliding abrasion		X		
	sliding	*(symbol)*	sliding abrasion (three body abrasion)		X		
	rolling	*(symbol)*	rolling abrasion (three body abrasion)		X		
Solid — fluid with particles	flow	*(symbol)*	particle erosion (erosion wear)		X	X	X
Solid — gas with particles	flow	*(symbol)*	fluid erosion (erosion wear)		X	X	X
	impact	*(symbol)*	impact particle wear		X	X	X
Solid — fluid	flow oscillation	*(symbol)*	material cavitation, cavitation erosion			X	X
	impact	*(symbol)*	drop erosion			X	X

the kinematics. This is in analogy both to the classification used in the field of the strength of materials, as mentioned above, and to the common classification of friction, such as sliding friction, rolling friction, etc. During the occurrence of the different types of wear, in the interface of the wear couple, one or more different wear mechanisms described in the foregoing sections may act.

5. Friction and wear as system properties

The discussions in the foregoing sections have shown that friction and wear are not intrinsic material properties but depend on so many influencing factors that, in any given situation, the whole "tribological system" must be considered [48]. Thus, measured quantities of friction and wear, e.g. friction coefficient or wear rate, depend on the following basic groups of parameters.

(a) The "structure" of the tribological system, i.e. the material components of the system and the tribologically relevant properties of the system's components.

(b) The operating variables, including load (or stress), kinematics, temperature, operating duration, etc.

(c) The tribological interactions between the system components.

The procedure of a systematic analysis of tribological parameters is illustrated in the simplified diagram of Fig. 10.

Consider a technical entity in which friction and wear processes occur. As indicated in the upper central part of Fig. 10, the material components and substances directly participating in the friction and wear processes have first to be

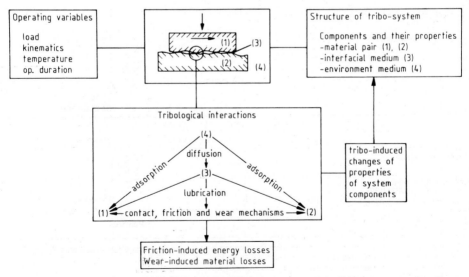

Fig. 10. Basic parameter groups of tribological systems.

TABLE 4

Data sheet for the description of tribological parameters

General description of the wear process:	Date: Sheet No:	Inspector:

I Technical function of the tribological system

II Operating variables

Form of motion	Sliding ☐	Rolling ☐	Impact ☐	Flow ☐	Superimposition:

Development of motion	Operating duration t_B in		Remarks:		

Load [a] F_N in N	F_N at t_0	F_N at t_B	Velocity [b] v in m/s	v at t_0	v at t_B	Temperature [c] T in °C	T at t_0	T at t_B

F_N ↑ ▭ Time t in →

v ↑ ▭ Time t in →

T ↑ ▭ Time t in →

III Structure of tribological system

	Elements	Body (1)	Counter body (2)		Elements	Interfacial medium (3)	Surrounding med. (4)
Properties [d]	Designation Material			Properties [d]	Designation Substance		
Volume properties Dimensions				Physical condition			
Chemical composition, structure				Chemical composition, chemical structure			
Physical data for the material [e]				Physical data for the material [e]			
Mechanical properties, hardness				Viscosity data			
Surface properties Surface roughness data				Other properties			
Physical-chemical surface data [f]							

Tribological contact area A_0			Friction condition in accordance with DIN 50 281:	**Coefficient of friction**	Friction diagram
Wear track ratio [g] ε in %	ε (1)	ε (2)			▭
Wear mechanisms and interactions:			Remarks:		Time t in → or distance s in

IV Wear characteristics

	Body (1)	Counter body (2)		Wear diagram
Appearance of wear (description, picture)			**Amount of wear** ↑	▭
Total amount of wear in accordance with DIN 50 321				Time t in → or distance s in
Mean velocity of wear $\dfrac{\Delta l}{\Delta t}$ in $\dfrac{\mu m}{h}$				
Mean wear/ distance ratio $\dfrac{\Delta l}{\Delta s}$ in $\dfrac{\mu m}{km}$			Remarks:	

[a] The nominal pressure on the surface can be obtained from the load F_N and the tribological contact area A_0, $p_0 = F_N/A_0$
[b] v is the relative velocity between the body (1) and the counter body (2)
[c] T is the macroscopic average temperature of the tribological system
[d] Where present, wear-dependent changes in properties should also be indicated
[e] Tribologically relevant quantities are, for example, density, coefficient of expansion, thermal conductivity, volume, pressure
[f] Properties of the outmost atomic layers of (1) and (2), e.g. thickness and composition of oxide layers
[g] Ratio of geometrical tribological contact area A_0 to the total wear track area covered A_1

identified and conceptually separated from the other parts of the technical entity considered. These components, designated as (1)–(4) in Fig. 10, are part of the structure of the tribological system. In addition to the structural elements, the operating variables have to be identified (see Fig. 10, left). Through the action of the operating variables on the structural elements of the system, tribological interactions occur which may lead to tribo-induced changes of properties of system components as well as to friction-induced energy losses and wear-induced material losses.

To assist the systematic analysis of relevant tribological parameters in a given friction and wear situation, a methodology and a data sheet have been worked out [48] (see Table 4). The use of such data sheets may be helpful for both the planning of friction and wear tests (especially for the problem of laboratory simulation tests) and the documentation of the obtained results of tribological investigations.

List of symbols

A_r	real area of contact
A_{0r}	area of contact (static)
c	coefficient
d	depth of penetration by wear particles (Fig. 3)
d	asperity contact diameter (Fig. 8)
E	elastic modulus
E'	composite elastic modulus
f	friction coefficient
f_a	friction coefficient (adhesion component)
f_{abr}	abrasive model factor ($f_{abr} = 1$ for micro-cutting)
f_d	friction coefficient (deformation component)
f_p	friction coefficient (ploughing component)
F_F	friction force
F_N	normal load
H	hardness
k	Archard wear coefficient
k''	velocity factor of oxidation
K_{Ic}	fracture toughness
K_1	relaxation of normal and shear stress
K_2	texture factor ($K_2 = 1$ for f.c.c. metals)
n	work hardening factor
p_y	yield pressure
r	radius of wear particle
s	sliding distance
t	time
T	macroscopic average temperature
v	relative velocity between body 1 and body 2
w	width of wear particle indentation

W_V wear volume
W_{12} surface energy
$\bar{\epsilon}_1$ strain to failure in one loading cycle
η line distribution of asperities
α junction growth factor
β tip radius of asperity
ϑ angle between the regions of plastically deformed contact
δ_c critical crack opening factor
γ particle size constant
γ_1 surface energy of body 1
γ_{12} interfacial energy between 1 and 2
γ_2 surface energy of body 2
λ proportion factor of plastically supported load
ψ plasticity index
ρ friction angle at abrasive/material interface
ρ thickness of tribochemical reaction layer
σ^* deviation of asperity heights
σ_{12} interfacial tensile strength
τ_c shear stress for dislocation movement
τ_{max} ultimate shear strength
τ_s average interfacial shear strength
τ_{s12} interfacial shear strength
θ slope of asperity
ξ critical thickness of tribochemical reaction layer

References

1 Lubrication (Tribology) Education and Research. A Report on the Present Position and Industry's Needs, Her Majesty's Stationery Office, London, 1966.
2 M.B. Peterson and W.O. Winer (Eds.), Wear Control Handbook, American Society of Mechanical Engineers, New York, 1980, pp. 1143–1303.
3 D. Dowson, History of Tribology, Longman, London, 1979.
4 H. Czichos, Umschau, 71 (1971) 116 (in German).
5 N.P. Suh and H.C. Sin, Wear, 69 (1981) 91.
6 D. Tabor, Trans. ASME, 103 (1981) 169.
7 T.R. Thomas (Ed.), Rough Surfaces, Longmans, London, 1982.
8 D. Tabor, in J.M. Blakely (Ed.), Surface Physics of Materials, Academic Press, New York, 1975, pp. 475–529.
9 H. Czichos, J. Phys. D, 5 (1972) 1890.
10 J. Ferrante and J.R. Smith, Surf. Sci., 38 (1973) 77.
11 J. Ferrante and J.R. Smith, Solid State Commun., 20 (1976) 393.
12 E. Rabinowicz, Friction and Wear of Materials, Wiley, New York, 1965.
13 U. Marx and H.G. Feller, Metallurgie, 33 (1979) 380 (in German).
14 L. Gümbel, Friction and Lubrication in Mechanical Engineering, Krayn, Berlin, 1925 (in German).
15 D. Landheer and J.H. Zaat, Wear, 27 (1974) 129.

16 K.-H. Zum Gahr, Abrasive Wear of Metallic Materials, VDI-Fortschr. Ber., Reihe 5, Nr. 57, VDI-Verlag, Düsseldorf, 1981 (in German).
17 H.C. Sin, N. Saka and N.P. Suh, Wear, 55 (1979) 163.
18 A.P. Green, J. Mech. Phys. Solids, 2 (1955) 197.
19 H. Drescher, VDI Z., 101 (1959) 697 (in German).
20 J.M. Challen and P.L.B. Oxley, Wear, 53 (1979) 229.
21 P. Heilmann and D.A. Rigney, Wear, 72 (1981) 195.
22 B.J. Briscoe, in K.L. Mittal (Ed.), Physicochemical Aspects of Polymer Surfaces, Plenum Press, New York, 1983, pp. 387–412.
23 J.T. Burwell, Wear, 1 (1959) 119.
24 B.J. Briscoe, Tribol. Int., 14 (1981) 231.
25 D.A. Rigney (Ed.), Fundamentals of Friction and Wear of Materials, American Society for Metals, Metals Park, 1980.
26 K.-H. Habig, Wear and Hardness of Materials, Hanser Verlag, Munich, 1980 (in German).
27 H. Czichos (Ed.), Friction and Wear of Materials, Components and Constructions, Expert-Verlag, Grafenau, 1982 (in German).
28 N.P. Suh, in P. Senholzi (Ed.), Tribological Technology, Vol. I, Martinus Nijhoff, The Hague, 1982, pp. 37–208.
29 D. Tabor, Proc. Int. Conf. Wear Mater., American Society of Mechanical Engineers, New York, 1979, p. 1.
30 O.R. Lang, Wear, 43 (1977) 25.
31 D. Kuhlmann-Wilsdorf, in D.A. Rigney (Ed.), Fundamentals of Friction and Wear of Materials, American Society for Metals, Metals Park, 1980, pp. 119–186.
32 J. Halling, Wear, 34 (1975) 239.
33 N.P. Suh, Wear, 25 (1973) 111.
34 J.P. Hirth and D.A. Rigney, Wear, 39 (1976) 133.
35 M. Füchsel, Organ Fortschr. Eisenbahnwes., 84 (1929) 413 (in German).
36 M.M. Krushov, Wear, 28 (1974) 69.
37 E. Hornbogen, Wear, 33 (1975) 251.
38 K.-H. Zum Gahr, Z. Metallkd., 73 (1982) 267 (in German).
39 T.F.J. Quinn, ASLE Trans., 10 (1967) 158.
40 K. Razavizadeh and T.S. Eyre, Wear, 79 (1982) 325.
41 D.H. Buckley, Surface Effects in Adhesion, Friction, Wear, and Lubrication, (Tribology Series, Vol. 5), Elsevier, Amsterdam, 1981.
42 D.H. Buckley, Wear, 20 (1972) 89.
43 N. Ohmae, T. Okuyama and T. Tsukizoe, Tribol. Int., 13 (1980) 177.
44 D.H. Buckley and R.L. Johnson, Wear, 11 (1968) 405.
45 K.-H. Habig, Materialpruefung, 10 (1968) 417 (in German).
46 J.F. Archard, in M.B. Peterson and W.O. Winer (Eds.), Wear Control Handbook, American Society of Mechanical Engineers, New York, 1980, pp. 35–80.
47 German Standard DIN 50 320, Wear. Terms, Systems Analysis of Wear Processes, Classification of the Field of Wear, Beuth-Verlag, Berlin, 1979 (available in English).
48 H. Czichos, Tribology. A Systems Approach to the Science and Technology of Friction, Lubrication and Wear, Elsevier, Amsterdam, 1978.

Chapter 2

Interfacial Friction of Polymer Composites.
General Fundamental Principles

BRIAN J. BRISCOE

Imperial College, London (Gt. Britain)

"I don't believe in demons"
E.M. Rogers, 1965

Contents

26

Abstract

This chapter reviews the processes which are generally regarded as being responsible for the frictional work dissipated when two solid bodies, one of which is a polymeric composite are moved over one another while in contact. Two major processes are envisaged as separate entities; the ploughing or deformation mode and the interface or adhesive mode. The distinction between these two mechanisms is based on the thickness of the interface region undergoing deformation and is imprecise, but nevertheless serves as a useful approximation. The characteristics of the dissipation processes in the two components are recognised as being rather different and these differences are considered first in the context of pure polymers. An examination of the changes produced by the inclusion of secondary phases is then discussed in terms of the two prime dissipation processes, Coulombic and Schallamach mechanisms, which do not fit easily into the two main categories are considered separately and their potential contributions are discussed.

1. Introduction

1.1. Scientific explanation vs. demons

E.M. Rogers' text "Physics for the Inquiring Mind" [1] contains many interesting problems, the first of which is this. "How do you know it is friction, not demons, that brings a rolling ball back to rest? Suggest experiments to support your view." At first sight, this appears to be a joke; it is not. The problem raises a number of seminal questions about the basis of scientific explanations and the laws derived from them. Rogers does not generally offer detailed model answers to his problems, but in this case he does. The answer is in two parts, but unfortunately it is too long to record in full here. The first part is in the form of a dialogue between the reader and a neighbour, Faustus, who argues for demons. The reader begins by stating that "I don't believe in demons". Faustus is clever and whatever properties the reader ascribes to friction he conveys in some special characteristic of demons. For example, demons push against the ball to provide the static friction and the dynamic friction is a consequence of the crushing of their bones. Lubrication reduces friction because it drowns demons and so on. Friction, argues Faustus, is about the sociology of demons.

The second part of Rogers' model answer is more serious. The case against demons is that they are arbitrary. Scientific explanations of friction must fit into a self-consistent body of knowledge, which spans other subjects. In science, we need the minimum number of basic ideas and physical laws. So it is with the fundamental principles of polymer friction, which is about the work done during the relative movements between and within molecules in interfacial regions. It is a part of the science of material deformation and our explanations must be part of this science.

The explanation of interactions between solid bodies is unusually difficult. An apt, and final quotation, favoured by Tabor and attributed to Pauli is "God made solids – but surfaces were the work of the Devil" * [2]. To aid our discussion of friction, we will follow the classical precedent embodied in the "two-term non-interacting model" popularised by Bowden and Tabor [3,4]. This is an important simplification, without which this difficult problem would be unmanageable. Bowden and Tabor originally attempted to rationalise ductile metal friction and their early scientific explanations referenced friction behaviour to continuum deformation studies in bulk materials using ideas based on plastic flow. Later, particularly Tabor, began to rationalise viscoeleastic evescent grooving in elastomers, a study which focuses on viscoelastic losses. The idea of two ranges of surface deformation (the non-interacting two-term model) is now commonly adopted and it is particularly valuable for organic polymers.

1.2. The two-term non-interacting friction model of friction

This idea can be introduced in a number of ways. The simplest is shown in Fig. 1. When a rigid asperity is slid over an organic polymer, experience tells us that, to a first order, the frictional work is often dissipated separately in two discrete zones. The interface zone may be about 100 nm thick, whilst the cohesive zone depth is comparable with the contact length.

The virtues of this division for organic systems are apparent when we review Table 1, bearing in mind that we will attempt to describe friction by analogy with bulk deformation behaviour. The deformation characteristics of polymeric matrices are very sensitive to temperature, hydrostatic stress, strain, rate of strain, time, history and, to some extent, to the chemistry of the environment. There are very clear and major differences in the conditions which exist in each zone; for example, the rates of strain may differ by several orders of magnitude. This attempt to define the magnitude of the important contact conditions is extremely important when comparing experimental studies of friction and bulk deformation. Interestingly, this aspect was not of great significance in the original interpretation of metallic friction in terms of bulk deformation, where the response is less sensitive to the deformation variables.

Another way of introducing the distinction between interfacial or cohesive work is to use the original ideal that interfacial work was done on the system by stresses

---

* The reader will realise that, in tribology, we have two interacting surfaces.

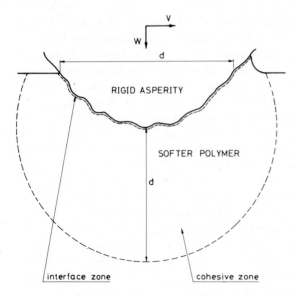

Fig. 1. A schematic illustration of the two-term model of friction. The two zones, the interface and cohesive zone, dissipate the frictional work in separate and non-interacting processes.

generated by the action of adhesive forces only. This is the adhesion model of friction which involves very localised interface shear; originally considered to involve plastic yielding of the softer body [2,3]. The large volume deformations do not involve adhesion at all in the simple model. The work is done by geometric engagements. This is the ploughing or deformation term. Experimental practice has often used rolling or efficient lubrication to suppress adhesion and hence obtain an unambiguous value for the deformation work term.

An important idea in the two term model is the simple additivity of the two contributions. In reality, this is never the case although it may be a good approxima-

TABLE 1

Notional deformation conditions in the two primary friction dissipation zones

Variable	Cohesive or ploughing zone [a]	Interfacial or adhesive zone
Temperature	Nearly ambient	Invariably high
Contact pressure	Close to ambient	Often high
Strain	Low	Sometimes very high
Strain rate	Moderate	Generally very high
Contact time	Similar	Similar

[a] The conditions developed in the cohesive zone are comparable with the conditions produced in certain bulk deformation experiments.

tion. We know, for example, interfacial adhesion plays an important role in plastic grooving from detailed studies on machining of solids [5]. Bearing this approximation in mind, we will review the fundamental ideas of both adhesive and ploughing friction in the context of polymeric composites. Much of the general principles are derived from the study of pure matrices and hence it is necessary to quote these data at some length. There have been many reviews of friction, several of which deal specifically with organic polymers; a selection is given in refs. 6–30 along with brief comments on their content. In these reviews, and indeed this article, there are invariable references to surface damage, wear, adhesion and lubrication.

2. Deformation or ploughing friction

2.1. General ideas

Ploughing friction arises from deformations which can be reproduced in some measure by conventional bulk deformation apparatus such as torsional pendulums. As we shall review this type of friction in terms of the bulk rheological properties of the polymer or composite, it is helpful to introduce friction in categories which reflect the nature of the deformation. Two broad cases are relevant here; small strain viscoelastic ploughing and large strain deformations which produce surface damage.

2.2. Friction in viscoelastic grooving

Much of the experimental work in this area has been devoted to the rolling, rather than the sliding, of rigid bodies over viscoelastic solids. This experimental trick often removes the adhesion component of the friction. The pictorial section of a cylinder or sphere rolling over a viscoelastic plane is shown in Fig. 2; this is a useful general case. As the roller moves over the surface of the polymer, energy is fed into the polymer ahead of the roller and some of it is resorted to the rear of the roller because of elastic recovery and urges it forward (Fig. 2). We can treat the friction problem in terms of the nett forces acting on the roller, as the figure depicts, or consider the work done on the system per unit rolling distance. The available analyses consider the latter. There are many studies of this process reported in the literature [3,4,10,14,20,31–37]. If the elastic work done on the contact per unit length is ϕ, the energy dissipated will be some fraction of this, $\alpha\phi$. A contact mechanical approach for a sphere of radius R on an elastic body gives

$$\phi = 0.17 \, W^{4/3} R^{-2/3} (1 - \nu^2)^{1/3} E^{-1/3} \tag{1}$$

where E is the real part of the Young's modulus, W is the normal load and ν is Poisson's ratio. The modulus E is a function of time and temperature. For a sinusoidal deformation, which is a good approximation, α is equivalent to $\pi \tan \delta$ where $\tan \delta$ is the loss tangent. Again, $\tan \delta$ is a function of time and temperature. So, the friction, f_d, is given by

$$f_d = KE^{-1/3} \tan \delta \tag{2}$$

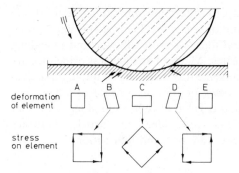

Fig. 2. The cohesive volume or ploughing term in viscoelastic grooving. The frictional work is dissipated beneath the contact by the deformations indicated. The figure denotes rolling, but the process is the same for sliding with efficient lubrication. The details of the surface deformation are not shown, but the size of the arrows indicate that the force at the front, restraining motion, is less than that at the rear, promoting motion. The extent of the difference in these forces, the nett friction, is a function of the way the polymer relaxes at the rear of the contact. This is governed by its viscoelasticity.

for a constant load in this geometry. K contains the constants given in eqn. (1). Actually, the problem is more complex than depicted. Each element beneath the contact is, on average, deformed three times more than suggested by eqn. (1). The important point, however, is that the important material parameters (excluding ν) are expressed in (2). A good test of eqn. (2) is shown in Fig. 3 for unfilled PTFE [36]. A notable feature of viscoelastic deformation friction is the relatively low frictional work; coefficients of friction are often of the order of 0.01 or less.

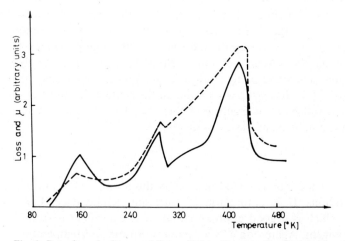

Fig. 3. Data from Ludema and Tabor [36] to illustrate the viscoelastic origin of rolling [eqn. (2)]. The measured rolling resistance is shown as is the quantity $E^{-1/3} \tan \delta$. E and $\tan \delta$ correspond to frequencies comparable with the contact frequency. ------, Coefficient of rolling friction.

We can now turn to the response expected in composite materials. A survey of the literature on the dynamic mechanical properties of composites indicates [27,38]

(i) The Young's modulus invariably increases with the incorporation of rigid fillers. These fillers may also increase the glass transition temperature by a small amount.

(ii) Rigid fillers have the effect of broadening the transitions seen in the viscoelastic spectrum. This effect is particularly pronounced with flake fillers such as graphite and mica.

(iii) The magnitude of tan δ at the maximum is often reduced, simply because the volume of the polymer is decreased. There are cases where new damping processes may be introduced by the presence of the filler.

(iv) Soft filler matrices, which are produced by polyblends, block and graft polymers, exhibit two sets of relaxation peaks characteristic of each phase. The sizes of the peaks, and hence the losses, are in proportion to the concentration of the phases. Similarly, the modulus will roughly scale with the composition.

The general conclusion is that the rolling friction of composites with hard filler phases will be decreased. The author knows of no major systematic study of this subject for composites, but White and Lin [39], for example, have found that theory and experiment are in general agreement for an automobile-type composite.

The practical significance of this dissipation mechanism is probably familiar to us all; it represents a significant fraction of the frictional work dissipated by well-lubricated rubber contacts [10]. Rolling and lubricated sliding have much in common. The grip provided by automobile types and certain types of footware require that the quantity $E^{-1/3} \tan \delta$ be optimised. Large values are required for lubricated traction but in dry conditions this may produce large amounts of heating and hence damage [27]. Wannop and Archard [40] have exemplified the problem of subsurface heating in rolling contacts for poly(methylmethacrylate). Under appropriate conditions, gross subsurface softening occurs and catastrophic spalling is produced.

So far, we have made no direct reference to the deformation conditions beneath the contact. The stress distribution in elastic contacts has been examined by several authors [41,42]. In the absence of interfacial shear stresses, the maximum shear stress is under the contact at a depth of about one-third of the contact diameter. The maximum heating rate occurs here [40]. The deformation frequency is approximately equal to the contact time; the relative velocity divided by the contact length. It could be three times this value. The strains are relatively modest and large hydrostatic stress fields are not developed.

2.3. Friction where cracking and tearing is involved

Several authors have investigated the consequences of exceeding the reversible elastic strain in contacts with either highly elastic solids [43,44] or glassy polymers [44–46]. A variety of tears and cracks are observed and their presence has implications as far as wear is concerned. The formation of these discrete damage sites is seen in the frictional character for certain contacts. Continuous motion is sometimes replaced by stick–slip where each stick–slip cycle represents a crack formation.

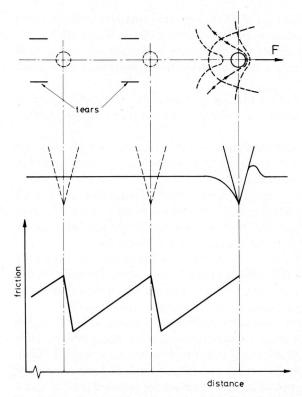

Fig. 4. Data abstracted from the work of Schallamach [11] and others. A sharp needle traverses the surface of an elastomer and, as a result, a series of discrete traction tears are produced. The figure shows how the tears are formed in tension at the rear of the contact and how these tears relax to give the observed damage. The middle figure is a section of the surface and the upper is the view down the axis of the needle. The needle jumps over the surface and, as it does so, it produces stick–slip motion (lower figure). The areas under each stick cycle are roughly proportional to the toughness of the rubber.

Schallamach's data on the friction and damage produced by needles on the surface of elastomers are a good example (Fig. 4). The area under each stick–slip event is, on the basis of limited data, proportional to the toughness of the polymer. A more quantitative example is given by Tabor and co-workers [15,34] for cones, Fig. 5, on well-lubricated rubber. The analogue of eqn. (1) for cones is

$$\phi = \frac{W}{\pi} \cot \theta \tag{3a}$$

where θ is the semi-apical angle of the cone. The load is fixed and, providing that no permanent deformation occurs, the friction, f_d, is given by

$$f_d = 2.3\alpha\theta \tag{3b}$$

Once tearing becomes apparent, additional work is done in this process and the

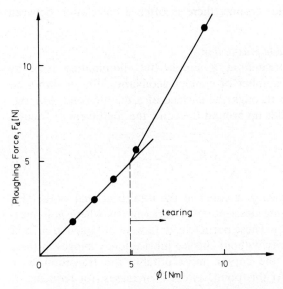

Fig. 5. The observed ploughing force, F_d, as a function of the energy input per unit distance ϕ on a well-lubricated rubber. The sliders are cones of varying semiapical angle, θ, under a normal load of 16 N. Up to $\phi = $ ca. 5 Nm, the gradient shown is 2.3α and smooth grooving occurs. Beyond a value of ϕ of 5 Nm, tearing occurs and there is a disproportionate increase in F_d. The parameters α and ϕ are defined in eqns. (3a) and (3b).

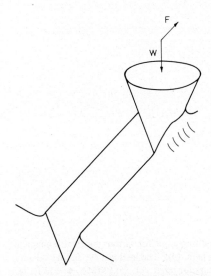

Fig. 6. The plastic ploughing caused by a rigid cone (after Archard).

friction is increased (Fig. 5). In this regime, there is often a correlation between friction and wear.

2.4. Friction which involves appreciable plastic flow

This problem was originally treated at a simple but illuminating level by Greenwood and Tabor [34] for a number of contact geometries. Fig. 6 shows an example where a rigid cone moves through the surface of a ductile solid. For this case, assuming that no material builds up around the cone, the coefficient of friction is given by

$$\mu = \frac{2}{\pi} \cot \theta \qquad (4)$$

This is the solution for eqn. (3) when ϕ is equal to the total frictional work. This relationship is a surprisingly accurate description of experiments where μ is measured as a function of θ (Fig. 7) [47]. These particular data were obtained at a fixed temperature, load and sliding velocity without surface lubrication. Changes in these variables affects certain parts of the curves, most notably the transition from ploughing to cutting (see Fig. 7). At this point, as $\cot \theta$ increases (for reducing θ, sharper cones), the increase in μ with $\cot \theta$ is slowed. For example, at higher loads this transition occurs at a larger critical cone angle. However, the central linear region remains essentially unchanged as load and velocity are varied. In contrast to elastomers, surface rupture, in this case cutting, requires less frictional work. This effect is thought to arise from a change in manner in which the deformed polymer in the prow regions is displaced by the cone.

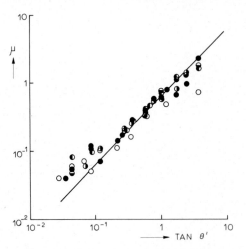

Fig. 7. The coefficient of friction, μ, as a function of θ', the cutting angle for various cones of semiapical angle θ. θ' is $(\pi/2) - \theta$. The solid line has a gradient of $2/\pi$ [eqn. (4)]. The load is 1 N; sliding velocity 4.2×10^{-5} m s^{-1} and the contact is unlubricated. Cutting occurs for $\tan \theta' > 1$. ○, PMMA; ●, PTFE; ◑, PC; ◔, PPR; ◐, EPOXY.

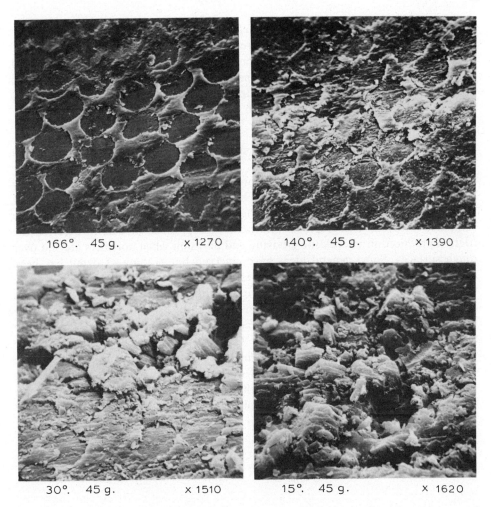

166°. 45 g.   × 1270   140°. 45 g.   × 1390

30°. 45 g.   × 1510   15°. 45 g.   × 1620

Fig. 8. Micrographs of the surface damage produced by various hard cones on the surface of an epoxy/I carbon fibre composite. The complex nature of the damage process is evident. Courtesy of Dr. J.K. Lancaster.

No comprehensive friction experiments of this kind have been carried out on composites, but one can imagine the type of microscopic behaviour that might occur in both particle- and fibre-filled matrices. It will be an extremely complex process where the indentor periodically engages regions of different mechanical properties. Figure 8 shows micrographs kindly provided by Dr. J. Lancaster where cones (Fig. 7) have been slid over the surface of a type I carbon-filled epoxy resin [48]. The fibres are normal to the surface, which had been previously polished in water. No friction data are available. The appearance of the damaged surface indicates that the cones engage with the fibres. The matrix is deformed and the fibres are bent, broken and even pulled out of the matrix. The frictional work will therefore be dissipated by many processes such as matrix flow, fibre rupture and bending and the parting of the fibre–matrix interface. The relative contributions from the various processes will be a function the size, orientation and density of the fibres as well as the mechanical properties of the fibre and the matrix. Interestingly, this type of friction process has features which are evident in the Coulomb model of friction (Sect. 4.2) [20].

2.5. Ploughing friction generated by rough surfaces and the relationship between friction and wear in abrasion

The characterisation of the microgeometry of rough surfaces using profilometry has advanced greatly in recent years [42,44–51]. It is therefore resonable to ask whether the ideas produced by the investigations into single model asperities can be combined with this topographical information to produce models to describe the friction produced in viscoelastic grooving and abrasion when polymers are slid over rough surfaces. Some general ideas have emerged but nothing of detail has been produced. Grosh's early work on elastomer friction on rough substrates was used by the author to calculate a characteristic length for the asperity contacts [52]. It is also evident that the greater the stress levels produced at the asperities, the greater will be the contribution that each interaction makes to the overall frictional work; more tearing or plastic work will be produced. Often, it is the changing geometry of the polymer surface by debris deposition which precludes any detailed analysis. The evaluation of the origins of the frictional work in model asperity contacts does not encourage a belief that we can predict the friction in multiple contacts. We may, however, conclude that there may be some correlation between the friction and the

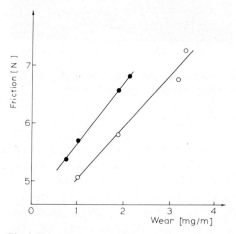

Fig. 9. Friction against wear rate during abrasion for a poly(ether–ether ketone) when the polymer is slid against range of abrasive papers under a constant load of 22 N. Closed circles are for silicon carbode papers and open circles for aluminium oxide papers. The data fall on two lines depending upon the type of particle used to construct the paper. The greater the wear, the higher is the friction.

rate of wear in these cases. This is sometimes seen in practice when unfilled ductile polymers are slid over abrasive papers. Figure 9 is an example of the type of correlation sometimes observed between abrasion rate and friction [53]. Similar correlations have been produced for composite systems.

3. Interfacial or adhesion friction

3.1. General notions: $f_a = A\tau$

Returning to Fig. 1 reintroduces the adhesion model of friction. The stresses are now transmitted by surface or adhesive forces. These forces operate at very short range, less than 0.1 nm, and are hence only effective at real areas of contact where the repulsive forces of the molecules in the interface interact. This is an important difference between the two modes of friction; in the ploughing component it is the geometry of the contact, which is related to the apparent area, which is important.

The adhesion component can be expressed in a deceptively simple equation [3,4,14]

$$f_a = A\tau \tag{5}$$

where f_a is the interfacial friction contribution, expressed either as a force or a work per unit sliding distance, and A is the real area of contact. The parameter τ is a specific shear stress or specific shear work [54–60]. It is common to regard τ and A as non-interacting parameters, although we know that this is not the case [5,61–63]. Like the separation of the adhesive and ploughing terms, this is a useful, but sometimes dangerous, expedient. The approach here will be to consider each term separately and then to review the problems associated with this method of analysis. It is convenient to discuss τ first.

3.2. The parameter τ; general characteristics

Figure 10 illustrates the significance of τ and also indicates the experimental technique whereby its numerical value can be obtained by investigating the frictional characteristics of their polymeric films. A number of ideas need to be introduced. These are (i) the significance of τ, (ii) its variation with contact parameters, and (iii) the interpretation of τ in terms of precedents in the study of bulk rheology.

3.2.1. The significance of τ

The measurement of friction alone in the contact configuration shown in Fig. 10(a) may provide an approximate numerical value for τ (see later) but it does not define the thickness of the interfacial region in which most of the friction work is dissipated. Nor does it provide information on the nature of the dissipation process. As we shall see, if our only interest is the magnitude of the friction, careful consideration of these points is not necessary. An interpretation of the significance of τ does, however, require a brief discussion of both points.

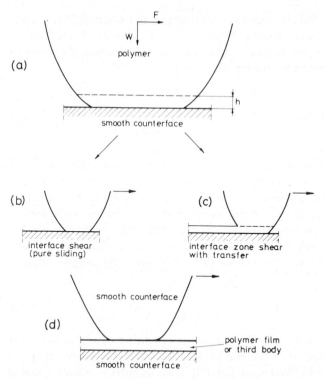

Fig. 10. Models of the interfacial or adhesion model of friction. In (a) is shown the general case where the energy is primarily dissipated in a zone of thickness h. The sections (b) and (c) show the case of pure interface sliding and interface zone shear, respectively. A sample friction experiment cannot distinguish (b) and (c), but transfer films indicate that (c) occurs (Table 3). The technique commonly used to measure τ, the interface shear stress is shown in (d).

Considering first the case of static friction: the force required to initiate motion. The picture given in Fig. 10(a) suggests that τ is a rupture stress or work for cohesive [Fig. 10(c)] or interfacial [Fig. 10(b)] mode II fracture. Some contacts, for example poly(methylmethacrylate) (PMMA), appear to fail at the interface. Other polymers, usually semi-crystalline polymers such as poly(ethylenes) and poly(tetrafluoroethylene) above their glass transitions, fail in a cohesive way and deposit transfer films on the counterface [14,15,64] (Table 3). Studies on the film thickness dependence of τ for organic coatings (see later) indicates that the shear work is almost completely dissipated in interface thicknesses of ca. 100 nm for all cases that have been studied.

Once sliding has commenced, the significance of τ is less certain. It could be argued that on rough surfaces the asperity contacts are continually being formed and lost in the way suggested by Barber [65]. τ then refers to a stress-modified interface region. If this is not the case and the contract region is in continuous shear, it is usual to consider τ as a measure of an "interfacial viscosity".

TABLE 2

τ_0(bulk)/τ_0(thin film) and α(bulk)/α(thin film) values for various polymers

Quantities in parentheses are arithmetic means. The overall average values of α(bulk)/α(thin film) and τ_0(bulk)/τ_0(thin film) are 0.9 and 11.3, respectively.

Polymer	$\dfrac{\tau_0 \text{(bulk)}}{\tau_0 \text{(thin film)}}$	$\dfrac{\alpha \text{(bulk)}}{\alpha \text{(thin film)}}$
PMMA	4.6–53.0 (28.8)	0.5–2.9 (1.7)
PS	10.0–31.0 (20.05)	0.4–1.7 (1.05)
PET	3.1–11.6 (7.35)	0.3–1.0 (0.65)
PC	3.5– 8.4 (5.95)	0.3–1.3 (0.8)
PP	3.2– 4.0 (3.6)	0.7–1.2 (0.95)
HDPE	7.0–14.0 (10.5)	0.3–0.7 (0.5)
PVAC	2.9　　(2.9)	1.5

As far as composites are concerned, the important feature is that the work is initially dissipated in a very thin interface region. Ultimately, the majority of this work is distributed throughout the system as heat. This is effectively a description of boundary lubrication where the weak layer is sheared between hard substrates. The numerical values of τ are usually measured by this technique [Fig. 10(d)]. The techniques for measuring τ have been described elsewhere; they involve measuring the friction of thin films on rigid substrates [58–60]. A measurement or estimate of the contact area is required and hence this usually necessitates the use of smooth elastic substrates. The next subsection reviews how τ changes with the significant contact parameters.

3.2.2. The variation of τ with contact parameters

There are many published studies on this topic [54–61]. The important variables are contact pressure [54], temperature [57], the rate of interfacial strain [66], effective contact time [66], and the chemistry of the environment [60]. A particular polymer may also show effects associated with its thermal or strain history [60].

The details are not important and it is sufficient to record that realistic values of τ can be obtained. As an example, we may examine the influence of contact pressure, P, on τ. Figure 11 gives experimental data obtained for a range of purified polymers. To a very good approximation in these cases

$$\tau = \tau_0 + \alpha P \tag{6}$$

where α and τ_0 are constant for a given polymer when other contact variables are fixed. Using eqn. (5) and noting that $P = W/A$, where W is the normal load for large loads

$$\mu = \frac{\tau_0}{P} + \alpha \tag{7}$$

μ being the coefficient of friction. Often, $\alpha > \tau_0/P$ and hence $\alpha \approx \mu$. For many

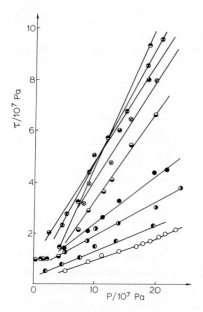

Fig. 11. The interface shear stress, τ, as a function of mean contact pressure, P, for various polymeric films measured using the configuration shown in Fig. 10(d). The temperature and sliding velocity are fixed. ◐, HDPE; ◑, LDPE; ○, PTFE; ●, PP; ◒, nylon 6; ◓, PET; ◔, PMMA; ◖, PS; ⊗, PVC.

polymers, α ranges from 0.1 to 0.4 and we have another explanation of Ammonton's laws of friction [20]. Equation (7) may be tested using the configurations shown in Fig. 10. Table 3 shows the calculated values of μ and data obtained from sliding polymers on themselves and on glass. The contact pressure has been equated with

TABLE 3

Values of the parameters in eqn. (7) and a comparison of calculated and measured coefficients of friction [a]

Polymer	P_0 (Pa)	τ_0 (Pa)	α	$\mu_c = (\tau_0/P_0)+\alpha$	μ_b	μ_g	Polymer transfer to glass 20°C
LDPE	1.53×10^7	6.0×10^6	0.14	0.53	0.52	0.42	Yes
HDPE	3.90×10^7	2.5×10^6	0.10	0.53	0.16	0.08	Yes
PTFE	2.27×10^7	1.0×10^6	0.08	0.12	0.16	0.13	Yes
PP	4.75×10^7	5.0×10^6	0.17	0.27	0.26	0.26	No
PMMA	2.68×10^8	1.0×10^7	0.36	0.39	0.36	0.41	No
PVC	1.01×10^8	-9.0×10^6	0.57	0.46	0.55	0.54	
PS	1.39×10	4.0×10	0.45	0.48	0.42	0.45	No

[a] $P_0 = 20$ s Vickers hardness; τ_0, α eqn. (7); μ_g = coefficient of friction against glass; μ_b = coefficient of friction against itself; μ_c = coefficient of friction calculated from eqn. (7).

the twenty second pyramidal hardness [59]. The agreement is good, which indicates the basic assumptions in the adhesion model of friction.

3.2.3. The interpretation of τ in terms of bulk rheological precedents

The study of the rheology of thin polymeric films provides a number of empirical equations to describe τ as a function of mean contact pressure, P, sliding velocity, V, contact time, t_c, and temperature, T. An empirical relationship which combines these variables is [56,57]

$$\tau = \tau_0' \ln(V/hv_1) + \tau_0'' \exp\{(Q/RT)\} + \alpha_0 \exp\{(V/dv_2)P\} \tag{8}$$

where τ_0', τ_0'', θ, α_0, v_1, v_2 are material constants. The parameters τ_0', τ_0'', and α_0 are not generally strong functions of the variables T, P, V or t_c. The functionality of v_1 and v_2 have not been fully explored, but the indications are that they are functions of these variables. The quantities h and d are related to the film thickness and the contact length. The terms V/h and V/d express the time variables as rate of strain and contact frequency. Equation (8) is a more explicit form of eqn. (6).

$$\tau = \tau_0 + \alpha P \tag{6}$$

The term $\alpha_0 \exp(V/dv_1)$ replaces α and accounts for viscoelastic retardation in compression [66]. τ_0 is expanded to $\tau_0'[\ln(V/hv_2) + \tau_0'' \exp(Q/RT)]$ to include the influence of the interfacial shear strain rate and the temperature [57,58].

The form of eqn. (8) is similar to the equations developed to describe the rheology or rupture of bulk polymers [67]. For example, eqn. (6) is a good description of the variation of the shear yield stress of bulk polymers as a function of hydrostatic stress. Table 2 shows a comparison of the values of α and τ_0 obtained in thin film shear, measured in the experiment depicted in Fig. 9, and the comparable values measured for bulk isotropic shear deformation under hydrostatic stress [68]. On average, τ_0 for the bulk is about ten times greater than for the corresponding film. The α values are similar. When making this type of comparison, the differences in deformation conditions between the two experiments should be borne in mind. In particular, the fact that the rates of strain and overall strains are very different (Table 1). The bulk deformation experiments were carried out at ca. 10^{-3} s^{-1} strain rate, whilst the film experiments correspond to nominal rates of strain of about 10^3 s^{-1}. The film experimental technique also produces a transient quasi hydrostatic stress.

A broad general conclusion that may be drawn from this comparison is that the processes responsible for the dissipation of the interfacial friction work in polymeric contacts are similar to those which exist in bulk flow. The conditions which exist in contacts are, however, not generally reproduced in macroscopic deformations. A similar conclusion is drawn from the observed variation of τ with t_c, T and V [60,66].

It is always desirable to offer microscopic interpretation of physical properties and some attempt has been made to do this with the quantity τ [17,57,60,69,70]. In view of the similarity between τ and the bulk analogue, it is obvious that these approaches will follow the precedents set in the analysis of bulk flow or yield. This is a difficult area, even the viscosity of simple fluids cannot be accurately predicted. Currently, there are several groups working on computer simulations of flow using molecular dynamic models [71]. A simpler but less discriminating approach is to use a modification of the Eyring stress-modified thermally activated microscopic flow model [5,17,60,69,70,72]. There have been several attempts to apply this model to thin polymeric films undergoing interface shear. The Eyring model can, for example, provide a simple but crude molecular interpretation of the term α_0 in eqn. (8) [α in eqn. (6)]. Amuzu et al. [69] have interpreted the friction of a series of poly(n-alkyl methacrylates) by considering the dimensions of the molecular side groups that might interact during shear using a modified Eyring model.

3.2.4. The real area of contact

3.2.4.1. General comments. The measurement or the accurate prediction of the real area of contact produced by the contacting of two rough solids is problematical, if not impossible. Direct measurement is impracticable and predictions are uncertain. The early analytical models used by Bowden and Tabor for plastic contacts [3,4,61] and the models developed later by Lodge and Howel [73] and Archard [74,75] for elastic contacts probably give realistic first-order estimates for pure phases. New approaches using bearing area curves offer interesting possibilities [76]. The problem is aggravated in multiphase systems. Bowden and Tabor and Archard were also concerned with metals where the material properties which define the contact area are relatively insensitive to time and temperature. With polymers, however, the extent of junction growth (or contraction for elastomers) does not seem as important as it is with metals [80–83].

3.2.4.2. Simple contact area models and their value in the computation of friction. The most effective approach is to use the Bowden and Tabor approximation which appears to be valid at high loads. Greenwood [84] has reviewed the range of contact stress produced at asperities and concludes that at high loads, where plastic deformation occurs, the contact stresses are close to the plastic flow stress, p_0. We can then compute the friction as

$$f = \frac{W}{P_0}\left(\tau_0 + \alpha P_0\right) \tag{9a}$$

$$= \frac{W\tau_0}{P_0} + \alpha W$$

Table 3 exemplifies this approach for the calculation of the coefficient of friction.

Equation (9a) also illustrates the influence of load on the frictional force. As W decreases, f decreases almost in proportion. The trend in the coefficient of friction, μ, is shown in Fig. 12 for a series of explosives [77]. At high loads, μ tends to α whilst at lower loads, the value of α becomes infinite. In the low load regime, the

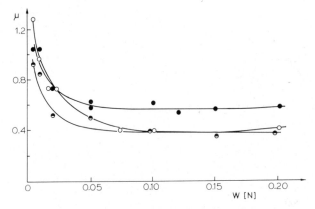

Fig. 12. The coefficient of friction in slow speed sliding of single crystals of explosives on smooth clean glass plates as a function of load. The data are generally reversible. Equation (7) explains the trend if P is an increasing function of load. ●, Cyclotetramethylene tetranitramine; ○, pentaerythritol tetranitrate; ◐, cyclotrimethylene trinitramine.

average contact pressure is significantly less than p_0. The asperities are deforming in an elastic manner. Actually, even at high loads, the asperities may rapidly achieve a state where they are deforming elastically but the average contact pressure is close to the plastic flow stress. This was Archard's view for metals. Multiple asperity elastic contacts were modelled by Archard and others to show that

$$A = K'W^{n'} \tag{9b}$$

where n' ranges from 2/3 to 1. If the number of asperities remains constant, then n' is 2/3 [75]. When the average contact area per asperity is constant, n is unity. The latter is the most common case. At high loads, K' is apparently close to p_0.

A number of studies [9,78–81] have examined the friction index in

$$f = K'W^n \tag{10}$$

In general, $n > n'$, which reflects the pressure dependence of τ [80,82]. These studies have usually assumed that the real area of contact is equal to the apparent contact area and applied the Hertz equation. A similar approach, which assumes a constant, but rather small, number of asperities which deform elastically, has also been used on starch compacts [83]. Figure 13 is an example where a poly(styrene) hemisphere is slid over a smooth glass plate at low speeds [82]. At lower loads, $n = 0.88$, but at high loads, n tends to unity. The approximate value of the load required to induce irreversible flow for a smooth contact is shown as W^*. If we take into account the pressure dependence of τ, we may compute n' by assuming that eqn. (6) applies. For these data, $n' > 2/3 \approx 0.88$. An unambiguous calculation of the numerical value of the friction from eqn. (6) or eqn. (7) is not possible; the fact that plastic flow is detected at the correct load for a smooth contact is only indirect evidence. The

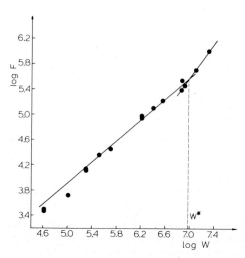

Fig. 13. The friction of a smooth poly(styrene) sphere on a clean glass plate as a function of load. The data fit eqn. (10) and the load index, n, is ca. 0.88 up to W^*, when gross plastic flow occurs, after which n is ca. 1. The result confirms an elastically deforming interface where the real area of contact is proportional to $W^{2/3}$. The fact that $n > 2/3$ is strong evidence that τ is a function of contact pressure. Temperature 20°C; sliding velocity 0.2 mm s^{-1}.

essence of the problem is that an underestimate of A provides an overestimate of τ. The product τA is thus rather insensitive to the magnitude of A; this fact may explain the quality of the agreement seen in Table 3.

The outline of the types of analysis which may be used to compute A, and hence f, given above can only be regarded as crude, but nevertheless interesting exercises. The whole problem of the appropriate time-dependent mechanical properties which are needed to compute A has been avoided. The comparison in Table 2 uses the 20 second pyramidal hardness to compute p_0. In the analysis cited for poly(styrene) in Fig. 13, values of the Young's modulus and the Poisson's ratio are required; the small strain values for 2 hertz were used. We have also neglected retardation in compression in τ. In addition, it was assumed that τ and A are not inter-related, i.e. there is no junction growth. In many practical systems, we would also have to consider the influence of the surface heating generated by the frictional work. The data shown in Fig. 14 indicate the nature of this problem [84]. In this experiment, the surface heating is generated by external means; a heated steel wedge is slid over the surface of the polymer at low speed (the wedge contact line is in the sliding direction). The polymer is a coating of ca. 2 mm on a steel plate. The same type of data may be obtained by studying the friction as a function of velocity at ambient temperature. These data are for poly(ether–ether ketone) but they are representative of the behaviour of many polymers in selected temperature ranges. The data may be rationalised as follows. For modest surface temperatures, local surface heating reduces τ [58], but the subsurface deformation characteristics of the polymer are unaffected and hence A remains unchanged; the friction decreases. At higher surface

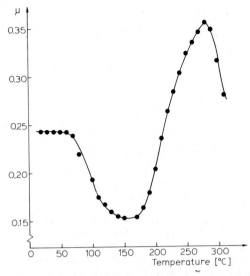

Fig. 14. The friction coefficient of a contact between a steel ball and a flat specimen of poly(ether–ether ketone). The contact is unlubricated; load 10 N; diameter of ball 6.5 mm; sliding velocity 0.45×10^{-3} m s^{-1}. The ball is heated to the temperature shown to simulate frictional heating.

temperatures, subsurface softening becomes important and the contact area begins to increase. A simple treatment predicts that $\tau(T) \times A(T)$ $(=f)$ would attain a constant value at high temperature. The data shown indicate that this occurs over a brief temperature range, but is followed by a rapid increase in frictional force; above 200°C, there is indirect evidence of appreciable junction growth at the asperities. We may therefore ascribe the rapid increase in friction to a disproportionate increase in the real contact area. The decrease in the friction at very high temperature is probably an artifact. At this point, the system has reached the stage where a very thin layer of polymer is being sheared between two hard substrates. The contact area is now defined by the mechanical and surface properties of the wedge and the steel substrate which supports the polymer. The contact pressures are much higher and eqn. (7) would predict a decrease in friction coefficient.

3.2.5. The prediction of interfacial friction in composite systems

3.2.5.1. General summary of ideas on interfacial friction. A number of ideas emerge from the study of interfacial friction in single component systems, viz.

(i) The friction may be regarded as the product of a real area of contact, A, and an interfacial rheological parameter which is designed as τ.

(ii) τ has the characteristics of a bulk flow parameter, but it is associated with a very thin interfacial region. The magnitude of τ depends on the properties of this interface region and the contact conditions.

(iii) The real area of contact is defined mainly by the surface topography of the bodies and the subsurface mechanical properties.

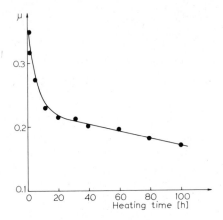

Fig. 15. The friction of a steel ball on a plate of 0.5% stearamide filled with low density poly(ethylene). Initially, after cleaning, the friction of the steel ball on the surface is high, ca. 0.35 coefficient of friction. With time (at 40°C) the friction slowly decreases until a value characteristic of an externally applied stearamide coating is obtained. Separate experiments show that the amide has diffused to the free surface to form a crystalline layer. The rate of friction decrease is a function of the heating temperature, the sample history, the amide nature and concentration, and the topography of the polymer surface.

There is potential for an extended amplication of these ideas in the context of composites, but we will restrict ourselves to three examples; self-lubricating poly(ethylenes), particle-filled PTFE, and polymer–solid lubricant systems.

3.2.5.2. A surface segregating self-lubricating system; amides–low density poly(ethylene).   It is a common commercial practice to incorporate small quantities of low molecular weight additives into organic polymers in the expectation that these materials will ultimately modify the polymer surface [85–90]. Waxes are added to rubbers in order to produce a surface layer to suppress ozone attack [90]. Amides are often incorporated into poly(ethylene) to reduce both the friction and autoadhesion [89].

The data presented in Fig. 15 introduce the important features [90]. The low density poly(ethylene) was fabricated with ca. 0.5% of oleamide in the melt and then quenched. The surface was then cleaned by abrasion in water and a steel hemisphere was slid over the surface and the friction recorded. The specimen was then heated at 40°C for various times and the friction measured after each heating period at ambient temperature. The frictional force decreases with time until an equilibrium value is attained. Separate studies indicate that the amide has diffused to the free surface to provide a lubricating layer [87,91,92]. The rate of the friction decrease is a function of temperature and the topography of the polymer surface. It appears that the amide forms a molecular dispersion in the melt but is precipitated from the matrix on cooling. The stored elastic strains surrounding these micro inclusions probably provides the driving force for free surface segregation [90].

A simple analysis based on eqn. (7) is adequate to explain the trend in the data. The friction work is dissipated almost entirely in the interface region, which is ultimately a thin oleamide layer. The values of τ_0 and α are significantly less for the amide than the poly(ethylene); each parameter differs by about a factor of four [56]. The contact pressure, P_0, is unchanged, however. We predict a change in friction coefficient of ca. four, which is close to the observed reduction. A treatment of the behaviour at shorter times and lower coverage may be made using the two-zone model of interfacial friction presented by Bowden and Tabor for metals [3,4,93]. The important experimental point is that a higher surface concentration is required on rough surfaces in order to ensure complete coverage of the asperities. The equilibrium surface concentration is a function of the bulk concentration and the appearance rate is a strong function of temperature.

These results stress the importance of the interface in polymeric systems. Small amounts of low molecular weight species can have a marked effect. This is also sometimes the case in what are termed immobile lubricants such as certain silicone fluids in poly(styrene) [94]. Similar effects are seen with the creeping lubrication provided by hydrocarbons in certain polymers [95]. The low molecular weight species may be mobile and act like the amide–poly(ethylene) system. However, at high molecular weights, the fluid separates in the matrix as microdrops of a few microns in diameter at ambient. This material is virtually immobile but may enter the interface region as pockets of oil are breached during the wear of the system. Also, depending upon the fabrication procedure, a surface excess of oil may be present on the free surface after moulding. We will not dicuss solid lubricants, such as MoS_2 and graphite under this heading, although the principles are similar to those involved with immobile fluid lubricants.

3.2.5.3. PTFE–particulate filler systems. PTFE is a popular matrix or filler in the formation of composites for bearing applications because it produces a low friction. The values of τ_0 and α for the oriented material are low; they are comparable with oleamide, which is a fraction of the molecular weight of PTFE. The low friction is accompanied by high transfer wear, but fillers can effectively attenuate this mode of wear. There are many subtleties in the tribology of PTFE, but they are not all relevant here [96]. The main aspect of importance as far as the friction is concerned is that, during continuous sliding in clean contacts, the surface regions of the polymer becomes highly oriented and, as a consequence, shows appreciable interfacial strain softening [14,15,64,97]. The wear debris formed by the transfer process provides a very effective lubricant. A stylised pictured of a PTFE matrix composite is shown in Fig. 16. Each asperity resembles the micro contacts depicted in Fig. 16. This contact cannot be modelled with any accuracy for obvious reasons. The micro contact geometry and surface concentration of the fillers is not known and, in any event, the filler particles will actually modify the counterface topography. We can apply eqn. (7) to each asperity region after estimating the appropriate value of p and then sum the contributions. The contact stress at filler contacts will be high, but their total area is small. In unfilled regions, the contact area is greater but the contact stress proportionally decreases. On average, the friction per contact area is roughly constant or falls in a narrow range. We do not expect, therefore, that the friction will be a strong function of the nature or concentration of the filler particles

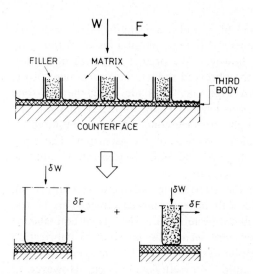

Fig. 16. A schematic model of the action of a third body at a composite/counterface interface. The frictional work is dissipated mainly in the third body. The specific frictional work is τ and hence each unit of contact δA contributes to the friction $\delta F = \tau \delta A$. τ is a function of contact pressure and temperature. Considering contact pressure only, from eqn. (6)

$$\delta F = \delta A \left(\tau_0 + \alpha \frac{\delta W}{\delta A} \right)$$

$$\Sigma \delta F = \tau_0 \Sigma \delta A + \alpha \Sigma \delta W$$

$$F = \tau_0 A + \alpha W$$

and

$$\mu = \frac{\tau_0}{P_m} + \alpha$$

and P_m is the mean contact stress. For PTFE, $\alpha \approx 0.08$ and $\tau_0 \approx 10^6$ Pa (Table 3) and for virgin PTFE, μ is ca. 0.12 when $P_m = P_0 \approx 2.3 \times 10^7$ Pa. Table 4 shows data for filled systems. The incorporation of hard fillers on hard substrates will increase P_m somewhat, although the load transmitted to the particles will be controlled by the value of P_0 for the PTFE. The minimum value of μ is $\alpha \approx 0.08$. This model assumes that the interfacial rheology of the third body is the same as undergraded oriented PTFE.

or indeed the counterface. This is found to be the case and any variations could reasonably be attributed to topographical changes on the counterface. Table 4 gives examples. The friction is generally an increasing function of counterface surface roughness beyond 0.2 μm c.l.a. for PTFE [64,98]. Higher surface roughness will certainly produce some cutting, which is more energy intensive as well as possibly inhibiting interfacial shear softening [64]. The data listed in Table 4 were obtained at rather high sliding speeds (5 m s^{-1}) and some account should be taken of thermal heating in the interface and subsurface (Sect. 3.2.4.2). The variation of μ for PTFE with temperature is well documented, but its variation with interface shear strain

TABLE 4A

The friction coefficients of various graphite–PTFE based composites measured in a pin or disc configuration. Also tabulated are the wear rates, thermal conductivity, and an indication of the abrasive nature of the filler

Wear path radius, 80 mm; velocity, 5 m s^{-1}; normal load, 40 N (35.0×10^4 N m^{-2} apparent contact pressure); bulk counterface temperature, $21 \pm 2°C$; counterface, mild steel (0.02 μm CLA); k_{50} = thermal conductivity at ca. 50°C.

Sample	Wt.% filler	Wear rate (mg h^{-1})	Coeff. of friction ± 0.03 (average value)	Counter roughness/μm CLA		k_{50} (W m^{-1} K^{-1})
				Initial	Final	
PTFE		3×10^2	0.12	0.02	0.02	0.26
Oleophillic	10	32.0	0.14	0.02	0.03	0.30
	20	9.1	0.14	0.02	0.04	0.40
	30	29.3	0.11	0.02	0.04	0.52
Polar	10	7.3	0.15	0.02	0.04	0.27
	20	1.5	0.14	0.03	0.05	0.30
	30	5.8	0.14	0.02	0.04	0.33
Nuclear	10	15.9	0.10	0.02	0.06	0.33
	20	1.0	0.12	0.02	0.05	0.39
	30	2.8	0.12	0.02	0.06	0.53

TABLE 4B

Friction coefficient for three PTFE composites on various substrates
Sliding conditions as for Table 4A.

Composite	Counterface	H_v	μ ± 0.02	Wear rate (mg h^{-1})	Counterface roughness/μm CLA	
					Initial	Final
10% ca. 2 μm alumina	Lead bronze	111	0.1	1.16	0.33	0.25
	Mild steel	118	0.09	2.95	0.31	0.29
	18/8 Stainless steel En 58 M Crotorite V	165	0.09	2.52	0.32	0.35
	10% Al bronze	168	0.09	2.92	0.36	0.27
	Grey cast iron	205	0.09	2.54	0.28	0.26
	FV 520 Stainless steel (martenistic)	330	0.09	4.62	0.42	0.32
	Glass		0.09		2.88	0.050.17
10% Mica	Lead bronze	113	0.09	0.35	0.32	0.29
	Mild steel	121	0.11	2.59	0.25	0.25
	18/8 Stainless steel En 58M Crotovite	167	0.09	5.83	0.31	0.28
	10% Al bronze	167	0.08	2.89	0.35	0.28
	Grey cast iron	203	0.10	1.67	0.28	0.26
Glass			0.09	1.09	0.05	0.09
5% Oleophillic graphite + 5% alumina	Mild steel	125	0.10	52.7	0.29	0.25

TABLE 4C

Friction coefficients for 10% glass-filled PTFE on mild steel. Also shown are wear rates and the initial and final values of the counterface roughness
Sliding conditions as for Table 4A.

Composite	Average coeff. of friction ±0.02	Wear rate (mg h^{-1})	Counterface roughness/μm CLA	
			Initial	Final
4–44 μm diam.	0.12	0.60	0.02	0.18
glass spheres	0.13	1.08	0.27	0.10
	0.12	0.96	0.56	0.17
	0.10	0.69	0.70	0.38
	0.12	1.00	0.94	0.33
	0.11	1.10	1.44	0.30
200–250 μm	0.10	2.96	0.03	0.38
diam. glass	0.09	2.43	0.46	0.23
spheres	0.10	3.52	0.63	0.21
	0.10	4.63	0.92	0.34
	0.11	3.80	1.24	0.38
350–400 μm	0.10	4.60	0.08	0.13
diam. glass	0.10	6.85	0.20	0.33
spheres	0.10	8.31	0.43	>1.5
	0.11	8.34	0.72	>1.5
	0.14	9.21	0.20	>1.5

rate has not been studied in detail [58,64]. At present, it is thought that the friction coefficients given in Table 4 are rather high (by a factor of ca. two) for a model which involves only localised surface heating. If we suppose that some bulk heating occurs and, as a consequence, p is reduced by about two to three, a reasonable agreement is obtained. This would require a temperature increase in the subsurface of perhaps 20°C, which is reasonable. We should also note that certain fillers will marginally improve the thermal conductivity of the matrix. Oleophilic graphite at 30 wt.% will double the thermal conductivity of PTFE [27,99] (Table 4). This change in conductivity may be critical for certain systems such as PTFE where the mechanical properties which control A are highly temperature-dependent. The filler particles themselves will generally improve the strength and creep resistance of the matrix and hence have the effect of minimising the contact area.

3.2.5.4. Polymers filled with PTFE. PTFE is an interesting filler for high temperature thermoplastic polymers and strong cross-linked resins. It acts much like the silicone fluid in poly(styrene). As the matrix wears, the PTFE transfers to the two ports of the contacting surfaces. Several authors term this layer the "third body" or "la troisième corps". The epoxy-impregnated PTFE–carbon fibre weaves used in airframe bearings seem to rely for their action upon this layer [100]. Figure 17 shows data for a series of PEEK–PTFE composites; the presence of relatively small amounts of PTFE in the PEEK is sufficient to effect a significant reduction in the

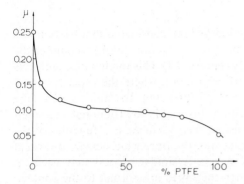

Fig. 17. The friction coefficient of a PTFE/poly(ether–ether ketone) composite as a function of weight percent of PTFE. Data obtained using a steel ball (6.25 mm diam.) sliding over a plate at 20°C under a load of 10 N. A relatively modest bulk concentration of PTFE is sufficient to ensure a surface excess of PTFE and hence to attenuate the friction of the polyketone.

frictional force. The analysis presented previously is appropriate as a first-order rationalisation of these data.

3.2.5.5. Concluding comments on interfacial friction in composites. The discussions given above are essentially an interpretation of the model of boundary lubrication developed by Bowden and Tabor for lubricated metals. The classical calcium stearate is replaced by a "third body" and one of the metals by a composite. In some cases, for example with amides or PTFE, the surface rheology may be defined with some confidence. There are unknowns such as surface temperature and several problems in the definition of contact area and average contact pressures. Similarly, the appropriate mechanical properties of the matrix are uncertain. However, a crude model based on the classical metal boundary lubrication picture provides a rationalisation of data with a limited predictive capacity. Unfortunately, there are systems where even this limited exercise is not possible. This is the case in brakes, for example, where very significant surface chemical degradation occurs [101]. Here, we do not even known the composition of the third body, let alone its rheology. In this area, it is probably easier to account for wear than friction. Lui and Rhee [102], for example, has used a modification of the Quinn oxidative wear model [103] to describe the high temperature wear of brake compounds. In this field, friction equations abound [104], but none has a strong physical basis.

4. Other models of friction

4.1. Introduction

In a general discussion of friction in organic polymers, it is appropriate to mention two friction models which are not really included in the two-term non-interacting model. They are the Coulombic and Schallamach wave friction processes.

4.2. Coulombic friction

Coulomb, and, indeed, Leonado da Vinci before him, considered that friction was due to rigid asperity engagements. Dowson has provided a fascinating account of the history of friction up to the early part of this century [20]. This ratchet mechanism is not popular today save in the area of interfibre friction where the asperity size is often comparable with the diameter of the fibres [105,106]. In these systems, the sliding motion is highly discontinuous in nature. Invariably the frictional character- istics of natural fibres, which contain cuticles on their surfaces, is a function of the sliding direction. In one direction, the cuticles trap the fibre which move towards it but in the other direction, it simply slides over it. A similar process may occur in fibre-reinforced composites, particularly when the fibres are normal to the contact. This point was introduced earlier (Sect. 2.4). Coulomb, who was a marine engineer, developed his ideas from the study of the friction of wood on metals for the control of marine slideways. The pictures produced by Coulomb for his ratchet model recognised the fibrilar nature of wood and, indeed, they resemble some of the sections used to illustrate the structure of fibrous composites. There are no realistic models for Coulombic friction which can be usefully used for composite interfaces; the old models are not dissipative, for example [20]. Nevertheless, the basic idea is appealing, not only for fibre composites but also for particulate systems.

4.3. Schallamach friction

There was a brief period spanning about ten years and beginning in about 1970 when several authors actively studied what have now become known as Schallamach waves or waves of detachment [107–110]. Schallamach observed that, with certain elastomers, when they were slid over glass, true sliding did not occur. Instead, relative motion was achieved by the formation and propagation of macroscopic dislocations. Later studies showed that the frictional work could be accounted for by the nett peeling working work required to propagate the dislocation [108,109]. If the work required to peel and form a unit of contact area at a velocity v is $\Lambda(v)$, then the rate of friction work is

$$V = \Lambda(v) As$$

where V is the imposed velocity, s the wave frequency and A the contact area [112]. These rubbers were particularly soft and the apparent contact area is a good approximation to the real area of contact [111]. The quantity $\Lambda(v)$ was measured in a conventional peel test at a peel velocity v and in a Schallamach wave experiment where the average wave velocity was v. The agreement between the two measure- ments was good [108,109].

At present we do not know how general is the phenomenon of Schallamach waves. Simply theory suggests that the material must be able to accommodate a unit strain reversibly and also generate a traction of the order of the Young's modulus [109]. This would restrict the process to quite smooth elastomers sliding on clean substrates. There is no reason why this should not occur with some soft elastomeric composites.

The prediction of the absolute value of the Schallamach friction is not possible at present, even if $\Lambda(v)$ is available.

5. Lubricated systems

5.1. Introduction

The regimes of lubrication have been described by several authors [16,20,72,113]. In this section, we review briefly the boundary lubrication of polymers where the lubricant film thickness is predominantly of the order of molecular dimensions. There are several good reviews of the hydrodynamic and elastohydrodynamic lubrication of polymers, particularly elastomers [10,113,114]. Two topics will be discussed; lubricant plastisation and the importance of wetting forces.

5.2. Lubricant plastisation

Rubenstein [115] recognised that some of the efficiency of nylon 6:6 could be attributed to the fact that the *surface* of the polymer was plasticised. The effect of plastisation is like the increasing of temperature (Sect. 3.2.4.2); the polymer can have too much of a good thing. Extensive plastisation in the bulk causes an excessive increase in contact area. This trend (friction increase at long exposures) was observed by Rubenstein although Cohen and Tabor [116] and Amuzu and Briscoe [117] did not detect large increases in friction. Table 5 shows data for the influence of water on the friction of nylon 6 [117]. The quantities τ_0, α and p_0 were measured separately and used to predict the values of the coefficient of friction using eqn. (7) (Sect. 3.2.2). p_0 decreases by ca. 70%, but τ decreases by about 4. Clearly, local surface plastisation is the major effect, but longer exposures (greater than 15 h to water and vapour) may reverse this balance.

Another example of the influence of fluid plastisation is shown in Fig. 18 for various fluids in contact with a sample of PEEK [84]. Dodecane provides some lubrication but dodecane–5% decanoic acid not only increases the friction but also

TABLE 5

Variation of hardness, shear strength and coefficient of friction of nylon 6 with relative humidity P_0, μ_g and μ_b, as defined for Table 2; $\tau(P_0)/P_0$ is the calculated friction coefficient based on eqn. (7).

R.H. (%)	H$_2$O content (%)	P_0 (Pa)	$\tau(P_0)$ (Pa)	$\dfrac{\tau(P_0)}{P_0}$	μ_b	μ_g
ca. 0	ca. 0	1.82×10^8	9.30×10^7	0.51	0.48	0.46
32.7	0.5	1.61×10^8	6.60×10^7	0.41	0.42	0.44
59.0	0.30	1.49×10^8	4.90×10^7	0.33	0.30	0.31
84.2	0.44	1.34×10^8	4.00×10^7	0.30	0.29	0.29
ca. 100	1.00	1.15×10^8	2.50×10^7	0.22	0.20	0.26

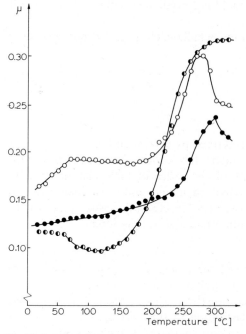

Fig. 18. Lubricated friction of a steel wedge on a poly(ether–ether ketone) plate. Sliding conditions as for Fig. 14; sliding direction in the axis of the wedge. ●, Dodecane; ○, dodecane + 5% dodecanoic acid; ◑, dodecane + 5% dodecylamine. Data obtained by heating the wedge only. Experiments in oxygen-free nitrogen with small quantities of lubricant in the contact.

reduces the critical scuffing temperature. The base is a more effective additive, at low temperatures at least. Separate microhardness studies show that certain additives have the facility to plastise PEEK, particularly at high temperatures; decanoic acid is an example.

5.3. Wetting forces in friction

The surface free energy of a polymer is generally not a prime variable in the consideration of friction. The deformation models ignore it and, while the adhesion model acknowledges its existence, it does not enter directly into the theorems. The adhesion has to be enough. An exception is in the study of Schallamach waves where it scales $\Lambda(v)$ [118,119] and hence probably the friction [120].

Wetting forces may, however, influence the entry condition in fluid lubrication, poor wetting giving starved lubrication [121,122]. Polymer–fluid interactions may also attenuate the auto-adhesion between contacts and hence reduce the area of contact. This is the case in monofilament contacts. Often, the fluid does not lubricate in the true sense, it simply appears to reduce the contact area. This has been demonstrated for a range of poly(ethylene terephthelate) monofilaments filled with titanium in contact with fluids of different surface tensions [123].

6. Conclusions

Frictional work is dissipated in solid contacts, which are in relative motion, through the mechanical deformation of one or both of the solids. Sometimes, the friction process creates a material such as the "third body" then the friction may be defined by the properties of that material. Ultimately, virtually all of the frictional work is dissipated throughout the system as heat.

The first problem in the study of friction is to judge how the moving contact transmits the work into the system. To a first order, this is done via adhesive forces at the areas of real contact or by non-adhesive geometric engagements. These interfacial deformations then dissipate the work by processes which are familiar in bulk deformations. Although the general types of these interfacial deformations may be reproduced in bulk studies, some of the conditions created in contacts are rarely reproduced in bulk studies. For example, the deformation times are usually significantly shorter in duration in contact zones. The second problem is to make realistic comparisons between bulk and interfacial deformations where these pronounced disparities in deformation variables exist. The molecular dynamic interpretation of bulk and interfacial deformations share a common root. The initial work is done mainly against short range intermolecular forces. For organic polymers, these are primarily van der Waals' forces. There may also be some chemical work dissipated and one can envisage entropic contributions.

In composite systems, the same arguments apply, but additional factors need to be considered, particularly when the friction is dissipated in thin reconstructed interfacial zones. The prospect of Coulombic or "ratchet" friction is an example. However, our understanding of the details of the friction processes in virgin polymeric systems is not complete and hence we can only describe the response of composite matrices in general terms. An exception is, perhaps, the case of visco-elastic rolling friction. While the case against demons and their peculiar sociology is good, even in the context of the friction of composites, it could be better. We have acceptable scientific explanations but only a limited predictive capacity.

List of symbols

d	a contact length
f	frictional force
f_d, f_a	frictional force for deformation friction and adhesive friction, respectively
h	film thickness
K_{50}	thermal conductivity at 50°C
n, n'	load indices for friction and contact area
s	frequency of wave propagation
v	wave velocity
A	real area of contact
E	Young's Modulus

H_{v}	Vicker's 20 s hardness
K, K'	material constants
P	contact pressure (mean)
P_0	contact pressure during plastic flow
Q	"activation energy"
R	radius of curvature
V	relative velocity of sliding or rolling
V_1, V_2	characteristic velocities associated with the shear rate and contact frequency dependence of τ
W	normal load
W^*	critical normal load required to induce plastic flow
α, α_0	pressure coefficients of τ
α	loss parameter in deformation friction
δ	loss angle
θ	semi included cone angle
μ	friction coefficient (f/w)
μ_{g}, μ_{v}	friction coefficient on glass and mutual value, respectively
ν	Poisson's ratio
τ	interface shear stress
τ_0, τ_0', τ_0''	intrinsic interface shear stresses
ϕ	a deformation work parameter
Λ	nett adhesive rupture energy

Acknowledgements

The author is grateful to several colleagues for allowing the presentation of unpublished data; they are Dr. J. Amuzu, Dr. M. Adams, Dr. J. Lancaster, Dr. T. Stolarski, Mr. P. Evans and Mr. N. Lin.

References

1 E.M. Rogers, Physics for the Inquiring Mind, Princeton University Press, Princeton, NJ, 1960, pp. 108, 343.
2 D. Tabor, Phys. Bull., 29 (1978) 521.
3 F.P. Bowden and D. Tabor, Friction and Lubrication of Solids, Clarendon Press, Oxford, 1950.
4 F.P. Bowden and D. Tabor, Friction and Lubrication of Solids, Part II, Clarendon Press, Oxford, 1964.
5 K.L. Johnson, in D. Dowson, C.M. Taylor, M. Godet and D. Berthe (Eds.), Friction and Traction, Westbury house, IPC Press, Guildford, 1981.
6 D. Dowson, C.M. Taylor and D. Berthe (Eds.), Friction and Traction, Westbury House, IPC Press, Guildford, 1981. A collection of reviews and papers devoted only to friction and traction. Probably the best recent account of the subject in one volume.
7 J.J. Bickerman, Rev. Macromol. Chem., C11(1) (1974) 137. The late J.J. Bickerman had interesting and sometimes controversial ideas on friction and this reviews his early approach.

8 L.H. Lee (Ed.), Advances in Polymer Friction and Wear, Plenum Press, New York, 1974. An important, but now rather dated, collection of reviews and research papers. Many illuminating discussions.

9 R.P. Steijn, Met. Eng. Q., May (1967) 9. A full account of early literature with the best compilation of load indices at present.

10 D.F. Moore, Friction and Lubrication of Elastomers, Pergamon Press, Oxford, 1972. A complete account, but no mention of recent work on fatigue and Schallamach waves. Also D.F. Moore and J. Geyer, Wear, 30 (1974) 25.

11 A. Schallamach, Wear, 1 (1958) 384. An early account of the friction and abrasion of rubbers.

12 J.K. Lancaster, in A.D. Jenkins (Ed.), Polymer Science. A Material Science Handbook, North-Holland, Amsterdam, 1972, p. 960. A very comprehensive review to the 1977 edition of Friction and Wear.

13 A.D. Roberts, Tribol. Int., April (1975) 75. Review of friction of elastomers.

14 B. Briscoe, in K.W. Allen (Ed.), Adhesion 5, Elsevier, London, 1981. A review of the role of adhesion in friction and wear.

15 B. Briscoe and D. Tabor, in D. Clark and W.J. Feast (Eds.), Polymer Surfaces, Wiley, Chichester, 1978. General outlines of the mechanisms of polymer friction and wear.

16 N.P. Suh and N. Saka (Eds.), Fundametals of Tribology, MIT Press, Cambridge, M, 1980. Many reviews, some specifically on friction of organic solids.

17 J.M. Georges (Ed.), Microscopic Aspects of Adhesion and Lubrication, Tribology Series, Vol. 7, Elsevier, Amsterdam, 1982. Contains several papers on the tribology of polymeric contacts.

18 H. Czichos, Tribology. A Systems Approach to the Science and Technology of Friction, Lubrication and Wear, Tribology Series, Vol. 1, Elsevier, Amsterdam, 1978. A broad treatment of tribology with discussion of polymer friction.

19 K. Mittal (Ed.), Physicochemical Aspects of Polymer Surfaces, Plenum Press, New York, 1983. Contains about six papers on polymer tribology, some of which deal with friction.

20 D. Dowson, History of Tribology, Longmans, London, 1979. A very comprehensive and interesting treatment of the early friction theories including rolling. still a good current reference for many aspects of lubrication.

21 I.V. Kragelsky, M.N. Dobychin and U.S. Kombalou, Friction and Wear Calculation Methods, Pergamon Press, Oxford, 1982. The most recent Soviet text which naturally emphasis Soviet work, but also covers most aspects of friction.

22 B. Briscoe, Chem. Ind. (London), July (1982) 467.

23 Wear of Materials, 1975, 1977, 1979, 1981, 1983, various editions, American Society of Mechanical Engineers, New York, appropriate year. Five edited volumes of a major international conference. Limited amount of material on friction of organic systems, but nevertheless a very significant reference source.

24 M.J. Schick (Ed.), Surface Characteristics of Fibres and Textiles, Dekker, New York, 1977. Several papers on fibre friction and lubrication.

25 G.M. Bartenev and V.V. Lavrentev, Friction and Wear of Polymers, Tribology Series 6, Elsevier, Amsterdam, 1981. Text book covering general aspects of friction and wear of polymers.

26 V.A. Belyi, A.L. Sviridyonok, M.I. Petrokovetso and V.G. Savkin, Friction and Wear in Polymer-based Materials, Pergamon Press, New York, 1981. Mainly a record of the studies carried out by Acad. Belyi's group in Gomel.

27 M.O.W. Richardson (Ed.), Polymer Engineering Composites, Applied Science Publishers, London, 1977. Useful general text but also contains three reviews of tribological subjects.

28 Speciality Bearing, Tribol. Int., 15(5) (1982). Several experienced authors highlight important features of polymer composite tribology: little on friction per se.

29 B. Briscoe, Philos. Mag., A43 (1981) 511. In the festschrift to mark Professor David Tabor's retirement. Deals with the adhesion model of friction for polymers.

30 J.K. Lancaster, in Tribology in the 80's, NASA Conference Publication 2300, NASA, Cleveland, 1984.

31 D. Tabor, Proc. R. Soc. London Ser. A, 229 (1955) 198.

32 D.G. Flom and A.M. Bueche, J. Appl. Phys. 30 (1959) 1725.
33 W.D. May, E.L. Morris and D. Attack, J. Appl. Phys., 30 (1959) 1713.
34 J.A. Greenwood and D. Tabor, Proc. Phys. Soc., 71 (1958) 989.
35 J.A. Greenwood, H. Minshall and D. Tabor, Proc. R. Soc. London Ser. A, 259 (1961) 480.
36 K.C. Ludema and D. Tabor, Wear, 9 (1966) 329.
37 W.O. Yandell, Wear, 17 (1971) 229.
38 L.E. Nielsen, Mechanical Properties of Polymers and Composites, Dekker, New York, 1974.
39 J.L. White and Y.M. Lin, J. Appl. Polym. Sci., 17 (1973) 3273. See also R. Bond, G.F. Morton and L.H. Krol, Polymer, 25 (1984) 132.
40 G.L. Wannop and J.F. Archard, Proc. Inst. Mech. Eng. london, 187 (1973) 615.
41 K.L. Johnson, Proc. Inst. Mech. Eng. London, 196 (1982) 363.
42 D.R. Thomas (Ed.), Rough Surfaces, Longmans, London, 1982.
43 A. Schallamach, J. Polym. Sci., 9(5) (1952) 385.
44 B. Bethune, J. Mater. Sci., 11 (1976) 199.
45 B. Lamy, Tribol. Int., 17(1) (1984) 35.
46 M. Coulon and W. Lenne, Polym. Test., 2 (1981) 199.
47 B. Briscoe, P. Evans and J. Lancaster, J. Mater. Sci., in press.
48 J. Lancaster, private communication, 1984.
49 N.S. Eiss, in Tribology in the 80's, Proc. Int. Conference, NASA conference Publication 2300, NASA, Cleveland, 1984.
50 D.J. Whitehouse, in N.P. Suh and N. Saka (Eds.), Fundamentals of Tribology, MIT Press, Cambridge, MA, 1980, p. 252.
51 N.G. Guy (Ed.), Surface Topography in Engineering, BHRA Fluid Engineering Series, Vol. 3, Cotswold Press, Oxford, 1977.
52 K.A. Grosh, Proc. R. Soc. (London) Ser. A, 274 (1963) 21.
53 C. Bird and D. Moran, unpublished work. See B. Briscoe, Chem. Ind. (London), July (1982) 467.
54 B. Briscoe and D. Tabor, J. Adhes., 9 (1978) 145.
55 B. Briscoe and D.C.B. Evans, Proc. R. Soc. (London) Ser. A, 380 (1982) 389.
56 B. Briscoe and D. Tabor, ACS Prepr., 21(1) (1976) 10.
57 B. Briscoe and A.C. Smith, Reviews on the Deformation Behaviour of Materials III, 3 (1980) 151.
58 B. Briscoe, B. Scruton and R.F. Willis, Proc. R. Soc. (London) Ser. A, 333 (1973) 99.
59 J.A.K. Amuzu, B. Briscoe and D. Tabor, Trans. Am. Soc. Civ. Eng., 20(40) (1977) 354.
60 B. Briscoe and A.C. Smith, J. Appl. Polym. Sci., 28 (1983) 3827.
61 D. Tabor, in E. Matijevic (Ed.), Surface and Colloid Science, Vol. 5, Wiley, New York, 1972.
62 E. Eissner and J.S. Courtney-Pratt, Proc. R. Soc. (London) Ser. A, 238 (1957) 529.
63 A.R. Savkoor and G.A.D. Briggs, Proc. R. Soc. (London) Ser. A, 356 (1977) 103.
64 C.M. Pooley and D. Tabor, Proc. R. Soc. (London) Ser. A, 329 (1972) 251.
65 J.R. Barber, Proc. R. Soc. (London) Ser. A, 312 (1980) 871.
66 B. Briscoe and A.C. Smith, J. Phys. D., 15 (1982) 579.
67 I.M. Ward, J. Mater. Sci., 6 (1971) 1397.
68 B. Briscoe and A.C. Smith, Polymer, 22 (1981) 158.
69 J.A.K. Amuzu, B. Briscoe and D. Tabor, ASLE Trans., 20(2) (1977) 152.
70 S. Rabinowitz, I.M. Ward and J.S.C. Parry, J. Mater. Sci., 5 (1970) 909.
71 D.M. Heyes, J. Chem. Soc. Faraday Trans. 2, 79 (1983) 611.
72 B. Briscoe and D. Tabor, in S. Eicke and G. Parfitt (Eds.), Interfacial Phenomena in Apolar Media, Dekker, New York, 1984.
73 A.S. Lodge and H.G. Howell, Proc. Phys. Soc. London, Sect. B, 67 (1954) 89.
74 J.F. Archard, Nature (London) 172 (1953) 918.
75 J.F. Archard, Proc. R. Soc. (London) Ser. A, 243 (1957) 190.
76 J.H. Warren and N.S. Eiss, Jr., Wear of Materials 1977, American Society of Mechanical Engineers, New York, 1977.
77 J. Amuzu, B. Briscoe and M. Chaudri, J. Phys. D, 9 (1976) 133.
78 B. Lincoln, Nature (London), 172 (1953) 169.

79 D. Tabor and D.E. Wynne Williams, Wear, 4 (1961) 391.

80 N. Adams, J. Appl. Polym. Sci., 7 (1963) 2075.

81 M.W. Pascoe and D. Tabor, Proc. R. Soc. (London) Ser. A, 235 (1956) 210.

82 M. Adams, J. Amuzu and B. Briscoe, J. Phys. D, in press.

83 B. Briscoe, M. Fernando and A.C. Smith, unpublished data.

84 B. Briscoe, T. Stolarski and S. Davis, Tribol. Int., July (1984).

85 A.L. McKenna, Fatty Amides, Witco Chemical Corp., Memphis, TN, 1982.

86 L. Mascia, Role of Additives in Plastics, Edward Arnold, Salisbury, 1984.

87 A.J.G. Allen, J. Colloid Sci., 14 (1959) 206.

88 R.C. Bowes, N.L. Jarvis and W.A. Zisman, Ind. Eng. Chem. Prod. Res. Dev., 4 (1965) 86.

89 B. Briscoe, V. Mustafer and D. Tabor, Wear, 19 (1972) 399.

90 S.H. Nah and A. Thomas, J. Polym. Sci., Polym. Phys. Ed., 18 (1980) 511.

91 D. Allan, B. Briscoe and D. Tabor, Wear, 25 (1973) 393.

92 J. Klein and B. Briscoe, Proc. R. Soc. (London) Ser. A, 365 (1979) 53.

93 W.L. Skelcher, in J.M. Georges (Ed.), Microscopic Aspects of Adhesion and Lubrication, Tribology Series, Vol. 7, Elsevier, Amsterdam, 1972. Also W.J. Skelcher, T.F.J. Quinn and J.K. Lancaster, Trans. Am. Soc. Civ. Eng., 25 (1982) 391.

94 M.P.L. Hill, P.L. Millard and M.J. Owen, in L.H. Lee (Ed.), Advances in Polymer Friction and Wear, Plenum Press, New York, 1974.

95 Various authors in D. Dowson, M. Godet and C.M. Taylor (Eds.), Wear of Non-Metallic Materials, Mechanical Engineering Publications, London, 1978.

96 B. Briscoe, in Polymer and Wear and its Control, American Chemical Society, New York, 1985.

97 B. Briscoe and T. Stolarski, Nature (London), 281 (1979) 206.

98 B. Briscoe and M.D. Steward, Materials Performance and Conservation, Institution of Mech. Engineers, London, 1978.

99 M.D. Steward, Ph.D. Thesis, Cambridge University, 1976.

100 J.K. Lancaster, Am. Soc. Test. Mater. Spec. Techn. Publ. 769, 1982.

101 B. Briscoe, H. Lin and T. Stolarski, Wear, in press.

102 T. Lui and S.K. Rhee, in Wear of Materials 1977, American Society of Mechanical Engineers, New York, 1977.

103 T.F.J. Quinn, ASLE Trans., 10 (1967) 158.

104 See, for example, several papers in Wear of Materials 1977, American Society of Mechanical Engineers, New York, 1977.

105 K.R. Makinson, Text. Res. J., 38 (1967) 763.

106 G.F. Flanagan, Text. Res. J., 36 (1966) 55.

107 A. Schallamach, Wear, 17 (1971) 301.

108 A.D. Roberts and A.R. Thomas, Wear, 33 (1975) 45.

109 G.A.D. Briggs and B. Briscoe, Philos. Mag., A38 (1978) 387.

110 M. Barquins, Wear, 91 (1983) 103.

111 K.N.G. Fuller and D. Tabor, Proc. R. Soc. (London) Ser. A, 345 (1975) 327.

112 G.A.D. Briggs and B. Briscoe, Wear, 35 (1975) 357.

113 Several authors in Tribology in the 80's, NASA Conference Publication, NASA, Cleveland, 1983.

114 A.D. Roberts, Eng. Mater. Des., 55 (1969) 22.

115 C. Rubenstein, J. Appl. Phys., 32 (1961) 1445.

116 S.C. Cohen and D. Tabor, Proc. R. Soc. (London) Ser. A, 291 (1966) 186.

117 J.A.K. Amuzu and B. Briscoe, unpublished data.

118 A.N. Gent and J. Schultz, Proc. Int. Rubber Congr., Brighton, I.R.I., London, 1972, paper C1.

119 E.H. Andrews and A.J. Kinloch, Proc. R. Soc. (London) Ser. A, 332 (1972) 401.

120 G.A.D. Briggs, Ph.D. Thesis, Cambridge University, 1975.

121 R.B. Lewis, American Society of Automotive Engineering Inc., National Combined Fuels and Lubrication Meeting, Houston, Texas, 1969.

122 T. Stolarski, private communication, 1984.

123 See, for example, M.J. Adams, B. Briscoe and S.L. Kremmtzer, in K. Mittal (Ed.), Physicochemical Aspects of Polymer Surfaces, Plenum Press, New York, 1983.

Chapter 3

Friction and Wear of Materials with Heterogeneous Microstructures

ERHARD HORNBOGEN

Institut für Werkstoffe, Ruhr-Universität, D-4630 Bochum (F.R.G.)

Contents

Abstract

Some general aspects of unlubricated tribological properties of materials which are composed of more than one phase or other microstructural components are discussed. Friction and wear are due to dissipation of energy and matter at the heterogeneous surface microstructure.

Comparable tribological properties are defined, i.e. properties which do or do not contain the external compressive stress, bulk hardness, or microhardness. The roles of work hardening of the surface and of fracture mechanical properties are then discussed.

Attempts were made to derive quantitative models for wear rates, w, as functions of volume fractions f_α, f_β, properties of phases α, β, interfaces $\alpha\beta$, and different types of microstructures. For anisotropic structures, tensors have to be used for a complete description of tribological properties. This is of concern for many composite materials. Materials exposed to sliding and abrasion can be subdivided into structures with one (isotropic), three, and six components of the tensor of the wear rate w_{ij}. For erosion, the number of components can reach nine, or eighteen if the direction of loading has to be considered.

Finally, the relation between friction and wear is discussed briefly, as well as the optimisation of these two groups of tribological properties in the design of composite materials with desired properties.

1. Introduction

The term "heterogeneous" is used for materials which are composed of two or more phases or other descrete microstructural components [1,2]. The latter could apply, for example, for tempered martensite or for a thermoplastic polymer with additives. Thus, microstructural components may consist of more than one phase, for example in the form of an ultra-fine dispersion. A wide variety of materials is included in this definition. Many of them can form elements of useful tribological systems. Their microstructure may originate from [3]

(a) solidification or solid state reactions, using heterogeneous thermodynamic equilibria,

(b) sintering reactions, which can combine even non-equilibrium phases and volume fractions, and

(c) all methods to produce more artificial morphologies of composite materials such as impregnation, up to the structural sophistication of integrated circuits.

A systematic approach is attempted to some tribological aspects which all heterogeneous materials have in common. The discussion is focussed on quantitative relations with microstructural parameters. They form a sometimes neglected part of the "intrinsic material properties" which, in turn, belong to a complete tribological system (Fig. 1(b) and (c) show examples for two different combinations of microstructure of material A and an abrasive material B (grinding paper). Changes in microstructure can produce great differences of friction and especially of wear behavior, even if macrohardness and volume fraction of phases are equal. It is the purpose of this paper to pay attention to such microstructural aspects, especially of systems which consist of combinations of soft and hard microstructural components such as, for example, enforced polymers or sintered hard metals.

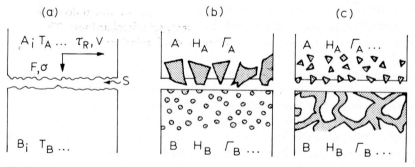

Fig. 1. Microstructural aspects of tribological systems. Materials, A, B; hardness, H_A, H_B; microstructures, Γ_A, Γ_B; temperatures, T_A, T_B.

2. Types of microstructure and anisotropy

A description of heterogeneous microstructures requires information on the nature of the phases (or microstructural components α, β,...), their volume fractions f_α, f_β

$$f_\alpha + f_\beta = 1 \tag{1}$$

and the type of microstructure [1,2]. We distinguish as microstructural elements grain boundaries $\alpha\alpha$, $\beta\beta$, phase boundaries $\alpha\beta$ and use them to define different microstructures [3]. Dispersion-, duplex-, and net- or cell-structures may be isotropic (Fig. 2). Suitable for their characterization are the densities of boundaries ρ_b, for example the interfaces $\rho_{\alpha\beta}$. A boundary density ρ_b is defined by the sum of the interfacial area over the specimen volume

$$\rho_b = \frac{\Sigma A_b}{V} \tag{2}$$

An ideal dispersion structure of β in α is defined by the fact that no $\beta\beta$-boundaries exist, because β exists as isolated particles only.

$$\rho_{\beta\beta} = 0 \tag{3a}$$

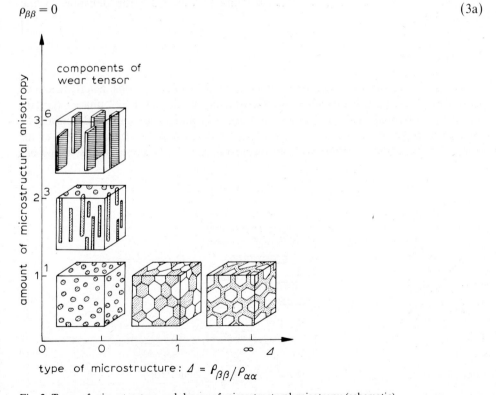

Fig. 2. Types of microstructure and degree of microstructural anisotropy (schematic).

The degree of dispersion is proportional to the density of interfaces $\rho_{\alpha\beta}$ for a certain volume fraction f_β (of cubes)

$$\rho_{\alpha\beta} = \frac{6f_\beta}{d_\beta} \tag{4}$$

Formation of a three-dimensional net- or cell-structure is associated with the formation of $\beta\beta$-boundaries and with percolation of β.

$$\rho_{\beta\beta} > 0 \tag{3b}$$

An ideal duplex structure consists of a mixture of an equal number of α- and β-grains of equal size and implies percolation of α as well as β. A duplex parameter Δ can be defined

$$\Delta = \frac{\rho_{\alpha\alpha}}{\rho_{\beta\beta}} = 1 \tag{5}$$

The average grain size, S_b, for this structure follows from the densities of the boundaries.

$$S_b = \left(\rho_{\alpha\alpha} + \rho_{\beta\beta} + 2\,\rho_{\alpha\beta} \right)^{-1} \tag{6}$$

All types of microstructure can be produced in anisotropic versions. Microstructural anisotropy must be distinguished from anisotropy at the level of "phase" (crystal anisotropy, oriented macromolecules) (Fig. 3). Any anisotropy can be described by three principle directions $a_1 \equiv a$, $a_2 \equiv b$, $a_3 \equiv c$. The difference between two, or all three of them leads to tensors with an increasing number of components. Uni-axial

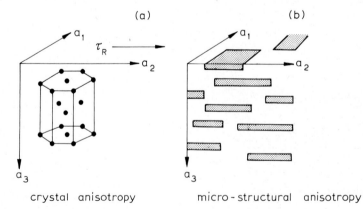

crystal anisotropy   micro-structural anisotropy

Fig. 3. Anisotropy based on (a) phase and (b) microstructure.

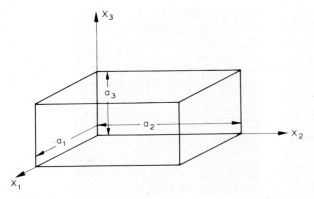

Fig. 4. Designation of materials dimensions a_i and directions of sliding or erosion x_j [compare Fig. 19(a)].

fiber composites and lamellates show bi-axial anisotropy; oriented ribbons, for example, imply tri-axially (Figs. 3, 4).

A complete description, for example, of microstructural anisotropy of a dispersion requires the following parameters:

(a) size and shape of the particles (β in α),
(b) spacings and local distribution, including order,
(c) orientation functions of non-spherical particles.

Microstructural anisotropy implies that the volume fractions of the phases must not be identical with the fractions found in the surface areas (Fig. 5). For known shapes and orientations of particles, these effective surface area fractions can be determined

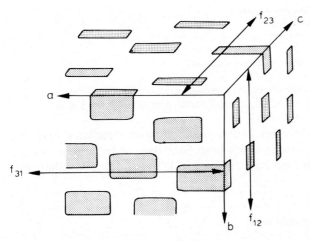

Fig. 5. Designation of surface fractions of the microstructural component f_{ij} for anisotropic microstructures.

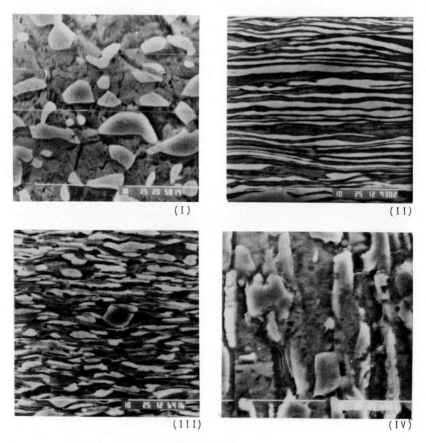

I	80% cold rolled + 100 h/680°C/H_2O, isotropic
II, III, IV	structure I + 90% cold rolled, anisotropic
II	plane I, III plane 2, IV plane 3.
x_1	transverse direction
x_2	rolling direction
x_3	short-transverse direction (coordinate system: Fig. 4)

Fig. 6. Examples of anisotropic microstructures. (a) Lamellar structure: Fe–5 wt.% Ni–0.1 wt.% C, ferrite + martensite, $f_\alpha \approx f_\beta \approx 0.5$, 90% rolling of duplex structure (compare Fig. 2).

by geometrical calculation. Oriented fiber composites and lamellates may, for example, be considered as structures which have been derived by uni- or bi-axial distortions of originally isotropic structures (for example by extrusion or rolling, Fig. 6).

3. Formal description of friction and wear

The frictional force is caused by the dissipation and storage of energy U during sliding in a direction x [4,5]. The total energy can be related to the nominal area,

 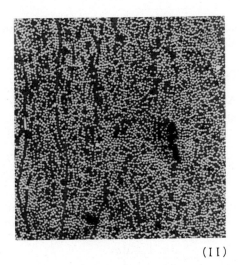

(I) (II)

Fibers parallel to direction x_2 (Fig. 4)
volume fraction $f = 0.5$
diameter $d = 8 \mu$m
length $1 < l < 3$ mm
I plane 1 or 3
II plane 2

Fig. 6. (b) Fiber structure: carbon fibre reinforced EP, 8 μm diam., 1–3 mm fibers, $f_\alpha \approx f_\beta \approx 0.5$.

A_0, of the sliding surface. What fraction of A_0 does form asperities depends on the morphology of the surfaces, on macrohardness (defined as HB or $HV = F/A'$, where $A' = A/S$) and pressure $\sigma = F/A_0$. The reactions which give rise to friction take place only in a fraction of the surface $A/A_0 = \sigma s/H \leqslant 1$. Microstructural details (fraction and shape of microstructural components in the surface) are ignored for a continuum approach. The systems factor, s, is explained after eqn. (8).

$$\frac{\mathrm{d}U}{\mathrm{d}x}\frac{1}{A_0} = \frac{\mathrm{d}u}{\mathrm{d}x} = \left| \frac{\mathrm{J}}{\mathrm{m}^3} \right| \tag{7}$$

This leads to the coefficient of friction μ.

$$\mu = \frac{F_R}{F} = \frac{\tau_R}{\sigma} = \frac{\mathrm{d}u}{\mathrm{d}x\sigma} = \frac{\mathrm{d}\gamma As}{\mathrm{d}xA_0\sigma} = \frac{\mathrm{d}\gamma s}{\mathrm{d}xH} \tag{8}$$

The specific energy, γ, is the dissipation in the effective asperity area $A < A_0$. The dimensionless factor s takes care of the surface morphology of the sliding partners and, if necessary, of other external system conditions such as environment. For adhesion, γ_{ad} is determined by the reaction of surfaces γ_{AO}, γ_{BO} to interfaces γ_{AB} of all microstructural components.

$$\gamma_{AO} + \gamma_{BO} - \gamma_{AB} = \gamma_{ad} \tag{9}$$

A certain share of the frictional energy may cause wear by either the direct or indirect actions of the stresses τ_R and σ.

Wear is separation of matter, M, from the surface of a material.

$$w_w = \frac{dM}{dx}\frac{1}{A_0} = \frac{dm}{dx} = \left|\frac{g}{m^3}\right| \tag{10}$$

It is useful to relate the gravimetric wear rate, w_w, to the density, ρ, to obtain the volumetric rate, w_v, and therefore the dimensional change of the material.

$$w_v = \frac{w_w}{\rho} = \frac{dm}{dx}\frac{1}{\rho} = \frac{da}{dx} \tag{11}$$

where m is the amount of matter lost in the nominal area A_0.

4. The coefficient of wear and the role of macrohardness

The wear coefficient, k, is defined as the probability of wear in the fraction of the surface which is interacting [A/A_0, cf. eqn. (8)] [6].

$$w \equiv w_v = \frac{da}{dx} = k\frac{A}{A_0} = k\frac{\sigma}{H} \tag{12}$$

Unlike the coefficient of friction, μ, hardness is not contained in k [Table 1, eqn.

TABLE 1

Comparability of tribological properties

Symbol	Units	Containing		
		σ	H	
$d\gamma/dx$	$J\,m^{-3}$	–	–	Energy dissipation in asperity area A
$k_v = k$	1	–	–	Volumetric wear coefficient
k_w	$g\,m^{-3}$	–	–	Gravimetric wear coefficient
μ	1	–	+	Coefficient of friction
$w_v/\sigma = k_v/H$	Pa^{-1}	–	+	Wear rate per nominal area
$w_w/\sigma = k_w/H$	$g\,m^{-3}\,Pa^{-1}$	–	+	Pressure $\sigma = F/A_0$
du/dx	$J\,m^{-3} = Pa$	+	+	Energy dissipation per normal area A_0; $u_i = U/A_0$
w_v	1	+	+	Volumetric wear rate
w_w	$g\,m^{-3}$	+	+	Gravimetric wear rate

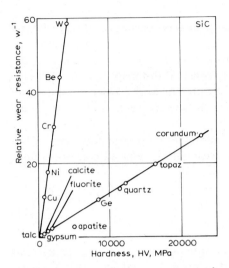

Fig. 7. Proportionality of wear resistance w^{-1} and bulk hardness implies k = const. for tough metals and brittle ceramics.

(8)]. The two "coefficients", therefore, are not comparable tribological properties. The coefficient of friction characterizes dissipation of energy related to the *nominal area* A_0; the coefficient of wear describes a probability for separation of matter in

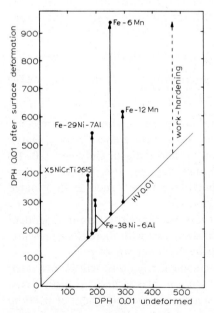

Fig. 8. Work-hardening ability of alloys exposed to abrasive friction, maximum for structures susceptible to stress-induced martensitic transformation. Fe–6 wt.% Mn–0.1 wt.% C.

the *effective area* A. Therefore, μ and k/H are related properties, as are $d\gamma/dx$ and k, but not μ and k [eqn. (8)].

A plot of wear resistance, w^{-1}, against hardness for constant pressure σ provides evidence for a unique wear mechanism, i.e. $k = \text{const.}$, if different materials fall on a straight line (Fig. 7) [7]. In this connection, the different work-hardening abilities of materials require attention. The hardness in the surface after action of the frictional shear stress is effective [eqns. (8) and (12)] [8,9].

In some alloys (metastable austenites and many thermo-plastic polymers), work-hardening, ΔH, will be considerable and the wear rate is reduced correspondingly (Fig. 8) [9].

$$(w - \Delta w)^{-1} \sim H_0 + \Delta H \tag{13}$$

If a proportionality between w^{-1} and H is still not found, a change in wear mechanism, i.e. of k must be associated with the change in toughness. This is often a change in the mechanism of microcracking.

5. Components of the coefficient of friction

In addition to adhesion [eqn. (9)], several other processes are able to either dissipate energy as heat or cracks, or store energy, for example as structural defects or electrical charges [eqn. (8)] [10–13].

$$\mu = \Sigma \frac{d\gamma}{dx}\frac{s}{H} = \left(\frac{d\gamma_{ad}}{dx} + \frac{d\gamma_{el}}{dx} + \frac{d\gamma_{d}}{dx} + \frac{d\gamma_{t}}{dx} + \frac{d\gamma_{f}}{dx} + \frac{d\gamma_{ch}}{dx} \right) \frac{s}{H} \tag{14}$$

In different types of material or microstructural components, some of the mechanisms may dominate: adhesion for materials with high surface energies, γ_{ad}, elastic deformation for rubbers, γ_{el}, plastic deformation in thermoplastic polymers [10–14] and metals, γ_{d}, crack propagation in all brittle materials, γ_{f}, and, finally, friction-activated chemical reaction, γ_{ch}, for example between partners such as fluorinated polymers and alloys containing alkaline metal atoms.

It is well known that the coefficient of friction is often not pressure-independent [as indicated by eqn. (14)]. The value of μ often decreases with increasing pressure, σ, for many polymers. The various terms are different functions of pressure. Some, such as plastic deformation, fracture and tribo-chemical reactions, will be activated only above a certain threshold pressure, σ_c. Adhesion, elastic and, in most cases, plastic deformation will always contribute to the macroscopic coefficient of friction. Composite materials provide opportunities to tailor-make coefficients of friction by selecting microstructural components which contribute favorable combinations such

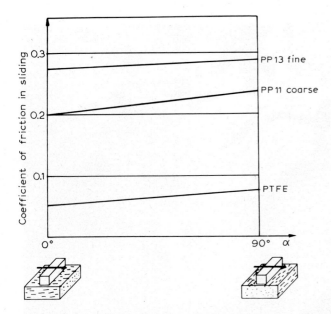

Fig. 9. Dry friction of thermoplastic polymers as a function of the angle, α, between the direction of sliding and molecular orientation. Isotactic PP with 5 vol.% (11) and 20 vol.% (13) admixture of atactic PP.

as high hardness and low adhesion. Examples are metal–PTFE composites for a low coefficient of friction (Fig. 9) [15,16].

6. Components of the wear coefficient, k, and the role of fracture

The wear coefficient expresses the probability of decohesion of matter in the asperity area A [eqn. (12)]. The large number of special wear mechanisms shall be subdivided into those which are determined by plastic deformation, i.e. the resistance to τ_R (ploughing, plastic chip formation) and crack formation mainly caused by tensile stresses [10]. It depends on the special wear system whether stable (fatigue, thermal fatigue, stress corrosion cracking) or unstable crack propagation determines the fracture mechanisms.

In a first attempt to consider micro-cracking, it was claimed that, for abrasion of brittle materials (i.e. for very low values of G_{IC} or K_{IC}), fracture mechanical properties can determine the wear rate, w [17].

$$w = \frac{s^*}{G_{IC}} \frac{\sigma}{H} = \frac{Es^*}{K_{IC}^2} \frac{\sigma}{H} = k_f \frac{\sigma}{H} \tag{15}$$

A careful microscopic inspection of wear mechanics indicates, however, that, in many cases, plastic deformation and immediate fracture occur simultaneously [18,19]. Brittle fracture dominating as wear mechanism is found for abrasion of extremely brittle materials only. Usually, both cooperate in an additive way (Fig. 10). This can

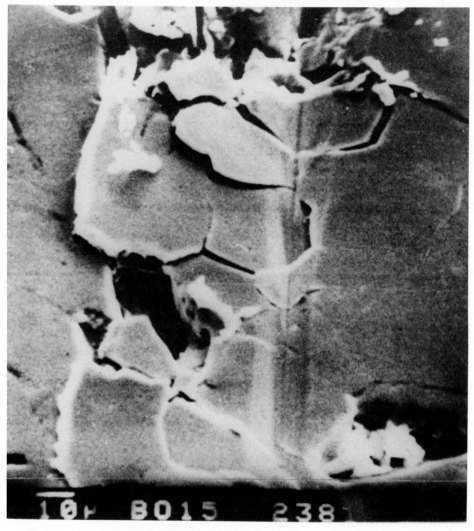

Fig. 10. Wear by microcracking of scratched grey cast iron. 12 vol.% graphite in pearlitic matrix induces cracking.

be expressed by terms of the wear coefficient for plastic deformation, k_d, and immediate fracture, k_f. For many composites, it is useful to add a third term to take care of separation along insufficiently bonded interfaces (k_i) [20].

$$w = (k_d + k_f) \frac{\sigma}{H} \tag{16a}$$

$$w = (k_d + k_f + k_i) \frac{\sigma}{H} \tag{16b}$$

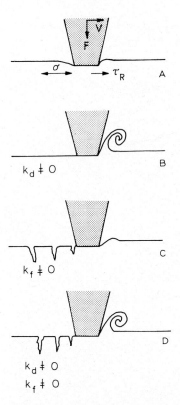

Fig. 11. Components of the wear coefficient (schematic). k_d = plastic deformation, ploughing or chip formation; k_f = microcracking.

The limiting cases between $k_d = 0$, $k_f = 0$, and $k_d \neq 0$, $k_f \neq 0$ are shown schematically in Fig. 11.

This approach implies an understanding of the fact that, in some materials, wear resistance can decrease with increasing hardness. This may be so because the change in hardness, dH/H, is overcompensated by a decrease in fracture energy $(-dG_{IC}/G_{IC})$ [Fig. 12, eqn. (15)] during a hardening process.

$$w = \left(k_d + \frac{s^*}{G_{IC}}\right)\frac{\sigma}{H} \tag{17}$$

For brittle materials ($G_{IC} \rightarrow 0$), the second term becomes dominant. This implies that the wear rate (for abrasive loading conditions) is inverse proportional to the critical microcrack length for the spontaneous crack propagation a_c.

For a composite material, a change in volume fraction, f_β, of a hard and brittle

74

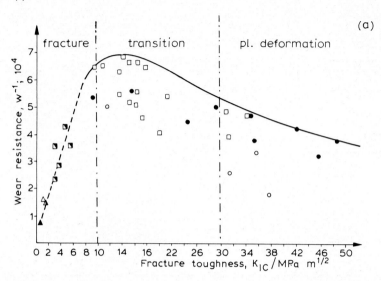

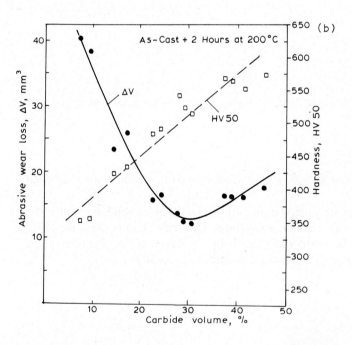

Fig. 12. Effect of microcracking on abrasive wear. (a) Wear resistance of metals and ceramic glasses as a function of fracture toughness. K_{IC} controls w^{-1} for brittle materials (left part of diagram). On the right, plastic deformation and work hardening influence the wear resistance. (b) Wear rate, w, of austenitic white cast irons as a function of volume fraction of the carbide M_7C_3. Increasing wear at $f_\beta > 0.3$ due to microcracking of carbide. (c) Prerequisite for increasing wear with increasing hardness, H (schematic).

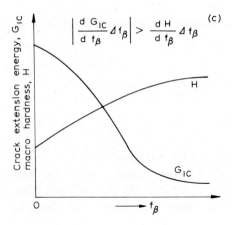

Fig. 12 (continued).

component β may lower fracture energy more than it raises its hardness (Fig. 12).

$$\left[\frac{\mathrm{d}H}{\mathrm{d}f_\beta}\frac{1}{H}\right] \lessgtr \left[-\frac{\mathrm{d}G_{IC}}{\mathrm{d}f_\beta}\frac{1}{G_{IC}}\right] \tag{18}$$

so that the wear resistance decreases with volume fraction f_β (Fig. 13).

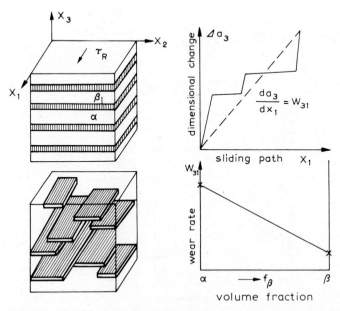

Fig. 13. Derivation of quantitative models for the wear rate. Sequential removal, Δa_3, of α and β by sliding in directions x_1 and x_2 (schematic) (lamellar structures).

Fracture mechanisms come into play usually above a certain threshold stress σ_c. Equation (16a) has been modified accordingly.

$$w = w_d + w_f = [k_d + k_f \varphi] \frac{\sigma}{H} \tag{19a}$$

$$\varphi = 1 - \frac{\sigma_c}{\sigma} \tag{19b}$$

where the microcracking factor $\varphi = 0$ for $\sigma < \sigma_c$, and $\varphi > 0$ for $\sigma > \sigma_c$.

Above a volume fraction $f_{\beta c}$, the type of microstructure may change, e.g. from dispersion to net (microstructural transformation; see Fig. 18). As a consequence, the fracture toughness and wear resistance may change discontinuously above $f_{\beta c}$ (see Fig. 18).

7. Isotropic heterogeneous microstructures

Heterogeneous structures are characterized by the volume fractions and properties of the phases, of the boundaries, and of the type of microstructure (Fig. 2) [1–3]. For the derivation of their behavior, a sequential or a simultaneous arrangement of microstructural components in the plane of sliding are the limiting situations

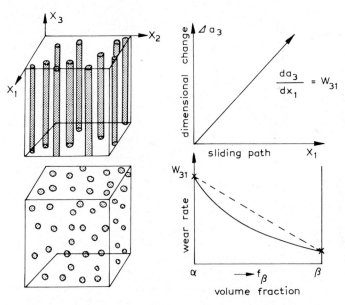

Fig. 14. Simultaneous removal Δa_3 of α and β by sliding in direction x_1 and x_2 (schematic) (fiber structure, dispersion structure).

[21–24]. These are represented by the anisotropic structural species lamellate and uni-axially aligned fiber composite (Figs. 13 and 14).

Wear of a lamellate in direction x_3 must occur simply by the layer-by-layer removal of the individual components with the partial rates w_α and w_β. The macroscopic wear rate w occurs in the case of good bonding (Fig. 13).

$$w = w_\alpha f_\alpha + w_\beta f_\beta = \left(\frac{k_\alpha f_\alpha}{H_\alpha} + \frac{k_\beta f_\beta}{H_\beta} \right) \sigma \tag{20a}$$

For equivalent wear mechanisms in both phases, the relation may be $k_\alpha = k_\beta = k$ (compare Fig. 12). Then

$$w = \sigma k \left(\frac{f_\alpha}{H_\alpha} + \frac{f_\beta}{H_\beta} \right) \tag{20b}$$

It becomes evident that the bulk wear resistance is not relatable to an average bulk hardness as it would follow from the rule of mixtures.

$$H = f_\alpha H_\alpha + f_\beta H_\beta \tag{21}$$

Assuming that the macrohardness is determined by the rule of mixtures [eqn. (21)], a relation is derived which implies that wear resistance of the phase mixture w^{-1} is proportional to its bulk hardness [eqn. (22)].

$$w = \frac{k\sigma}{H} = \frac{k\sigma}{H_\alpha f_\alpha + H_\beta f_\beta} \tag{22a}$$

$$w^{-1} = \frac{H}{k\sigma} = \frac{1}{k\sigma} \left(H_\alpha f_\alpha + H_\beta f_\beta \right) \tag{22b}$$

$$w^{-1} = f_\alpha w_\alpha^{-1} + f_\beta w_\beta^{-1} \tag{22c}$$

$$w = \frac{w_\alpha w_\beta}{f_\alpha w_\beta + f_\beta w_\alpha} \tag{22d}$$

This relation is found for many heterogeneous materials, e.g. for abrasion of fine dispersions of a hard phase in a soft matrix (Fig. 15). Prerequisites are that the width of the abrasive groove, S_g, is much larger than the particle size, d_β, and spacing S_β (Fig. 16), and perfect bonding between α and β.

Consideration of the type of microstructure and the microstructural dimensions shown in Fig. 16(a)

$$S_g \gg S_\beta \tag{23a}$$

$$S_g \gg d_\beta \tag{23b}$$

are useful to explain deviations from the bulk wear rates given by eqn. (22).

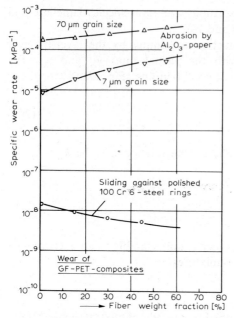

Fig. 15. Wear rate as a function of the volume fraction, f_β, of hard microstructural component. Effect of the wear system: wear is even increased by hard fibers for abrasion due to decohesion of the fiber–matrix interfaces (high k_i, see Fig. 19).

The following special cases are of interest for composite materials:

(a) *A material which contains one phase, the hardness of which is equivalent to or higher than that of the abrasive partner* $H_\beta > H_{ab}$. If this phase is dispersed finely in a soft matrix so that the particles can be shifted during (plastic) ploughing together with the matrix, eqn. (22) is valid. If, however, the dimensions of the hard phase become larger, or if it is anchored in the matrix as a net or cell structure, a supporting effect will impede wear of the soft matrix. The wear resistance will rise up to a value which is determined by the harder phase alone $[k_\beta \ll k$ from eqn. (22b)] (Fig. 17) [21].

$$w_\beta^{-1} > w^{-1} \approx w_\beta^{-1} f_\beta = \frac{f_\beta H_\beta}{k_\beta \sigma} \tag{24a}$$

For dispersion structures, this effect is related to the groove width. It is, therefore, also a function of the pressure, σ. High resistance should be expected in this case below a critical pressure σ_c which produces grooves of a diameter, S_g, less than the particle size, d_β, $d_\beta > S_g$ for $\sigma < \sigma_c$ [eqn. (23)].

$$w^{-1} = \frac{f_\beta H_\beta}{k_\beta (\sigma < \sigma_c)} \tag{24b}$$

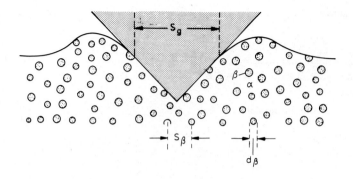

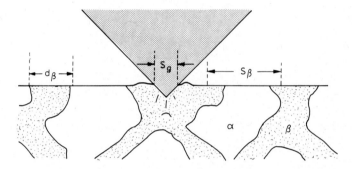

Fig. 16. Effect of the ratio of groove width, S_g, and particle spacing S_β: fine dispersions, coarse net (schematic).

(b) *A material in which one component increases its probability of fracture with increasing volume fractions, particle size, or type of structure.* This situation can be expressed by an increase of k_β to $k_\beta^+ = k_{\beta d} + k_{\beta f}$, for example by the onset of brittle fracture, above a pressure σ_c, or a particle size $d_{\beta c}$ (Fig. 18). The ratio of critical crack size, a_c to particle size, $d_{\beta c}$, ($a_c < d_{\beta c}$) can explain such a transition [eqns. (15)–(17)].

$$\frac{1}{w} = \frac{f_\alpha}{w_\alpha} + \frac{f_\beta}{w_\beta} = \frac{1}{(\sigma < \sigma_c)}\left(\frac{f_\alpha H_\alpha}{k_\alpha} + \frac{f_\beta H_\beta}{k_\beta^+}\right) \tag{25}$$

An additional probability for wear due to separation of interfaces can be explained in a corresponding way [eqn. (16b)].

8. Anisotropic structures

An additional feature of anisotropic structures is the directionality of friction and wear properties [25–27]. Tribological tensors are required for a full description of such materials [28]. An understanding of the components of the tribological tensors

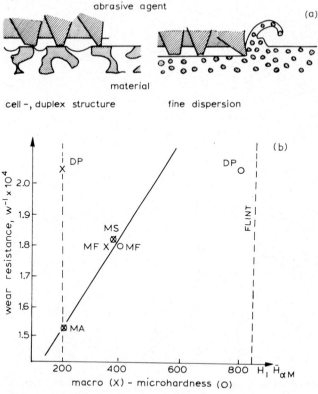

Fig. 17. (a) Influence of the degree of dispersion of very hard particles β on the mechanism of wear. Hardness of the abrasive $H_{ab} \approx H_\alpha$. (b) Comparison of the wear resistance of a very fine dispersion (MA, tempered martensite) and a coarse dispersion (DP, dual-phase structure) of equal bulk hardness. HSLA steel: 0.11 wt.% C, 1.54 wt.% Mn, 0.50 wt.% Si, 0.15 wt.% Mo. MS = martensite; MF = martensite + ferrite.

can be based on the mechanisms which have been described earlier in this paper. They will differ in different surfaces of one material. Anisotropy can originate from the levels of phase [10] and of microstructure (Fig. 3). The subject of this paper is microstructural anisotropy only.

Using the coordinate system of Fig. 4 and eqns. (7) and (11) for the dissipation of energy du/dx and volumetric wear rate w, these properties can be expressed for anisotropic microstructures. The specific energy, Δu, used for a sliding distance x is

$$\Delta u_i = \frac{du_i}{dx_j} x_j \tag{26a}$$

where u_i is the energy per unit area in the surface normal to i. The sliding direction is j. For example, $u_1 = U/a_2 a_3$ or $u_2 = U/a_1 a_3$, with $i = 1, 2, 3$, and $j = 1, 2, 3$ as

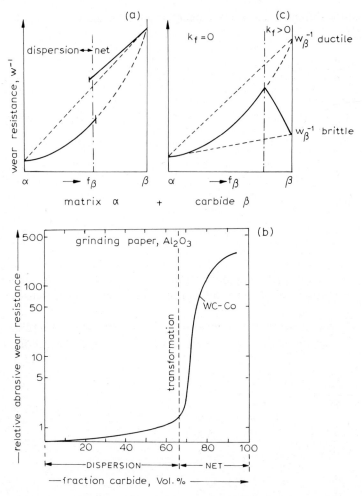

Fig. 18. Consequences of microstructural transformation on wear resistance (schematic). (a) Increased resistance due to transition dispersion → net. (b) Wear resistance of sintered hard metals. (c) Ductile → brittle transition inducing a decreasing wear resistance above a certain volume fraction [see Fig. 12(b)].

indicated in Fig. 4. Δu_i and du_i/dx_j are average values of the energy used, for example, during the formation of a groove by abrasion. This energy is also directly related to the geometry of the microstructure. The dissipation of frictional energy possesses a fine structure with the path of sliding (Fig. 5). This structure is of an origin different from vibrations caused by stick–slip in homogeneous materials. The tensor of the frictional energy has nine components, or eighteen if the sign of the sliding direction has to be considered: $\pm x_j$. The difference between the coefficients

du_i/dx_j originate from the differences in surface microstructure in the two principle directions of the sliding.

$$\Delta u_1 = \frac{\delta u_1}{\delta x_1} x_1 + \frac{\delta u_2}{\delta x_1} x_1 + \frac{\delta u_3}{\delta x_1} x_1$$

$$\Delta u_2 = \frac{\delta u_1}{\delta x_2} x_2 + \frac{\delta u_2}{\delta x_2} x_2 + \frac{\delta u_3}{\delta x_2} x_2 \tag{26b}$$

$$\Delta u_3 = \frac{\delta u_1}{\delta x_3} x_3 + \frac{\delta u_2}{\delta x_3} x_3 + \frac{\delta u_3}{\delta x_3} x_3$$

For sliding and abrasion

$$\frac{\delta u_1}{\delta x_1} = \frac{\delta u_2}{\delta x_2} = \frac{\delta u_3}{\delta x_3} = 0 \tag{26c}$$

For erosive loading conditions, these components must not become zero. The particular coefficients of friction can be derived from the components of this tensor using eqn. (8).

A similar form is suitable to describe the anisotropic volumetric wear rate [28]. Δa_i is the dimensional change in the directions $i = 1, 2, 3$ by sliding in the directions x_j, $j = 1, 2, 3$.

$$\Delta a_1 = \frac{\delta a_1}{\delta x_1} x_1 + \frac{\delta a_1}{\delta x_2} x_2 + \frac{\delta a_1}{\delta x_3} x_3$$

$$\Delta a_2 = \frac{\delta a_2}{\delta x_1} x_1 + \frac{\delta a_2}{\delta x_2} x_2 + \frac{\delta a_2}{\delta x_3} x_3 \tag{27a}$$

$$\Delta a_3 = \frac{\delta a_3}{\delta x_1} x_1 + \frac{\delta a_3}{\delta x_2} x_2 + \frac{\delta a_3}{\delta x_3} x_3$$

For sliding and abrasion, three components must become zero

$$\frac{\delta a_1}{\delta x_1} = \frac{\delta a_2}{\delta x_2} = \frac{\delta a_3}{\delta x_3} = 0 \tag{27b}$$

Erosive loading conditions may provide components perpendicular to the surface. Consequently, the components given in eqn. (27b) may not be equal to zero under these circumstances (Fig. 19).

The general formulation of the wear rate is

$$w_{ij} = \frac{da_i}{dx_j} \tag{27c}$$

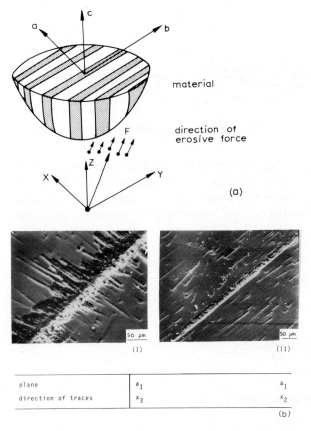

Fig. 19. (a) Conditions for the most general case of the tensor. Anisotropic material: $a = a_1$, $b = a_2$, $c = a_3$, and three directions of erosive load $x = x_1$, $y = x_2$, $z = x_3$. (b) Anisotropic abrasive wear of a fiber-composite scratched with a diamond needle parallel and perpendicular to the fibers (compare Fig. 6(b), Table 2).

For sliding and abrasion, the six components of the wear tensor are

$$w_{ij} = \begin{pmatrix} 0 & w_{12} & w_{13} \\ w_{21} & 0 & w_{23} \\ w_{31} & w_{32} & 0 \end{pmatrix} \tag{28}$$

Composite materials can be subdivided in three classes with increasing asymmetry and consequently 1, 3 and 6 principle wear rates (Fig. 2).

(a) $w_{12} \neq w_{13} \neq w_{21} \neq w_{23} \neq w_{31} \neq w_{32}$

(b) $w_{12} = w_{21} \neq w_{13} = w_{23} \neq w_{31} = w_{32}$ (29)

(c) $w_{12} = w_{13} = w_{21} = w_{23} = w_{31} = w_{32}$

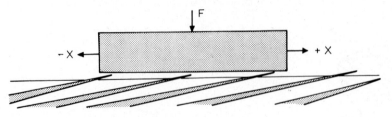

Fig. 20. Microstructure for which friction and wear depend on the direction of sliding $\pm x$ (schematic).

The symmetrical structures (a) have been discussed already. Uni-axial anisotropy (b) implies 3 components of the abrasive wear rates. It is represented by lamellates and uni-axial fiber enforcement. The number of components is doubled if wear depends on the direction of sliding. This is the case for fish-scale-like structures (Fig. 20 and Table 2).

Relations which control the bulk wear have been obtained from the discussion of the different types of isotropic structure (Chap. 7). They must be valid for particular surfaces of the anisotropic materials. A lamellate can be described by the components (cf. Fig. 13)

$$w_{31} = w_{32} = w_\alpha f_\alpha + w_\beta f_\beta \tag{30}$$

$w_{23} = w_{13} =$ perpendicular to the lamellae

$w_{21} = w_{12} =$ parallel to the lamellae

The reason for the difference represented by

$$w_{23} = w_{13} \neq w_{21} = w_{12}$$

may be due to the fact that a perpendicular passage of lamellae (or fibers) will favor

TABLE 2
Friction and wear of CF–EP ($\sigma = 2.22$ MPa) as a function of the direction of sliding
N, P and AP indicate directions normal, parallel and perpendicular (antiparallel) to the direction of fiber axes, respectively.

	Abrasion by 70 μm Al$_2$O$_3$ grain paper			Abrasion by 7 μm Al$_2$O$_3$ grain paper			Wear by diamond needle		
	N	AP	P	N	AP	P	N	AP	P
μ [1]	0.69	0.64	0.58	0.64	0.50	0.46	0.62	0.54 [a]	0.27 [a]
$\dfrac{w}{\sigma}$ [10^4 MPa^{-1}]	3.2	6.1	4.5	0.35	0.32	0.32			

[a] See Figs. 3–5 and 19(b).

separation of the $\alpha-\beta$ interface but a parallel passage will not (Fig. 14). The description of the components follows eqns. (22)–(25).

An important implication of microstructural anisotropy is that the direct relations to volume fractions become invalid [for example, eqn. (23), Fig. 5]. Geometric relations exist between volume fraction f_β and, for example, $f_{\beta 31}$. The index 31 indicates the fraction of β-phase encountered during sliding in the plane perpendicular to a_3 in a direction x_1. Equation (23) becomes [eqns. (26) and (27)]

$$w_{31}^{-1} = \frac{f_{\alpha 31}}{w_\alpha} + \frac{f_{\beta 31}}{w_\beta} \tag{31}$$

The wear resistance of composites can be optimized by adjusting the direction and plane of sliding to the optimum orientation of the material if the complete wear tensor is known.

Systematic experimental work on the effects of the type of microstructure and microstructural anisotropy on friction and wear is still in an early stage [17–20]. Composite materials will provide a wide field for such work with a good perspective for improvement of wear systems. The general principles outlined here could form a framework for such developments. The formulations describe relatively simple structures. There is a long way to go for a full and quantitative understanding of more complex or hybrid composite structures.

9. Relations between friction and wear

In practical systems, the relation between friction and wear is a complex matter. Many experienced researchers claim that not even qualitative rules exist. However, from a fundamental point of view, the situation is quite straightforward. The energy dissipated by friction induces stresses and structural changes in the surface [eqns. (7) and (8)]. A certain fraction of this energy is used for elastic and plastic deformation and for fracture. A relation should follow for the fraction of frictional energy which is utilized for the mechanisms which provide formation of wear particles. A correlation between friction and wear should be based on the equivalent tribological properties (Table 1), namely μ and w/σ [eqn. (8)].

$$\mu = \frac{\tau_R}{\sigma} = \frac{d\gamma}{dx} \frac{1}{H} \tag{32}$$

$$\frac{w}{\sigma} = k\frac{1}{H} = \frac{dm}{dx\rho} \frac{1}{\sigma} \tag{33}$$

$$k = \mu p \tag{34}$$

where p is the fraction of frictional energy which is used to produce wear ($0 \leqslant p \leqslant p_{max} = k_{max}/\mu$). p_{max} is determined by the hypothetical situation that the total

TABLE 3
Combinations of desired tribological properties

μ	$\dfrac{w}{\sigma}$	Examples for application
Min.	Min.	Bearings, pipings
Max.	Min.	Brakes, tires
Min.	Max.	Machinability, grindability
Max.	Max.	None ? ?

energy produces wear. $p = 0$ for damage-free adhesion–decohesion at the original interface, $d\gamma_{ad}/dx$, or for elastic deformation, $d\gamma_{el}/dx$, but not for direct crack formation, $d\gamma_f/dx$, and plastic deformation, $d\gamma_d/dx$, if it induces internal stresses and defects which prepare the structure for cracking [eqn. (14)].

Composite materials provide a chance to minimize friction and wear [15,16]. They should be composed of a component which leads to minimum frictional force ($d\gamma/dx \to$ min, symmetrical polymers) and a structure which requires a high energy for decohesion (tough metals) $p \to 0$. The third requirement for a successful structural design is always a high hardness H [eqn. (34)], provided, for example, by a net structure of a hard phase.

There are various requirements for optimum combinations of friction and wear in engineering applications. The principle cases are listed in Table 3. Optimum combination of these properties is usually not achieved by individual materials, but by composites which combine suitable materials in well-defined microstructural morphologies.

List of symbols

A	asperity area (m^2)
A_0	nominal area (m^2)
A_b	area of boundaries, interfaces (m^2)
a	dimension of material (m)
a_i	dimension in direction i (m)
Δa	change in dimension (m)
a_c	critical crack length (m)
d_β	particle diameter (m)
E	elastic modulus (Pa)
F	compressive force (N)
F_i	frictional force (N)
f_α, f_β	volume fraction
$f_{\alpha j}$	fraction of phases in area ij
G_{1C}	crack extension energy (J m^{-2})
H, HB, HV	indentation hardness (Brinell, Vickers) (Pa)

H_α, H_β	microhardness of α, β (Pa)
$i = 1, 2, 3$	subscript: direction of removal of matter
$j = 1, 2, 3$	subscript: direction of sliding
K_{IC}	fracture toughness (Pa m$^{1/2}$)
k	wear coefficient
k_α, k_β	partial wear coefficient of α, β
k_d, k_f, k_i	wear coefficient (deformation, fracture, interface)
$k_\beta^+ = k_{\beta d} + k_{\beta f}$	wear coefficient of phase β
γ_{AO}, γ_{BO}	surface energy of material A, B (J m^{-2})
γ_{AB}	interfacial energy
	dissipation of energy in surface:
γ_{ad}	adhesion (J m^{-2})
γ_d	plastic deformation (J m^{-2})
γ_{el}	elastic deformation (J m^{-2})
γ_f	fracture (J m^{-2})
γ_{ch}	tribochemical reaction (J m^{-2})
γ_t	phase transformation (J m^{-2})
M	removed matter (g)
$m = M/A_0$	matter per nominal area (g m^{-2})
$m = M/A$	matter per asperity area (g m^{-2})
μ	coefficient of friction
p	fraction of energy used for wear
$\rho = M/V$	density (g m^{-3})
$\rho_{\alpha\alpha}$, $\rho_{\alpha\beta}$, ρ_b	density of boundaries, interfaces (m^{-1})
S_b	spacing of boundaries (m)
S_g	width of groove (m)
s	tribo-system factor
s^*	system factor for onset of microcracking (J m^{-2})
$\sigma = F/A_0$	compressive stress (Pa)
σ_c	critical stress for the formation of cracks (Pa)
$\tau_R = F/A_0$	frictional shear stress (Pa)
U	dissipated energy (J)
u	energy per nominal area (J m^{-2})
u_{ij}	tensor of dissipated energy (J m^{-2})
$w = w_v$	volumetric wear rate
w_w	gravimetric wear rate (g m^{-3})
w_d	wear rate caused by plastic deformation
w_f	wear rate caused by fracture
w_α, w_β	wear rate of α, β
w_{ij}	wear tensor
w^{-1}	wear resistance
x	path of sliding (m)
x_j	sliding in direction j (m)

Acknowledgements

The support of our work on friction and wear of composite materials by the Land Nordrhein–Westfalen is gratefully acknowledged.

Thanks are due to my colleagues who contributed experimental results especially micrographs: U. Herold-Schmidt [Fig. 6(a)], K.-H. Zum Gahr [Fig. 11], and K. Friedrich [Fig. 6(b), Table 2].

References

1 S.A. Saltikow, Stereometrische Metallographie, VEB Dt. Verlag Grundstoffenindustrie, Leipzig, 1974 (translated from Russian).
2 E.E. Underwood, Quantitative Stereology, Addison-Wesley, Reading, MA, 1974.
3 E. Hornbogen, Acta Metall., 32 (1984) 615.
4 F.P. Bowden and D. Tabor, Friction. An Introduction to Tribology, Anchor Press, New York, 1973.
5 D. Dowson, History of Tribology, Longmans, London, 1978.
6 J.F. Archard, J. Appl. Phys., 24 (1953) 981.
7 M.M. Krushchov and M.A. Babichev, Research on the Wear of Metals, Moscow, 1960, Chap. 18; Frict. Wear Mach. (USSR), 19 (1965) 1.
8 K.H. Habig, Verschleiss und Härte von Werkstoffen, Hanser, München, 1980.
9 H.M. Bauschke, E. Hornbogen and K.H. Zum Gahr, Z. Metallkd., 72 (1981) 1.
10 B.J. Briscoe and D. Tabor, Br. Polym. J., 10 (1978) 74.
11 D.W. van Krevelen and P.J. Hoftyzer, Properties of Polymers, Elsevier, Amsterdam, 1976, p. 170.
12 J.K. Lancaster, Wear, 14 (1966) 223.
13 E. Hornbogen and K. Schäfer, in D.A. Rigney (Ed.), Fundamentals of Friction and Wear of Materials, American Society for Metals, Metals Park, OH, 1981, pp. 409–438.
14 G.C. Pratt, in M.O.W. Richardson (Ed.), Polymer Composites, Applied Science Publishers, London, 1977, pp. 237–261.
15 J.P. Giltrow, Composites, 3 (1973) 55.
16 J.M. Thorp, Tribology, 4 (1982) 69.
17 E. Hornbogen, Wear, 33 (1975) 251.
18 K.H. Zum Gahr, Z. Metallkd., 67 (1976) 678.
19 A.R. Rosenfield, Br. Polym. J., 10 (1978) 221.
20 K.H. Zum Gahr, in K.H. Zum Gahr (Ed.), Reibung und Verschleiss, DGM, Oberursel, 1983, pp. 135–156.
21 J. Becker, E. Hornbogen and K. Rittner, Prakt. Metallogr., 15 (1983) 342.
22 S.V. Prasad and P.D. Calvert, J. Mater. Sci., 15 (1980) 1746.
23 H.M. Hawthorne, Proceedings of the International Conference on Wear of Materials, ASME, New York, 1983, pp. 576–582.
24 T. Tsukizoe and N. Ohmae, Proceedings of the IV International Conference on Composite Materials, Milan, Italy, 1980.
25 K. Friedrich, Fortschr. Ber. VDI Z., 18 (15) (1984).
25 K. Tanaka and S. Rawakami, Wear, 79 (1982) 221.
27 N.H. Sung and N.P. Suh, Wear, 53 (1979) 129.
28 T.P. Harrington and R.W. Mann, J. Mater. Sci., 19 (1984) 761.

Chapter 4

Tribological Properties of Selected Polymeric Matrix Composites against Steel Surfaces

JOHN M. THORP

Coromandel (New Zealand)

Contents

Abstract

A unique tri-pin-on-disc tribometer was used to determine the tribological properties of commercial polymeric bearing materials, comprising various nylons, a silica-filled ultra-high molecular weight polyethylene (UHMWPE) and polyurethane elastomers.

At moderate to heavy loads, coefficients of friction (μ) against dry and paraffin oil-lubricated steel were independent of normal force and are presented as a function of sliding speed. Values of μ for all the test materials except UHMWPE were high without lubrication.

Fatigue (frictional) wear and abrasive (cutting) wear were determined using discs of steel gauze and abrasive paper, respectively. Wear relationships are given, enabling comparisons to be made of fatigue wear on rough steel (gauze) in non-transfer film conditions, and of cutting wear on an abrasive counterface in transfer film conditions.

The wear tests largely endorsed manufacturers' recommendations regarding applications for the test materials. Friction test results indicated, however, that some of the polyurethanes would be unsuitable as dry bearing materials. The silica-filled UHMWPE proved outstanding, with low friction against unlubricated steel, good fatigue wear resistance and excellent abrasion resistance.

1. Introduction

1.1. Properties pertaining to the tribological application of polymers

1.1.1. Performance requirements of tribological components in modern industry

The introduction to industry of new technologies and automation, with ever faster production rates and more arduous and often hostile operating conditions, has made increasing demand on the performance and reliability of tribological components, such as bearings, gears, seals, chutes and wear guides. Further, it is no longer economic to halt production for regular maintenance periods. Consequently, tribological components must be able to function in dry conditions or incorporate a lubrication-for-life system. The bearing material must suffer little wear, yet tolerate abrasion, be inert to corrosive environments and accommodate shock loading or bending to some extent.

Such advances in industry have only been made possible by the development of components manufactured from polymeric composites. The latter are not only lighter in weight, more chemically inert, and in general operate more quietly than metal components but, owing to the molecular structure and viscoelastic nature of polymers, possess unique tribological properties. A polymeric surface is able to deflect elastically to some extent on impact, thereby being able to tolerate some bending and shock loading. Abrasive or wear particles impinging on a polymeric surface tend to be repelled or may be accommodated within a soft material.

1.1.2. Applications requiring abrasion-resistant materials

Tribological applications where a polymer composite is indicated thus include vanes and gears in pumps handling industrial fluids, sewage and abrasive-contaminated water; roll neck bearings in steel mills subject to heat, shock-loading, water and mill scale; chute liners abraded by coke, coal and mineral ores; guides in bottle-handling plant; bushes and seals in agricultural and mining equipment; and sluice gate bearings to name a few examples.

1.1.3. Dry bearing applications

Manufacturers of polymeric bearing materials often stress the advantage of the inherent low friction (or inherent lubricity) possessed by polymers sliding against an unlubricated metal (usually steel) counterface. On the contrary, most polymers exhibit a high coefficient of friction (μ) when sliding against unlubricated steel under a wide range of sliding speeds, loads and temperatures. Tests performed in the author's laboratory * gave values of μ ranging between 0.3 and 0.5 for a number of commonly used polymeric bearing materials sliding against steel, whereas 0.4 is given as a representative value of μ for most polymers [1]. The high friction is often accompanied by lumpy transfer of polymer on to the counterface and high wear.

---

* Physics and Engineering Laboratory, D.S.I.R., Lower Hutt, New Zealand.

Nevertheless, in the event of failure of the lubrication system, polymeric materials may slide against a metal counterface without the catastrophic cold welding or seizure which occurs with many metal-to-metal sliding contacts.

The frictional properties of polytetrafluoroethylene (PTFE) and ultra-high molecular weight polyethylene (UHMWPE) are exceptional, relatively low values of μ (0.1 and below) being achieved on repeated sliding against unlubricated steel accompanied by the transfer of highly oriented polymer fibres. Thus, after a running-in period, during which an oriented transfer film is established on the steel counterface, the coefficient of friction decreases to the inherent low value obtained when these materials, with smooth oriented molecular profiles, slide against themselves.

Accordingly, PTFE and PE composites are of value as dry bearings in repeated sliding conditions, i.e. as plain bearings, bushes, gears and seals. Applications include those

(1) where maintenance is spasmodic or nil (bushes and seals in domestic appliances, toys and instruments),

(2) where lubrication is sparse (aircraft control linkage bearings),

(3) where lubrication is unacceptable because of product or environment contamination (plain bearings and gears in food, paper and textile industries),

(4) where lubrication is a problem (tribological components in inaccessible equipment, in hazardous conditions, in vacuum or space) and

(5) as a safeguard in the event of failure of the lubrication system (gears in trains).

1.2. Commercial polymeric bearing materials

1.2.1. Thermoplastics as bearing materials

The selection of the correct polymer or, more usually, a polymer composite for a specific tribological application is a complex problem for the industrial and design engineer.

There is currently a large, ever-increasing selection of commercial polymer-based materials available with over 60 different types sold in the U.K. alone [2] as dry bearing materials. These are based on relatively few polymer types, however, comprising polyamides (nylon 6, 6/6 and 11), polyacetals (homo- and co-polymers), polyimides (high temperature polymers), polycarbonate, ultra-high molecular weight polyethylene (UHMWPE) and the fluorocarbon polymers [primarily polytetrafluoroethylene (PTFE)]. These thermoplastics, with large flexible linear molecules, soften or melt at characteristic temperatures and, compared with metals, have relatively high elasticity. Though the polymer types used are few in number, different methods of fabrication (extrusion, injection moulding and casting) and various combinations of polymers, fillers, fibre reinforcements and stabilizers have resulted in a seemingly bewildering choice.

1.2.2. Fillers in polymer composites

Fillers (e.g. glass, carbon, asbestos, oxides and textile fibres) are incorporated within many polymers in order to improve their mechanical strength. Further, most polymers, with the exception of PTFE and PE, require the addition of solid lubricants [graphite, molybdenum disulphide (MoS_2), PTFE] in order to reduce friction. The solid lubricant particles transfer fairly readily to a metal counterface. In the case of lamellar solid lubricants, however, the percentage volume necessary to effect an optimum reduction in the coefficient of friction is sufficiently high (10 vol.% or above) to cause a deterioration in the mechanical strength of the material. At lower concentrations, PTFE is more effective and may even improve the mechanical properties of the composite [3]. Finally, bronze, silver or graphite powder fillers are used to improve the poor thermal conductivity possessed by polymers, thereby facilitating the dissipation of frictional heat. The effect of such fillers and reinforcing fibres on the friction and wear properties of the composite material is discussed elsewhere in this volume.

1.2.3. Thermosets in water-lubricated applications

Thermosetting resins, in which molecular side chains are cross-linked to form a rigid network, are used mainly in water-lubricated applications. The brittle resin must be reinforced, however, with fabric or fibres of cotton, asbestos or other. Phenolic resins reinforced with cloth have long been established as high-strength abrasion-resistant bearing materials for use as roll neck bearings in rolling mills and marine stern tube bushes. Under excessive heat, these materials char and without lubrication (or water cooling) can only operate at low loads and speeds. However, as with thermoplastics, performance is improved significantly by the incorporation of solid lubricants (graphite, MoS_2 and PTFE) or by impregnation with oil.

1.3. Availability and interpretation of design data

Most manufacturers of polymer-based bearing materials publish data relating to the physical and mechanical properties, friction and wear life of their products. However, the friction and wear data are often inadequate and the test conditions are generally poorly specified. Various test rigs and conditions are used with different contact geometries, heat flow, counterfaces, loads, sliding speeds and temperatures. Thus, results from different laboratories frequently differ and are difficult to apply to an industrial situation with, yet again, inevitable differences in surface, environmental and operating conditions. The latter are of particular importance in the case of polymers owing to the dependence of their tribological properties on pressure, sliding speed and surface temperature. Another factor of importance, yet frequently ignored, is whether conditions allow the formation of a stable transfer film of polymer (or filler) on the metal counterface.

With the ever-increasing use of polymeric composites in tribological applications in current and future developments in industry, space and medicine, the need for friction and wear data is well recognised and of vital concern. A Data Item based on manufacturers' results was published by the Engineering Sciences Data Unit [4] in 1976. The need for strictly comparative wear data on commercially available polymeric dry bearing materials prompted research projects by independent laboratories, including the National Centre of Tribology (U.K.) [5], as described by

Anderson [2], and the Physics and Engineering Laboratory (N.Z.), as described by the author [6,7].

Strictly comparative friction and wear data determined on an appropriate test rig under carefully controlled conditions are useful in quality control and in determining the relative merits of materials against a counterface, both dry and lubricated, over a wide range of loads and speeds. Provided the transfer film, contact geometry and heat flow conditions are not too different from those in the laboratory, the same order of merit could be expected in a real situation with similar operating conditions. Such friction and wear test results thus complement existing data and simplify the task of the engineer in the specification and selection of polymeric bearing materials.

In this chapter, tribological properties are presented of a selection of commercial polymeric composites both sent to the author's laboratory for testing as to their suitably in a particular application and because of common usage.

2. Experimental

2.1. The test rig

2.1.1. General

A variety of test geometries (including thrust washer, pin-on-disc and journal configurations) and a variety of test methods (including determination of limiting PV values, static and kinetic coefficients of friction, and wear rate under various conditions) have been used by manufacturers and others to provide engineering design data. In the case of wear tests, an accelerated test is often used to reduce the test time, although different wear mechanisms from those occurring in a practical situation are likely to be introduced by increasing the severity of the test conditions.

Test data published on polymer-based material systems are generally for unlubricated conditions. However, use of even a poor lubricant or sparse lubrication can have a significant effect on friction and wear behaviour. Indeed, some polymeric composites cannot be recommended for use as bearings without lubrication. The test rig should thus be designed to perform friction and wear tests on both dry and lubricated polymer/steel systems.

2.1.2. The PEL tri-pin-on-disc tribometer

The friction and wear tests performed by the author were made on a unique tri-pin-on-disc tribometer (Fig. 1) designed and built at PEL * as a versatile research and test machine with wide ranges of loads, sliding speeds and temperatures, but particularly to retain lubricating fluids up to high sliding speeds. Test pins can easily be machined from a wide range of materials, including polymer composites, whereas

---

* Physics and Engineering Laboratory, D.S.I.R., Lower Hutt, New Zealand.

Fig. 1. The tri-pin-on-disc tribometer. (From Thorp [9].)

the disc is usually of steel and the harder of the mating materials. Pin geometry can be varied (hemispherical, conical, cylindrical), but in the polymer test series cylindrical flat-ended pins were used in order to obtain conformity between the test specimen and counterface as quickly as possible, thereby reducing running-in periods.

In a conventional pin-on-disc tribometer, a stationary pin is loaded against a horizontal rotating disc. At sliding speeds typical of many industrial applications, however, centrifugal forces cause fluid lubricants to be flung off the rotating disc. Drip feed or disc-immersion methods are commonly used to replenish the lubricant. The former may still result in lubricant starvation in the contact area and presents problems of lubricant temperature control, whereas the latter requires a relatively large volume of fluid (uneconomic in the case of special lubricants), which is still agitated turbulently at high disc speeds.

The PEL tri-pin-on-disc tribometer (Fig. 1) overcomes this problem by rotating the pins instead of the disc, thereby causing comparatively little agitation of the fluid lubricant, which is retained on the stationary disc up to high (5 m s^{-1} or more) pin sliding speeds.

2.1.3. Design details

The tribometer design is described in detail elsewhere [8,9]. The main features are illustrated schematically in Fig. 2. Basically, three equi-spaced loaded cylindrical test pins are rotated on a stationary horizontal disc attached to the flat base of a disc-lubricant container (A, Fig. 2). (The Mark II version incorporates a built-in heating element which enables lubricant temperature control up to $300 \pm 0.5°C$. The

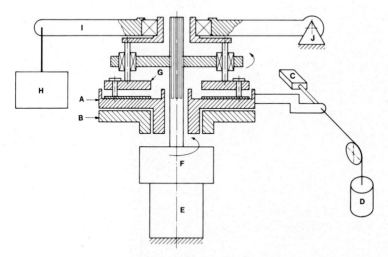

Fig. 2. A schematic drawing of the basic design of the tribometer. (From Thorp [8].) A, Disc-lubricant container and test disc; B, hydrostatic oil bearing; C, torque sensor; D, torque calibration weights; E, drive motor; F, gearbox; G, workhead containing the test pins; H, loading weights; I, loading arm; J, fulcrum.

tests described here were performed earlier on the Mark I rig, however, at a room temperature of about 20°C.)

A main feature of the design is that the disc-lubricant container is supported on a virtually frictionless hydrostatic oil bearing, B, which allows all the torque generated at the pin–disc interfaces to be sensed by the torque measuring system, C. The disc-lubricant container is restrained by a torque arm which engages a flexure-spring designed to provide a restoring torque of up to 50 Nm. The spring deflection (2 mm maximum) is sensed by a linear voltage displacement transducer and the output voltage, which is proportional to the torque, is recorded on a strip chart recorder. A built-in torque calibrating system (using two weights, D) enables a proportionality constant to be determined which is used to calculate the coefficient of friction of the test system.

The drive is from a servo-controlled 1 h.p. d.c. motor, E, which is coupled to a four speed gearbox, F, visible in Fig. 1 through the perspex front cover. A vertical splined shaft protrudes from the top of the gearbox and passes through the centre of the disc-lubricant container (and disc) and rotates the complex workhead, G, independently of the loading system, by means of a splined collet-drive spider.

The workhead G contains a lower circular pin-carrier plate which is attached to a gimbal to allow for equal distribution of the load on the three test pins. The pins are positioned in one of five available locations so that they travel on the same path on the disc, being one of five concentric circles. The applied normal force is variable, in increments of 10 N, up to 310 N. Loading is by means of weights, H, which are supported on one end of a loading arm, I. The other end of the arm is pivoted on the machine work-top at J. The normal force is transmitted to the workhead (and thence to the test pins) by means of a thrust ball bearing. The upper race of the latter is attached, at centre, to the underside of the loading arm. The lower race is mounted on top of the workhead (Fig. 3) on an upper circular thrust plate supported on three vertical pillars. These latter pass through linear ball bushes (one mounted in each of the three arms of the drive spider) and engage, at their lower ends, the gimbal/pin-carrier assembly. The low-friction ball bushes permit the workhead to be rotated yet allow the normal force to be transmitted almost entirely to the test pins, with little force (about 0.2% at maximum) being diverted through the three bearings. The latter also allow the workhead and loading arm to slide down as the test pins wear.

2.1.4. Calculation of the coefficient of friction

The kinetic coefficient of friction, μ, is calculated from the recorder deflection, x, the torque calibration constant k, the wear track radius, r, and the total normal load (force), comprising the sum of the zero load and the applied load. However, a more accurate value of μ is obtained from the slope of a plot of x against the applied load, L_a, which was found to be linear at constant sliding speed over a wide range of loads, even with polymeric materials. Then

$$\mu = \frac{k}{r} \frac{\Delta x}{\Delta L_a} \tag{1}$$

where $\Delta x / \Delta L_a$ is the slope of the linear plot.

Fig. 3. The workhead showing the drive-spider, gimbal and pin-carrier with the five pin locations visible. The lower race of the thrust ball bearing, which transmits the normal force, is mounted on the upper load-bearing plate. A steel test pin, its holding plate and retaining screw lie in front. (From Thorp [8].)

Tests have shown that, under carefully controlled sliding speed and lubricant (or dry) conditions, reproducible values of the coefficient of friction (generally within 3%) are given by a pin–disc sliding system by use of this stepwise loading procedure with the pins located on any one of the five available wear tracks [10].

2.1.5. Estimation of wear

Wear is measured by removing, cleaning and weighing the test pins at timed intervals. Marks on the shoulders of the test pins and the pin-carrier enable the pins to be returned to exactly the same positions (and orientations) each time.

2.2. Test materials

2.2.1. Polymeric test pin materials

A number of commercial thermoplastic bearing materials were tested, comprising extruded and cast nylon 6, a PTFE/PE-filled nylon 6/6, a silica-filled UHMWPE and PTFE. Three polyurethane-based elastomers were also tested.

The basic composition and typical physical properties (quoted from manufacturers' data) are given in Table 1.

TABLE 1

Typical physical properties at room temperature (from manufacturers' data) for the polymer-based bearing materials selected

Code no.	Base polymer	Filler	Colour	Specific gravity	Ultimate tensile strength (MN m⁻²)	Break elongation (%)	Hardness Rockwell	Shore D	PV limit (kNm⁻² ms⁻¹)	Limiting speed (m s⁻¹)
Thermoplastic										
100 [a] (A/B)	Nylon 6 (extruded)	Nil	White/ opaque	1.14	81.4	20	119		87.5 70	0.05 0.5
101	Nylon 6/6	PTFE, PE	Grey-black	1.18	72.4	10	116	86	525	2.5
102 [a] (A/B)	Nylon 6 (cast)	Nucleated with MoS$_2$ + 5% dye	Dark grey	1.16	82.7	30		85	105	1.5
103	UHMWPE	Silica sand	Grey-blue	0.96	38.6	334	70			
104	PTFE	Nil	White	2.1	6.9	100	58	52	42 63	0.05 0.5
Elastomeric										
200	Polyurethane	About 2% (unknown)	Black	1.24	34.5	219		68	595	
201	Polyurethane	Insignificant	Green	1.17	43.4	86		78	858	
202	Polyurethane	Insignificant	Cream	1.10	32.1	240		68	718	

[a] (A/B) samples from different batches.

Cylindrical flat-ended test pins (with a surface contact diameter of 3.5 mm for friction tests and 3.0 mm for wear tests) were machined from each of the polymeric test materials.

2.2.2. Disc materials

Friction tests were performed against a mild steel (0.03% C, 0.03% Si, 0.29% Mn) disc with a ground finish, representative of typical industrial counterfaces.

Wear tests performed on the smooth steel disc under normal load and sliding speed conditions proved too time consuming. The testing time was reduced, not by increasing the severity of the conditions (which could introduce atypical wear mechanisms), but by the use of abrasive disc surfaces, comprising stainless steel plain weave gauze (aperture size 300 μm) and silicon carbide-coated (grain size 400 grade) paper. Each abrasive disc was glued to the upper face of a chemotextile diamond polishing cloth with a self-adhesive backing. The latter facilitated the attachment and removal of the composite abrasive disc to the base of the disc container.

2.2.3. The lubricant

The lubricant used in friction tests against smooth steel was a non-additive, highly refined paraffin oil (density 0.881×10^3 kg m^{-3}, kinematic viscosity 76.0 cSt at 37.8°C and 8.0 cSt at 98.9°C, viscosity index 72).

3. Coefficients of friction given by selected polymeric materials against smooth steel

3.1. Procedure

Unlubricated tests were performed first to facilitate the cleaning procedure. The test pins were cleaned with a non-polar solvent using an ultrasonic cleaner. They were then dried in hot air and placed in the pin-carrier with forceps. The disc-lubricant container and steel disc were also solvent-cleaned and then handled with clean tissues.

An unused disc track was selected for each test material. After five polymers had been tested (i.e. after the five available disc tracks had been used) the disc surface was reground.

Test pins were initially run-in at a sliding speed of 0.1 m s^{-1} under a suitable load (which was varied between 2.0 and 6.8 MN m^{-2}) until an equilibrium steady chart recorder reading was observed. The load was then varied stepwise between the minimum load of 1.28 MN m^{-2}, corresponding to the weight of the pin-carrier/loading arm components with zero applied load, up to a maximum (depending on the material and sliding speed) as illustrated in Fig. 4.

Tests were repeated until reproducibility of the coefficients of friction (generally

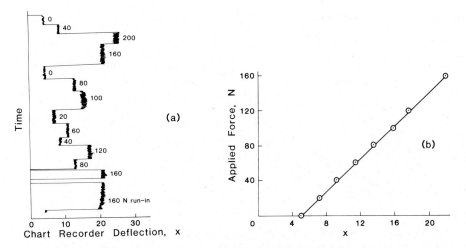

Fig. 4. (a) Chart recorder deflection, x, against time for various applied normal forces. (b) The mean value of x plotted against the applied normal force for nylon 6 pins sliding on a mild steel disc at 0.1 m s^{-1}.

to within 3%) was obtained, values of μ being calculated from the slopes of the plots of the recorder deflection against the applied load according to eqn. (1).

3.2. Results

3.2.1. Coefficient of friction as a function of load

Values of μ were found to be independent of the running-in load, although the running-in period was reduced at the higher loads which were largely used.

Following the running-in of the pins on a freshly ground steel disc, all the polymeric materials tested gave good linear plots of the chart recorder deflection against the applied load (force) over a wide range, as typically illustrated in Fig. 5. The frictional properties of the materials against smooth steel are thus presented in terms of the variation of μ with sliding speed, the values of μ being independent of normal force over the range tested and representing all but lightly loaded applications. The comparative values at a particular sliding speed should then be indicative of the relative merits of the polymeric test materials when sliding against steel of comparable surface roughness under comparable conditions, although the absolute values of μ presented here may not be those obtained in industrial or other laboratory test situations.

3.2.2. Reproducibility

Values of the coefficient of friction, μ, for duplicate (or more) tests made at 0.1 m s^{-1} under dry and lubricated conditions are given in Tables 2 and 3, respectively. The running-in load and load range over which the coefficient of friction was calculated are also tabulated. Generally, only two or three repeat tests (with new pins) were required to obtain values of μ reproducible to within a few per cent. The

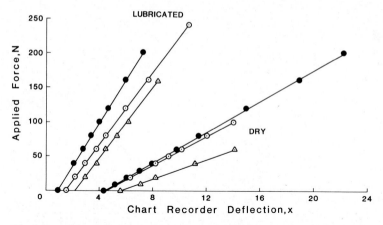

Fig. 5. Plot of the applied normal force against the chart recorder deflection (proportional to the frictional torque) for dry and lubricated nylon 6 (102) test pins sliding on mild steel at $\odot$, 0.01 m s^{-1}; $\bullet$, 0.10 m s^{-1}; and $\triangle$, 1.00 m s^{-1}.

TABLE 2

Coefficients of friction for unlubricated polymer test pins sliding at 0.10 m s^{-1} on a mild steel disc (from Thorp [7])

Material code no.	Run-in load (MN m^{-2})	Load range over which μ was calculated (MN m^{-2})	Coefficient of friction	
			μ	μ (mean)
100	1.97	1.97– 6.83	0.369	
	6.83	1.28– 6.83	0.378	
	6.83	3.36– 6.83	0.365	0.37
101	6.83	2.67– 7.52	0.354	
	6.83	1.28– 6.83	0.350	0.35
102	4.75	1.28– 8.21	0.320	
	4.05	1.28– 4.75	0.385	0.35
103	6.83	1.28–12.02	0.189	a
	6.83	1.28–11.68	0.169	a
	6.48	1.28– 9.60	0.153	a
	6.83	1.28–10.98	0.140	
	6.83	1.28–11.68	0.142	
	6.83	1.28–11.68	0.136	
	6.83	1.28– 6.83	0.136	0.14
200	4.05	1.28– 5.44	0.521	
	4.05	1.97– 4.05	0.514	0.52
201	4.75	1.97– 4.75	0.520	
	4.05	1.28– 4.05	0.505	0.51
202	4.75	1.28– 4.05	0.330	
	1.97	1.28– 4.75	0.362	
	4.05	1.28– 4.05	0.370	0.35

a Non-equilibrium value.

TABLE 3
Coefficients of friction for liquid paraffin-lubricated polymer test pins sliding at 0.10 m s^{-1} on a mild steel disc (from Thorp [7])

Material code no.	Run-in load (MN m^{-2})	Load range over which μ was calculated (MN m^{-2})	Coefficient of friction μ	μ (mean)
100	6.83	1.28– 6.83	0.109	
	6.83	1.28–11.68	0.104	0.11
101	6.83	1.28– 8.21	0.122	
	6.83	1.28– 8.21	0.130	0.13
102	4.05	1.28– 8.21	0.115	
	6.83	1.28– 8.21	0.109	0.11
103	6.83	1.28– 8.21	0.070	
	6.83	1.28– 8.21	0.065	0.07
200	6.83	1.28– 6.83	0.129	0.13
201	4.05	1.28– 4.05	0.170	0.17
202	4.05	1.28– 4.05	0.185	0.19

TABLE 4
Coefficients of friction for the dry and lubricated polymer/mild steel system as a function of sliding speed (from Thorp [7])

Material code no.	Dry μ			Lubricated μ			Comments
	$v = 0.01$ m s^{-1}	$v = 0.10$ m s^{-1}	$v = 1.00$ m s^{-1}	$v = 0.01$ m s^{-1}	$v = 0.10$ m s^{-1}	$v = 1.00$ m s^{-1}	
100	0.34	0.37	0.44	0.11	0.11	0.11	Fairly satisfactory dry at low speeds but should lubricate at higher speeds
101	0.35	0.35	0.35	0.12	0.13	0.14	Fairly satisfactory dry, better with lubrication
102	0.34	0.35	0.51	0.14	0.11	0.14	Fairly satisfactory dry at low speeds but should lubricate at higher speeds
103	0.12	0.14	0.18	0.06	0.07	0.13	Excellent dry
200	0.49	0.52	0.67	0.17	0.13	0.16	Should not run dry
201	0.47	0.51	0.49	0.20	0.17	0.06	Should not run dry
202	0.33	0.35	0.33	0.17	0.19	0.26	Fairly satisfactory dry, better with lubrication

polyethylene-based polymer took a longer period to establish a stable transfer film on the unlubricated steel disc, however, requiring some seven tests before equilibrium values could be ensured.

3.2.3. Values of μ for dry conditions

The results for dry conditions confirm the known high friction exhibited by many polymeric materials when sliding against steel under a wide range of conditions. Values of μ for the materials presented here range (with one exception) from 0.35 to 0.52 at a sliding speed of 0.10 m s^{-1} (see Table 2). The nylons (100–102) are all grouped at 0.35–0.37, μ apparently being little dependent on the method of manufacture (extruded or cast), the nylon type or filler at this speed. Benefits of the filler are illustrated (Table 4) at the higher speed (1.0 m s^{-1}), however, the PE/PTFE-filled nylon (101) then proving superior. One of the polyurethane-based elastomers (202) possesses (at 0.1 m s^{-1}) a coefficient of friction (0.35) of the same order of magnitude as the nylons, but the other two (200 and 201), with values of μ of 0.52 and 0.51, respectively, should clearly not be used as dry bearing materials. The inherent (dry) lubricity of the ultra-high molecular weight polyethylene (103) is clearly apparent (with an equilibrium value of μ of 0.14 at 0.10 m s^{-1}), as is also the transfer film mechanism which gives the least value of μ for this material after repeated sliding over the same area of steel counterface.

3.2.4. Values of μ with lubrication

Tables 3 and 4 show the significant reduction in friction (up to three-fold or more) obtained with most of the polymers on lubrication with paraffin oil. The polyethylene composite (103), which already has inherent lubricity, benefits least from lubrication yet with up to a two-fold reduction in μ nevertheless.

3.2.5. Variation of μ with sliding speed

The variation of the mean value of μ with sliding speed (0.01, 0.10 and 1.00 m s^{-1}) for both dry and lubricated conditions is shown in Table 4 together with comments as to the suitability of the material as a dry bearing over this speed (and load) range. In dry conditions, both nylon 6 samples (100 and 102) and one of the polyurethane-based elastomers (200) show significant increases in μ with increase in sliding speed, although three of the polymers (101, 201 and 202) show little or no change.

With lubrication, values of μ show less variation with speed in general owing to the more effective dissipation of frictional heat by the fluid lubricant. In one case (material 201), a significant decrease in μ was observed at the highest speed, indicating the onset of hydrodynamic lubrication. This material had a high rate of wear, resulting in an exceptionally smooth run-in contact surface.

3.3. Discussion

3.3.1. Comparison of the results with published values

As previously mentioned, values of μ may vary from one test rig to another owing to differences in contact geometry, transfer film formation, dissipation of frictional

TABLE 5

Comparison of tri-pin-on-disc coefficients of friction with manufacturers' data for unlubricated polymer/steel systems (from Thorp [7])

Material code no.	Manufacturers' data				μ	
	Test rig configuration	Load (MN m^{-2})	Speed (m s^{-1})	μ	Tri-pin-on-disc [a]	Literature values
100	Thrust washer	0.28	0.25	0.26	0.37	0.37–0.42
101	Thrust washer	1.73	0.10	0.17	0.35	
102					0.32	
103				0.11	0.14	0.15
200	Thrust washer	1.64	0.10	0.21 [b]	0.52	
201	Thrust washer	1.64	0.10	0.19 [b]	0.51	
202	Thrust washer	1.64	0.10	0.17 [b]	0.35	

[a] Load range, 1.3–7.5 MN m^{-2}; sliding speed, 0.1 m s^{-1}.
[b] Estimated from a brochure graph.

heat, surface topographies, running-in conditions (surface conformity) and surface contamination, even if the counterface, loads, sliding speeds and lubrication conditions are identical. This is clearly illustrated in Table 5, which compares the value of μ determined with the tri-pin-on-disc rig at 0.1 m s^{-1} with those quoted by the relevant manufacturers using a thrust washer configuration and similar loads and speeds. With the exception of the polyethylene, the manufacturers' values are considerably lower, yet literature values (shown in Table 5, column 7) are supportive of the values determined by the author.

Steijn [11], for example, reports values of 0.37–0.42 for coefficients of friction given by a steel sphere loaded on a slowly moving flat specimen of nylon 6. This compares reasonably well with 0.34–0.37 given for (the extruded) nylon 6 (100) against steel (see Table 4) for the slow speed range (0.01–0.10 m s^{-1}). *The Metals Reference Book* [12] gives a value of 0.4 for low sliding speeds. At higher speeds, Bilik and Donskikh [13] report a coefficient of friction of 0.37 for dry nylon 6 against steel using a roller test configuration with a slip speed of 0.5 m s^{-1} and a load of 0.2 MN m^{-2}. At a similar speed, the author's value would be of the order of 0.40. On the other hand, the manufacturer's data give the low value of 0.26 using a thrust washer configuration, a sliding speed of 0.25 m s^{-1} and a load of 0.28 MN m^{-2}.

Literature values of μ for the composite polymers sliding against steel are not available. However, the value reported [12] for unlubricated high density polyethylene (0.15) is within the range of values (0.12–0.18) obtained here for the overall (0.01–1.0 m s^{-1}) speed range at moderate to heavy loads. In this case, the manufacturer's value of 0.11 (for unspecified conditions) is in closer agreement.

3.3.2. Dependence of μ on load

In contrast to metals, the coefficient of friction given by a polymeric sliding system (at constant sliding speed) is generally dependent on load, decreasing with increase in load [11,14]. However, for the moderate to heavy load range used in these tests (i.e. from the minimum of 1.28 MN m^{-2} up to values close to the recommended limit for the test material) values of μ for both dry and lubricated conditions were virtually independent of load at each selected sliding speed. Nevertheless, higher values may be expected at lighter loads.

3.3.3. Dependence of μ on sliding speed

A plot of μ against sliding speed (or temperature) for a thermoplastic tends to go through a maximum in a similar manner to the curve given by a viscoelastic rubber [11]. Thus for most plastics μ initially increases with speed, thereby facilitating smooth sliding without stick–slip. The sliding speeds used in the tests (0.01, 0.10 and 1.00 m s^{-1}) were chosen to represent real situations, ranging from very low speed expansion bearings to sleeve bearings for medium speed applications. With most of the thermoplastic test materials, μ accordingly increased with increase in speed, although for the PTFE-filled nylon 6/6 it showed no change, presumably due to the effect of the transfer film of PTFE. Two of the elastomers (201, 202) appear to exhibit a maximum at 0.10 m s^{-1}.

3.3.4. Lubricated polymer friction

Surface adhesion forms the major contribution to polymer friction, as with metals. There is also a deformation contribution caused by the asperities of the mating surface ploughing through the viscoelastic polymer surface, which is important in the case of a hard (e.g. steel) counterface. The adhesion component to friction is virtually eliminated by use of a lubricant, as illustrated by the significant reduction in μ shown in Table 4. The deformation friction remaining is known to be related to the dynamic mechanical losses in the polymer and therefore will depend on the molecular structure and viscoelastic properties. This deformation contribution is relatively low, at least for the thermoplastic materials selected here.

3.3.5. The importance of the coefficient of friction in material selection

Apart from the strength of a polymeric material and its wear life against a particular counterface under specific conditions, the coefficient of friction is a prime factor in the selection of polymer composites for tribological applications, especially in unlubricated conditions. Excluding high friction braking uses, a low inherent (dry) coefficient of friction is generally a vital requirement, particularly for chute liners or conveyor wear strips and guides. In the selection of polymer composites for sleeve bearings and bushes, the coefficient of friction given by the material (against the appropriate counterface and under similar conditions) is indicative of the frictional heat that must be dissipated if degradation and failure of the bearing surface is to be avoided at all but the lowest sliding speeds. Consequently, only those materials possessing low frictional properties can be recommended for use as dry bearings unless sliding speeds are low (e.g. as for wheel bearings in lawn mowers and household appliances) or the frictional heat can be dissipated satisfactorily (e.g. as by use of a thin polymeric film on a steel backing, or by the incorporation of a solid lubricant either within the composite material or as a surface layer).

Published values of μ for polymer sliding systems are dependent on the test conditions, which may not represent those in the intended application. Nevertheless, the relative frictional performance of polymer composites given by a particular and appropriate test rig can be of assistance in material selection in situations where the conditions are not too different from the test conditions.

The data given in Table 4 for selected polymers and composites sliding on smooth steel under moderate to heavy loads cover a wide range of speeds. Throughout this speed range, silica-filled polyethylene (103) exhibited the least friction of the materials tested, thereby endorsing the manufacturer's recommendation for its use in dry running applications, particularly in conditions when repeated cycling allows surface films to be transferred to the metal counterface. The formation of such a beneficial transfer film is illustrated in Table 2, which shows (at 0.1 m s^{-1}) a decrease in μ for UHMWPE for the first three tests performed on the same disc track. After a further running period, a stable value of μ (0.14) is eventually attained.

All nylons exhibit values of μ that are two to three times greater than those given by the polyethylene composite at each selected speed. The PE/PTFE-filled nylon (101) proved to be superior, however, in that μ remained independent of sliding speed, at least up to 1.0 m s^{-1}. The unfilled nylons are commonly used as dry bearing materials in low speed applications, but at higher speeds lubrication or the incorporation of a solid lubricant (e.g. PTFE) within the nylon is indicated.

Two of the polyurethane-based materials (200 and 201), on the other hand, are clearly unsuitable in unlubricated conditions at even the lowest speed tested (0.01 m s^{-1}).

4. Coefficients of friction given by selected polymeric materials against an abrasive counterface

4.1. Procedure

The friction (torque arm deflection) was recorded at the beginning and at intervals during each of the abrasive wear tests. In the latter, polymeric test pins were rotated under constant load and at a sliding speed of 0.1 m s^{-1} on unlubricated steel gauze and abrasive paper discs. The coefficient of friction was calculated from

$$\mu = \frac{k}{r} \frac{x}{L_0 + L_a} \tag{2}$$

where k is the torque calibration constant, r is the radius of the wear track, x is the recorder deflection, L_0 is the normal force at zero applied load (associated with the weights of the pin-carrier and loading arm components) and L_a is the normal applied force.

Values of μ calculated in this manner contain small errors in x (associated with zeroing the recorder) and in L_0 but are good approximations for coefficients of friction greater than 0.1, as here. (These values of μ are thus less accurate than those determined with smooth steel using stepwise loading, which enables $x/(L_0 + L_a)$ to be replaced by the slope of the linear region in the plot of x against the applied normal force, L_a.)

The pin (polymeric test material) cleaning procedure was as before. The abrasive discs were of steel gauze and abrasive (silicon carbide-coated) paper, each supported on a chemotextile polishing cloth. The steel gauze was cleaned with solvent before being glued to its support cloth, whereas a new sheet of abrasive paper was used without additional cleaning.

4.2. Results

Typical friction traces recorded at the start and at intervals during wear tests on abrasive paper are shown in Fig. 6. Corresponding values of the coefficient of friction calculated at the start (μ_i) and at the end (μ_f) of each wear test were found to be remarkably reproducible and independent of the wear track diameter. Average values of μ_i and μ_f are given in Table 6 for abrasion of the polymeric materials by both steel gauze and abrasive paper. In all cases, the friction decreased during the initial running-in period. Once the pin/gauze system was fully run-in, however, the value of μ changed little with time, which infers that the transfer material remaining on the gauze tracks did not reach the level of the raised mesh intersections during each test duration. This was confirmed by observation and by the fact that the wear rate reached a constant value after the initial stabilisation period. The initial values of the coefficient of friction (μ_i) for abrasion by abrasive paper are higher than those given by steel gauze, but after ten minutes' wear (the test duration time with abrasive paper) values of μ_f reached similar values to those given by the metal gauze with four of the materials (101, 201, 102B and 200) and an even lower coefficient of friction with material 202, quoted by the manufacturer as a "low friction" polyurethane-based elastomer. The abrasive paper wear track was clogged with a thick transfer film in the latter case.

The percentage reduction in friction $[R_f = 100(\mu_i - \mu_f)/\mu_i]$ gives a measure of the transfer film deposited. The magnitude of values of R_f given by the abrasive paper wear tests (shown in Table 6, column 7) corresponds well with the estimate (shown in column 8) of the thickness of the stable deposit (i.e. not loose wear debris) observed on the abrasive paper tracks. The effect is less marked with steel gauze which is not clogged by polymer films or wear debris for a considerable time.

Comparison of the final (equilibrium or run-in) values of the coefficient of friction (μ_f), listed in Table 6, columns 2, 4, and 6, representing, respectively, the frictional performance of similarly loaded polymer test pins sliding at 0.10 m s^{-1} on smooth steel, steel gauze and abrasive paper, illustrates two types of behaviour.

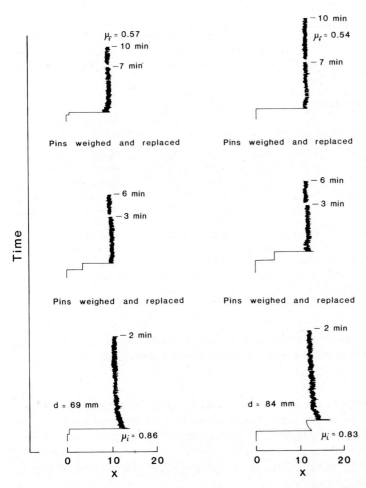

Fig. 6. Friction traces given by test pins of (cast) nylon 6 (102B) sliding on 69 and 84 mm diameter wear tracks on a SiC abrasive paper disc at timed intervals during a ten minute abrasion test. Sliding speed, 0.1 m s^{-1}; load per pin, 2.69 MN m^{-2}. (From Thorp [6].)

Excluding PTFE (104), which was only tested against steel gauze, the uppermost three polymers shown in Table 6 [the PTFE-filled nylon 6/6 (101) and poly-urethanes 202 and 201] built up heavy stable deposits on the rough counterfaces which reduced the coefficient of friction until a value even lower than that given by the same material sliding on smooth steel was reached. The transfer film on the latter is presumably less efficient than that on rough counterfaces. The remaining materials listed in Table 6 [the nylons 6 (102B, 100A), UHMWPE (103) and polyurethane 200] behaved conversely, with values of μ_f given by smooth steel being lower than those given by abrasive counterfaces. Two of these latter types of materials (the nylons 6) were observed to produce loose wear debris on abrasion. It would seem apparent

TABLE 6

Comparison of run-in values of the coefficient of friction given by the polymers sliding at 0.1 m s^{-1} on steel, μ_s, with initial, μ_i, and final, μ_f, values on fresh steel gauze and abrasive paper. The percentage reduction in μ_i [$R = 100(\mu_i - \mu_f)/\mu_i$] gives a measure of the effectiveness of the transfer film (from Thorp [6])

Material code no.	Steel μ_s	Steel gauze		Abrasive paper			
		μ_i	μ_f	μ_i	μ_f	R_f (%)	Deposit on track
104		0.48	0.33				
202	0.35	0.48	0.33	0.70	0.21	70	Heavy
101	0.35	0.42	0.29	0.63	0.26	59	Heavy
201	0.51	0.39	0.31	0.59	0.32	46	Heavy/medium [a]
102B	0.35	0.54	0.55	0.83	0.56	33	Medium
103	0.14	0.39	0.28	0.64	0.48	25	Medium
100B				0.83	0.64	23	Medium/small [a]
200	0.52	0.75	0.67	0.85	0.66	22	Medium
102A				0.68	0.53	22	Medium/small [a]
100A	0.37	0.47	0.36	0.72	0.67	7	Small [a]

[a] Loose debris on the abrasive paper track.

that the second type of material forms a more stable transfer film (if any) on smooth steel than on a rough counterface, which produces lumpy transfer debris.

4.3. Discussion

The surface roughness of the metal counterface is clearly an important factor in determining the stability of the wear transfer film, be it of polymer filaments (as with polyethylene), filler (as with the PTFE-filled nylon), or both. Thus the PTFE filler in the nylon 6/6 (101) appears to be entrapped more efficiently by a rougher surface. The polyethylene transfer film, on the other hand, is of greater benefit when deposited on a smooth steel surface.

5. Abrasive wear of selected polymeric materials

The acceptability of a polymeric composite for a tribological application, particularly in abrasive conditions, depends on a number of factors including the suitability of the physical and mechanical properties of the material, the ease of component fabrication, the relative cost, and most important the frictional behaviour and wear life of the material against the intended counterface under the required operating and environmental conditions. Of these factors, the wear life is probably the most significant and also the most difficult to determine. Clearly, for an expected wear life of some years, test time must be reduced to days, hours or even minutes, yet without changing the types of wear encountered in a real situation. Consequently, accelerated wear tests in which the severity of the conditions (loads, speeds, temperatures) are increased are unlikely to give a projected wear life representative of milder conditions.

Polymers wear abrasively by two extreme mechanisms [15–17]. One extreme is when the polymer surface is plastically deformed and micro-cut by hard sharp counterface asperities. The other extreme is when the polymer surface is elastically deformed and hence fatigued by rounded counterface asperities. In plastic deformation (abrasive) wear, furrows are made in the polymer surface in the direction of sliding. In elastic deformation (fatigue) wear, the furrows (or ridges) are perpendicular to the sliding direction. A combination of these two types of surface degradation and their interaction determines the overall wear of the material. The relative amounts of abrasive and fatigue wear thus depend on the sharpness of the abrading asperities and the elasticity of the polymeric surface.

The abrasion resistances of a polymeric material determined with both abrasive paper (sharp projections) and metal gauze (rounded wire mesh intersection asperities) will thus give a measure of the two extreme types of wear. The wear life of the material in a real situation can then be expected to lie within these extremes and may be estimated approximately from the topography of the metal counterface.

In abrasive wear tests performed on pin-on-disc type rigs, test pins are often made to traverse a spiral path on fresh abrasive paper in order to avoid the effects of transfer films and wear debris.

A circular path is generally used on metal gauze, however, since it does not become clogged with wear debris and transfer films and the abrasive capacity remains unchanged for some considerable time. Steel gauze results thus represent single traversal or non-transfer film conditions. Applications include wear strips, chute liners, and sleeve bearings and bushes in liquid environments which wash away wear debris.

In single traversal (non-transfer film) conditions, the wear volume of a polymer is proportional to the sliding distance. The specific wear rate, v_{sp}, given by the wear volume per unit area per unit sliding distance, varies with the nominal pressure, p, according to

$$v_{sp} = Kp^\alpha \tag{3}$$

where K is a proportionality constant representing the specific wear rate at unit pressure, and α is a parameter dependent on the properties of both the abrading surface and the polymer type which is specific to fatigue wear [15–17]. For wear on metal gauze $\alpha > 1$, whereas for abrasion by fresh abrasive paper $\alpha = 1$. (The specific wear rate is thus directly proportional to pressure for abrasion of a polymer by fresh abrasive paper.)

Ratner et al. and others [15–18] have found a correlation between the wear of polymers on metal gauze and single traversal wear on steel. Gauze wear tests thus provide useful data for non-transfer film applications of polymers against steel.

Neither metal gauze wear tests nor spiral path abrasive paper tests take into account the benefits which ensue when a polymer film or solid lubricant filler is transferred to a dry counterface during repeated traversals over the same path. The equilibrium wear rate finally attained in transfer-film conditions (as with a steel shaft rotating in a dry polymeric sleeve bearing) is proportional to the load [18], in contrast to results obtained with metal gauze.

Wear tests in the author's laboratory were performed using the tri-pin-on-disc tribometer in which three polymer test pins were cycled repeatedly on circular wear paths on steel gauze and abrasive paper discs. Unlubricated conditions and a constant intermediate sliding speed of 0.1 m s^{-1} were adopted throughout. Wear was measured by removing, cleaning and weighing the test pins at timed intervals. The numbered test pins were replaced in the pin-carrier in the same (similarly numbered) locations. Identical orientation of the pins was ensured by aligning marks on the pin shoulder and pin-holder. An optical stereo-microscope was used to check that the wear furrows on each pin were in the direction of sliding for abrasion by abrasive paper and perpendicular to the direction of sliding for fatigue wear by steel gauze. Loose wear debris was blown off the disc wear track before the test was continued.

The gauze wear results detailed in the following sections thus give the comparative fatigue wear of the selected polymers against steel in single traversal conditions. The abrasive paper results, on the other hand, give the comparative abrasive wear of the materials in conditions where transfer films reduce the abrasive capacity of the counterface.

5.1. Fatigue wear against stainless steel gauze

5.1.1. Procedure

The duration of each wear test ranged from 2 min for the least wear resistant material (a faulty polyurethane-based elastomer (201) with small cavities distributed throughout the material) to 1–2 h for the most wear resistant material, a cast nylon 6 (102B).

To determine the relationship between the specific wear rate and the nominal contact pressure, a series of wear tests was performed on steel gauze at various loads using a fixed sliding speed of 0.1 m s^{-1}. For each load, the wear test was continued until the plot of the wear (the average loss in weight per pin, ΔW) against running time, t, showed sufficient linearity (generally three or four points) to allow an accurate estimate of the wear rate ($\Delta W/t$) to be made from the slope.

To check the reproducibility, both new and used test pins were run on different fresh gauze tracks under the same load and on the same used track under different loads. In all such variations, the wear rate was found to be reproducible to within 1%, the only differences being in the duration of the running-in periods. (Thus worn pins sliding on a previously used track did not need a further running-in period, the linear plot of ΔW against t passing through the origin.)

Polytetrafluoroethylene (104) was selected for wear tests on steel gauze in which the sliding speed was varied stepwise from 0.02 m s^{-1} up to 1 m s^{-1}. A constant pressure of 2.22 MN m^{-2} was maintained throughout these speed tests.

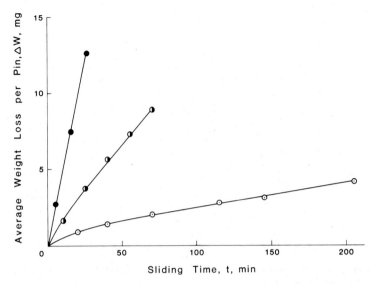

Fig. 7. Average weight loss per pin against sliding time for test pins of three nylon types rotating on a steel gauze disc. Sliding speed, 0.1 m s^{-1}; load per pin 5.05 MN m^{-2}. ⊙, Unfilled nylon 6 (100A); ◑, dyed (cast) nylon 6 (102B); ●, filled nylon 6.6 (101).

5.1.2. Results and discussion

5.1.2.1. Wear as a function of load. Typical plots of the average weight loss per pin, ΔW, against sliding time t on steel gauze are shown in Fig. 7 for the three nylons under a nominal contact pressure of 5.05 MN m^{-2} and at a fixed sliding speed of 0.1 m s^{-1}.

The specific wear rate, ν_{sp} (mm^3 m^{-2} m^{-1}), is calculated from

$$\nu_{sp} = \frac{s}{A \rho V} \tag{4}$$

where s is the linear mass wear rate ($\Delta W/t$) in mg min^{-1} for a specific load, determined (using the method of least squares) from the linear plot of ΔW against t, A is the pin contact area (m^2), ρ is the material density (mg mm^{-3}) and V is the sliding speed in m min^{-1}.

Typical plots of the wear ΔW as a function of load are illustrated in Fig. 8 for the gauze abrasion of the silica-filled UHMW polyethylene. Logarithmic plots of the specific wear rate ν_{sp} against the nominal contact pressure p (Fig. 9) were found to be linear for all the materials tested, thereby confirming the validity of eqn. (3). Values of the parameter K and exponent α, listed in Table 7, were obtained respectively from the intercept and slope of each plot according to

$$\log \nu_{sp} = \log K + \alpha \log p \tag{5}$$

with ν_{sp} in mm^3 m^{-2} m^{-1} and p in N m^{-2}.

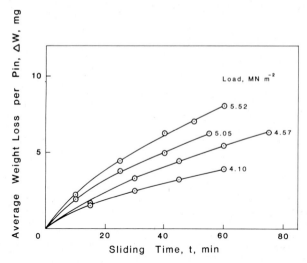

Fig. 8. Average weight loss per pin against sliding time for test pins of filled UHMWPE rotating on a steel gauze disc under various loads. Sliding speed, 0.1 m s^{-1}.

The value of K for a particular material depends on the units used for pressure and specific wear rate. The values of K previously published * [6] were calculated with p in kg(f) cm^{-2} and ν_{sp} in mm^3 cm^{-2} m^{-1} and are therefore related to the values given in Table 7 according to log K (mm^3 N^{-1} m^{-1}) = log K (mm^3 kg(f)$^{-1}$ m^{-1}) $-(5\alpha - 4)$, to a close approximation.

Unfilled nylon 6 (100A) shows a rapid increase in wear rate above a critical pressure of about 5.7 MN m^{-2} (see Fig. 9) which is attributed to thermal softening. The values of α and K listed in Table 7 are accordingly deduced from the linear region below the critical pressure. Similar sudden changes in wear rate with increase in load have been reported by Evans and Lancaster [19] for various polymers sliding on mild steel.

The pressure exponent α is specific to fatigue wear and depends on the polymer type and the properties of the abrading surface.

It has been established that α increases with increase in the number of cycles for abrasive degradation of a polymer surface [16,17]. Thus, α increases with increase in the degree of smoothness of the metal gauze (defined by n^2/r, where n is the number of grid meshes per cm^2 and r is the radius of the grid wire) and with increase in polymer molecular interaction forces.

---

* In publication, values of K submitted by the author in ref. 6 were mistakenly divided by 9.81 in an attempt to change the units from mm^3 kg(f)$^{-1}$ m^{-1} to mm^3 N^{-1} m^{-1}. The correct values of K [in mm^3 kg(f)$^{-1}$ m^{-1}] are obtained accordingly by multiplying those listed in Table 3 of ref. 6 by 9.81. Such values can then be related to those given in Table 7 (in mm^3 N^{-1} m^{-1}). Values of α obtained from the slopes of log p vs. log ν_{sp} plots are not affected by the units used.

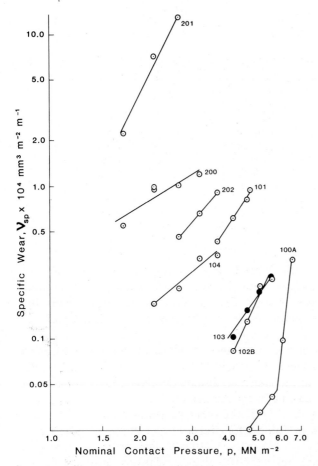

Fig. 9. Logarithmic plots of specific wear rate against nominal contact pressure for the wear of selected polymers against steel gauze. Sliding speed, 0.1 m s^{-1}. (From Thorp [6].)

Values of α listed in Table 7 range from 1.17 for polyurethane 200 up to 4.11 for polyurethane 201, illustrating the diversity of these polyurethane-based elastomers. Values for the thermoplastics lie within the range 1.60–4.07. Ratner and Farberova [17] report values ranging from 1 to 3 for rigid thermoplastics.

Table 8 compares values reported by Ratner and Farberova, α(R & F), for nylon 6, high density PE and PTFE with values determined by the author, α(T), for similar materials (nylon 6, silica-filled UHMWPE and PTFE). Values of α(T) are, on average, about 1.5 times greater than corresponding values of α(R & F), indicating that either a rougher gauze was used by the Soviet workers or their faster sliding speed (0.3 m s^{-1}) and hence greater frictional heat contributed to the differences.

The values of α reported for PTFE [α(T) = 1.6, α(R & F) = 1.2] are in agreement with those reported by Uchiyama and Tanaka (1.26 and 1.63 at 29 and 80°C,

116

TABLE 7

Values of the pressure exponent, α, and the wear factor, K, (from $\nu_{sp} = Kp^{\alpha}$) for polymeric materials abraded on steel gauze at a sliding space of 0.1 m s^{-1}

R is the linear correlation coefficient for the plot of log ν_{sp} against log p.

Material code no.	α	K (mm^3 N^{-1} m^{-1})	R
100A	2.67	4.32×10^{-16}	0.999
101	2.42	6.40×10^{-13}	0.992
102B	4.07	1.05×10^{-24}	0.972
103	2.70	1.67×10^{-15}	0.999
104	1.60	1.19×10^{-7}	0.971
200	1.17	3.02×10^{-4}	0.890
201	4.11	5.03×10^{-22}	0.993
202	2.22	2.59×10^{-11}	0.999

respectively) for 3 mm diameter pins of PTFE on a chromium-plated brass disc [20]. These low values of α are associated with the low intermolecular forces between surface PTFE molecules which contribute (beneficially) to the exceptionally low frictional properties and (disadvantageously) to the low wear resistance of this material.

Table 7 illustrates the superior wear resistance of the cast nylon 6 (102B) under the test conditions, with a wear factor K of only 1.05×10^{-24} mm^3 N^{-1} m^{-1}. (The manufacturer states that traces of molybdenum disulphide were used to nucleate the crystalline structure during the casting of this material.)

The wear resistances (at 0.1 m s^{-1}) of the other polymers relative to nylon 6 (102B) (i.e. ν_{sp}/ν_{102B}) are given in Table 9 for pressures of 1 N m^{-2}, 0.1 MN m^{-2} and 1 MN m^{-2}. The unfilled extruded nylon 6 (100A) rates second at the higher pressures, whereas the filled nylonm 6.6 (101) gives a relatively mediocre perfor-

TABLE 8

Comparison of values of $\alpha(T)$ [a] with $\alpha(R \& F)$ [b] for similar polymers abraded on steel gauze (from Thorp [6])

Material	$\alpha(T)$	$\alpha(R \& F)$	$\alpha(T)/\alpha(R \& F)$
Nylon 6/6 (filled)	2.4		
Nylon 6 (unfilled)	2.7	1.5	1.8
High density PE		2.1	
UHMWPE (filled)	2.7		1.3
PTFE	1.6	1.2	1.3
Babbitt metal		1.0	
Aluminium		2.8	

[a] $\alpha(T)$ From Thorp [6], obtained with a sliding speed of 0.1 m s^{-1}.
[b] $\alpha(R \& F)$ From Ratner and Farberova [17], using a speed of 0.3 m s^{-1}.

TABLE 9
Specific gauze wear rate relative to nylon 6 (102B) at pressures of 1 N m^{-2}, 0.1 MN m^{-2} and 1 MN m^{-2}
At each pressure the ranking order of the materials is listed, with 1 being most wear resistant and 8 being least wear resistant.

Material code no.	α	1 N m^{-2}		0.1 MN m^{-2}		1 MN m^{-2}	
		ν_{sp}/ν_{102B}	Ranking order	ν_{sp}/ν_{102B}	Ranking order	ν_{sp}/ν_{102B}	Ranking order
102B	4.1	1.0	1	1	1	1.0	1
100A	2.7	4.1×10^{8}	3	41	2	1.6	2
103	2.7	1.6×10^{9}	4	222	3	9.5	3
201	4.1	4.8×10^{2}	2	794	4	878	7
101	2.4	6.1×10^{11}	5	3300	5	73.5	4
202	2.2	2.5×10^{13}	6	13500	6	190	6
104	1.6	1.1×10^{17}	7	51300	7	174	5
200	1.2	2.9×10^{20}	8	962700	8	1226	8

mance owing to the weakening of the material strength by the filler. Two of the polyurethane elastomers and PTFE show the poorest overall performances over this pressure range and at this speed.

In general, a high value of α (Table 9, column 2) is associated with good (fatigue) wear resistance, whereas the least wear-resistant materials (200 and 104) have the lowest values of α (1.2 and 1.6, respectively). Material 201 would appear to be an exception, possessing a high α (4.1) and good wear resistance at 1 N m^{-2}, yet deteriorating rapidly with increase in pressure. This material was found to contain small cavities throughout. The high wear at the higher pressures can therefore be explained in terms of the collapse of the cavities and the disintegration of the material under pressure. Had the cavities not been present, this material could have been expected to possess high fatigue wear resistance on a par with the cast nylon (102B), as illustrated by the comparable values of α and K given in Table 7.

5.1.2.2. Wear as a function of speed. Polytetrafluorethylene (104) was selected for wear tests against steel gauze in which the sliding speed was increased stepwise from

TABLE 10
Stabilised average coefficient of friction, μ, and the specific wear rate of PTFE (104) on metal gauze as a function of sliding speed
Pressure = 2.216 MN m^{-2}.

v (m s^{-1})	a (mg min^{-1})	ν_{sp} (mm^3 m^{-2} m^{-1})	μ
0.02	0.033	1850	0.31
0.10	0.151	1700	0.32
0.20	0.323	1810	0.31
0.35	0.599	1920	0.31
1.00	0.395	3810	0.37

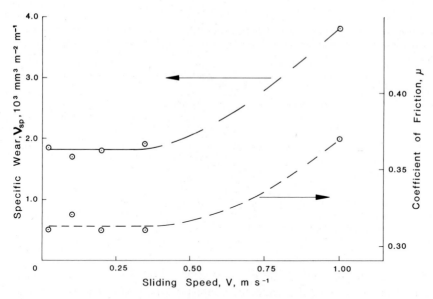

Fig. 10. Specific wear rate and stabilised coefficient of friction of PTFE (104) on steel gauze as a function of sliding speed. Nominal contact pressure per pin, 2.22 MN m^{-2}.

0.02 m s^{-1} up to 1.0 m s^{-1}. A constant pin (nominal) contact pressure of 2.22 MN m^{-2} was maintained throughout. The stabilised average coefficient of friction and the specific wear rate at 0.02, 0.1, 0.2, 0.35 and 1.0 m s^{-1} are listed in Table 10 and illustrated graphically in Fig. 10.

The specific wear rate is little changed for sliding speeds up to 0.35 m s^{-1}. At 1.0 m s^{-1}, however the wear rate has doubled. The variation of the coefficient of friction with sliding speed behaves similarly, as illustrated in Fig. 10, there being virtually no change in μ from 0.02 up to 0.35 m s^{-1}, but with a marked increase at 1.0 m s^{-1}. Clearly, the increase in friction and therefore in frictional heating above 0.35 m s^{-1} is associated with an increase in the specific wear rate.

Frictional heat thus appears to be effectively dissipated by steel gauze for pin sliding speeds up to 0.35 m s^{-1} at this load (2.22 MN m^{-2}). Excessive loading will also cause significant frictional heating, as illustrated in Fig. 9, which shows a sudden increase in wear rate for nylon 6 (100A) above a pressure of 5.7 MN m^{-2}. This was noted in Sect. 5.1.2 and explained in terms of thermal softening of the surface of the polymer.

5.1.3. Correlation with non-transfer film fatigue wear of polymers against steel

Polymer wear by metal gauze has been shown to correlate with polymer wear on a rough steel surface under single traversal conditions [15–18]. The fatigue wear processes are thus essentially the same, provided the asperities are rounded and do not cause plastic deformation and microcutting of the polymer surface. Accordingly, the relative wear resistances for abrasion by gauze (Table 9) should, in practice, apply to the fatigue wear of these materials against rough steel in conditions which do not allow a transfer film to build up on the counterface.

In agreement with these findings, it is known that materials 102B, 100A/B and 103 (i.e. nylons 6 and silica-filled UHMWPE) are used very successfully as wear strips, guides, chain conveyors and in other applications within this category. Thus, silica-filled UHMWPE is used as chute and hopper liners and as guides and star wheels in bottling plant. High density polyethylene is also used as the artificial acetabulum (socket) cemented into the hipbone in the Charnley hip prosthesis in which the ball of the femur is replaced by stainless steel. In this case, transfer film build up on the steel ball is hindered by the presence of synovial fluid. The use of nylon sheaves in place of steel is reported to increase wire rope life by up to 4.5 times [21].

The poor fatigue wear resistance of unfilled PTFE is well known and is endorsed in this work.

Polyurethane-based elastomers 200 and 202 also show poor wear resistance against steel gauze.

5.2. Abrasive wear against silicon carbide-coated paper

In many abrasive wear tests, a test pin is made to travel on an abrasive disc in a spiral path [22] or move helically over an abrasive cylinder [17]. The wear rate then refers to the test material sliding on a fresh abrasive path and the specific wear rate is found to be proportional to the pressure, i.e. $\alpha = 1$ in eqn. (3). The wear volume per unit load (normal force) per unit sliding distance is thus constant.

With repeated abrasion of metal or rubber over the same area of abrasive paper, the wear rate of the material first decreases and then becomes constant [17]. In general, there is no such stabilisation with plastics, the wear rate decreasing continuously. This is confirmed in the wear tests performed on the tri-pin-on-disc tribometer in which three test pins repeatedly traverse the same circular path on an abrasive paper disc.

For a comparable situation (the braking of a metal drum by asbestos-filled friction-polymer materials), Rhee [23] proposed an empirical wear relationship given by

$$\Delta W = kp^a v^b t^c \tag{6}$$

where k is the wear constant, p is the normal load, v is the sliding speed, t is the sliding time, and a, b and c are a set of parameters for the given friction pair. Under constant load/sliding speed conditions, eqn. (6) reduces to

$$\Delta W = Ht^c \tag{7}$$

at constant p and v, where $H(= kp^a v^b)$ is a constant dependent on the sliding speed and pressure.

A similar exponential equation was found to be valid for the polymer wear on abrasive paper in the tests outlined in the following section.

5.2.1. Procedure

A sliding speed of 0.1 m s^{-1} and a nominal contact pressure of 2.69 MN m^{-2} were maintained constant throughout the tests. The PV value (269 kN m^{-2} m s^{-1}) thus exceeded the recommended limit for unfilled nylon 6, but was well below values quoted (see Table 1) for filled nylon 6/6 and the polyurethane elastomers.

A new abrasive paper was used for each polymeric material, the test pins traversing up to 280 cycles on each of four, and mostly on five, of the circular paths available. The variation of the wear parameters with the path diameter, and hence with the number of cycles traversed by the pins per unit time, was thus determined.

For most of the materials, the total sliding time on each wear track was 10 min, during which the pins were removed (three or four times) at timed intervals, cleaned, weighed and replaced exactly as before. The abrasive wear was sufficient over this period to establish a basic wear equation.

With nylon 102B, the entire procedure was repeated with a new set of pins in order to check the reproducibility of the results, also with pins made from supposedly similar materials from different batches (100 A/B) and sources (102 A/B) to check quality control.

5.2.2. Results and discussion

Wear versus time curves for test pins of some of the polymers sliding on abrasive paper wear tracks of different diameter, d, are shown in Fig. 11. The plots include those given by the most abrasive wear-resistant material [UHMWPE (103)] and the least abrasive wear-resistant material [polyurethane (201)] to illustrate both the wide range of abrasive wear behaviour observed and, in contrast to gauze wear, the dependence of the wear on the abrasive track diameter. (Sets of overlapping curves given by the other materials tested have been omitted for clarity.)

Compared with the steel gauze wear tests, the test periods for abrasion on SiC paper were relatively short, mostly being 10 min. Nevertheless, during this time, the pin wear was equivalent to or exceeded that which occurred during 1–2 h abrasion on steel gauze. Test pins were often abraded down to the maximum allowable and in one case, with polyurethane material 201, the test pins wore down to the pin-carrier base plate in 2–3 min. Accordingly, the amount of pin wear occurring during the brief tests on abrasive paper discs was considered sufficient for the determination of the form of the wear–time relationship.

In contrast to steel gauze wear, when after a running-in period, plots of ΔW vs. t became linear, the curves for wear on abrasive paper followed an exponential relationship given by

$$\Delta W = Dt^c \tag{8}$$

(p, v, d constant) where ΔW is the average weight loss per pin (in mg), t is the sliding time (in min), and parameters D and c are specific to the polymeric material/abrasive system and dependent on the wear track diameter, d.

Equation (8) is of similar form to Rhee's empirical wear relationship given by eqn. (7) for constant sliding speed/load conditions. The exponential form can be

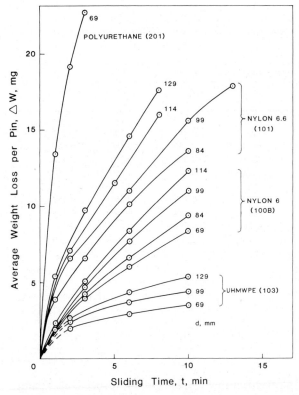

Fig. 11. Average weight loss per pin against sliding time for UHMW polyethylene (103), unfilled nylon 6 (100B), filled nylon 6.6 (101) and a polyurethane-based elastomer (201) sliding on circular tracks of different diameter (d) on a SiC-coated abrasive paper disc. Load per pin, 2.69 MN m^{-2}; sliding speed, 0.1 m s^{-1}.

explained in terms of progressive running-in of the pins accompanied by blunting and contamination of the abrasive track during repeated cycling.

The values of the parameters c and D, listed in Tables 11 and 12, respectively, were determined as a function of the wear track diameter from the slope (c) and intercept (log D) of each of the linear plots given by log ΔW against log t (Fig. 12) according to

$$\log \Delta W = c \log t + \log D \tag{9}$$

at constant p, v and d.

The parameters c and D specific to each polymeric test material were found to vary with the wear track diameter according to

$$c = Cd^{\beta} \tag{10}$$

TABLE 11

Values of the exponent c for each polymeric material as a function of the abrasive paper wear track diameter, d (from Thorp [6])

Material code no.	Mean ambient temp. (°C)	Wear path diameter (mm)				
		69	84	99	114	129
100A	20	0.722	0.756	0.753	0.727	0.783 [b]
100B	25	0.637	0.662	0.712	0.738	0.751
101	26	0.531 [b]	0.548	0.539	0.580	0.569
102A	20	0.470	0.536	0.586	0.535	0.564
102B	23	0.687	0.638	0.673	0.690	0.716
102B [a]	24	0.719	0.685	0.692	0.770	0.752
103	26	0.337	0.398 [b]	0.379	0.365	0.351
200	20	0.210	0.267	0.323	0.306	0.358
201	20	0.528	0.561 [b]	0.592 [b]	0.613	0.654
202	20	0.368	0.300	0.287	0.262	0.239

[a] Repeat test series.
[b] Calculated value.

and

$$D = Gd^\gamma \tag{11}$$

as illustrated by the logarithmic plots given in Figs. 13 and 14, respectively. Values

TABLE 12

Values of the wear factor, D (mg min^{-1}), for each polymeric material as a function of the wear track diameter, d (from Thorp [6])

Material code no.	Mean ambient temp. (°C)	Wear path diameter (mm)				
		69	84	99	114	129
100A	20	2.400	2.636	3.025	3.301	3.549 [b]
100B	25	1.934	2.041	2.135	2.248	2.456
101	26	3.283 [b]	3.774	4.428	4.673	5.318
102A	20	2.778	3.194	3.478	3.903	3.708
102B	23	2.077	2.363	2.388	3.078	3.007
102B [a]	24	2.331	2.453	2.457	2.942	3.031
103	26	1.609	1.769 [b]	1.852	2.057	2.367
200	20	4.891	5.724	6.034	5.800	5.977
201	20	13.357	15.930 [b]	18.463 [b]	20.767	23.367
202	20	5.179	6.128	7.311	7.608	8.637

[a] Repeat test series.
[b] Calculated value.

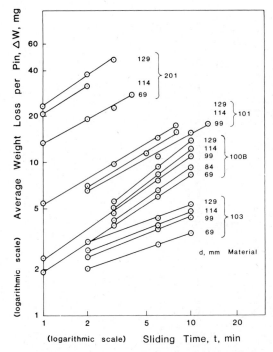

Fig. 12. Linear logarithmic plots of the average weight loss per pin against sliding time for materials 100B, 101, 103 and 201 sliding on wear tracks of different diameter on a SiC-coated abrasive paper disc. Load per pin, 2.69 MN m^{-2}; sliding speed 0.1 m s^{-1}.

of parameters C and β from eqn. (10) and G and γ from eqn. (11) are listed in Tables 13 and 14, respectively.

Substituting for D from eqn. (11) in eqn. (8) gives the wear volume, Δv, in terms of the abrasive wear track diameter and sliding time

$$\Delta v = G^1 d^\gamma t^c \tag{12}$$

where $\Delta v = \Delta W / \rho$ (mm^3), and $G^1 = G/\rho$ (mm^3 min^{-1} mm^{-1}), with ρ being the material density (mg mm^{-3}).

During the first cycle on any of the five circular paths available, the test pins slide on fresh abrasive paper. In following cycles, the wear behaviour changes as the abrasive path becomes progressively blunted and clogged with polymer and filler debris. It is more meaningful, therefore, to replace the sliding time by the number of cycles, N, traversed by the test pins on a circular abrasive path of diameter, d, according to

$$t = \frac{\pi}{V} dN \tag{13}$$

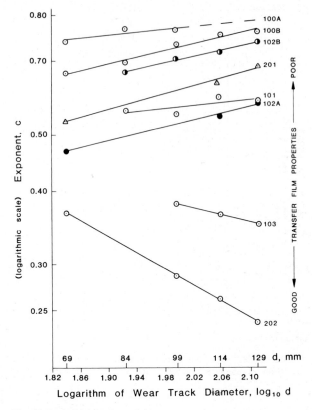

Fig. 13. Logarithmic plots of the exponent c against the wear track diameter for the abrasion of polymers on SiC-coated paper.

Then substituting for t in eqn. (12) gives

$$\Delta v = G^* d^{\gamma+c} N^c \tag{14}$$

where

$$G^* = G^1 \left(\frac{\pi}{V}\right)^c \tag{15}$$

Accordingly, eqn. (14) can be used to estimate the wear volume (per pin) of any of the test polymers after rotating N cycles on a SiC-coated abrasive paper wear path of any specified diameter.

The magnitude of the exponent c (see Table 11) appears to be related to the film transfer capability of the polymeric material. Thus, materials with low values of c (e.g. 103, 200 and 202) were found to build up stable transfer films on the abrasive paper tracks. For two of these materials (103 and 202), c decreased with increase in

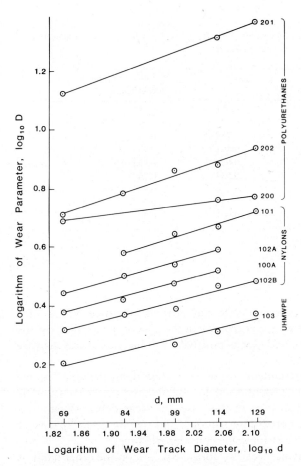

Fig. 14. Logarithmic plots of the wear parameter D against the wear track diameter for the abrasion of polymers on SiC-coated paper.

TABLE 13

Values of C and β (from $c = Cd^{\beta}$) from a linear plot of log c against log d for each material cycling on tracks of diameter d on an abrasive paper disc (from Thorp [6])

Material code no.	C (mm^{-1})	β	Linear correlation coefficient, R
100A	0.437	0.120	0.851
100B	0.191	0.284	0.987
101	0.310	0.127	0.701
102A	0.142	0.283	0.996
102B	0.201	0.261	0.994
103	1.476	−0.296	−0.999
200	0.008	0.790	0.945
201	0.130	0.330	0.994
202	6.737	−0.687	−0.999

TABLE 14

Values of G and γ (from $D = Gd^{\gamma}$) from a linear plot of log D against log d for each material cycling on tracks of diameter d on an abrasive paper disc (from Thorp [6])

Material code no.	G (mg min^{-1} mm^{-1})	γ	Linear correlation coefficient, R
100A	0.150	0.651	0.994
100B	0.552	0.296	0.998
101	0.132	0.759	0.987
102A	0.170	0.661	0.997
102B [a]	0.149	0.622	0.933
103	0.133	0.584	0.967
200	1.232	0.326	0.999
201	0.298	0.898	0.999
202	0.177	0.801	0.991

[a] Using combined results of two test runs.

the wear track diameter, i.e. the exponent β in eqn. (10) was negative. Values of c for the other test materials were higher, particularly in the case of the unfilled nylons which produced loose wear debris, and β was positive, i.e. c increased with increase in d.

For polymer wear by steel gauze under constant p, v conditions, the wear volume after running-in becomes directly proportional to sliding time (i.e. $c = 1$) and is independent of the wear track diameter. In this case, the wear debris is removed from the abrading asperities (wire mesh intersections) and the abrasive capacity of the gauze remains unchanged, at least for some considerable time.

The running-in process for polymer wear by abrasive paper, on the other hand, is continuous and the wear is affected firstly by the distance the pins travel on fresh abrasive paper (i.e. πd) and secondly by the number of cycles traversed in unit time (i.e. $N/t = V/\pi d$).

Initial wear of the spiral machine marks on the contact surfaces of newly machined pins and attainment of surface conformity with the disc counterface occurs increasingly with increase in the distance traversed by the pins on fresh abrasive paper, i.e. with increase in the wear path diameter.

With repeated cycling of the test pins over the wear path, however, blunting and contamination of the abrasive asperities by polymer transfer films increases with decrease in track diameter and hence with increase in rotational speed. This is illustrated by eqn. (10) with β negative, as observed for the two materials which gave the heaviest deposits on the abrasive paper tracks. In the case of the remaining test materials, with positive β the exponent c increases with increase in diameter, as illustrated in Fig. 13. These materials apparently do not deposit stable transfer films on abrasive paper and, accordingly, the exponent c, and thus the wear volume, increases with increase in the abrasive capacity of the wear path. With largely unclogged abrasive paper, the larger the wear path diameter the fewer the traversals in any fixed time and the greater the abrasive capacity of the wear track.

5.2.2.1. Reproducibility and test sample differences. The test series was repeated using a cast nylon 6 (102B) with test pins machined from the same rod sample. Values of the exponent c and wear parameter D were found to be reproducible to within about 5%, as illustrated in Tables 11 and 12.

Results for duplicate tests performed on supposedly similar materials but obtained from either different material batches (e.g. 100A/100B) or different sources (e.g. 102A/102B) were found to differ (see Tables 11 and 12) considerably. Changes in ambient temperature (being 3°C between tests on 102A and 102B, and 5°C between tests on 100A and 100B) are not considered sufficient to account for these differences.

Samples 102A and 102B (dyed cast nylon 6) were supplied before and after a take-over in which the factory was moved to a different location and equipment upgraded. Variations in performance are clearly due to differences in processing with new plant and in quality control.

Samples 100A and 100B (extruded nylon 6) were from the same source but with several months lapsing between deliveries. The wear performance in this case differs to an unacceptable degree, inferring poor quality control.

5.2.2.2. Polymer wear on a fresh abrasive path. The wear volume for a one cycle ($N = 1$) traversal on a fresh abrasive path of diameter d is given from eqn. (14) as

$$\Delta v_{N=1} = G^* d^{\gamma+c} \tag{16}$$

Substituting for G^* from eqn. (15) gives

$$\Delta v_{N=1} = G^1 (\pi d/V)^c d^\gamma \tag{17}$$

Running-in phenomena can be expected to affect one cycle wear behaviour for experimental values of d owing to the very short sliding times involved. Thus, the largest path ($d = 129$ mm) is traversed in 4 s. Wear volumes have thus been estimated for a hypothetical path ($d = 1910$ mm) which is traversed in 1 min for the sliding speed (0.1 m s^{-1}) used in the tests. Equation (17) then becomes identical to eqn. (12) with $t = 1$ (min), viz.

$$\Delta v_{t=1} = G^1 d^\gamma \tag{18}$$

since $\pi d/V = 1$ (min), with d in m and V in m min^{-1}.

These calculated wear volumes, wear volumes relative to material 103 (UHMWPE) and the ranking order of the polymers for such a hypothetical single traversal on abrasive paper are given in Table 15.

A reasonable correlation has been reported between the reciprocal of the work required to rupture the bulk polymer and the single traversal wear rate of the polymer sliding over a rough counterface under fixed load [18,24]. The rupture work

TABLE 15

Estimated wear volumes and wear relative to material 103 (UHMWPE) for one cycle on a (hypothetical) fresh abrasive paper track of 1910 mm diameter with a traversal time of 1 min (from Thorp [6])

Material code no.	$\Delta v_{t=1}$ (mm^3)	$\Delta v / \Delta v_{103}$	Ranking order (single traversal)
100B	4.5	0.4	1
103	11.4	1.0	2/3
200	11.7	1.0(3)	2/3
102B	14.1	1.2	4
100A	18.0	1.6	5
102A	21.6	1.9	6
101	34.6	3.0	7
202	68.3	6.0	8
201 [a]	225.1	19.8	9

[a] Faulty material with dispersed cavities.

is estimated from the product of the stress at rupture [ultimate tensile stress (uts)] and the associated strain (elongation) from a tensile test. Table 16 lists the reciprocal of the breaking work for each material (calculated from typical values of the uts and elongation given in Table 1) and, in column 3, the value relative to material 103. The estimated wear relative to that given by material 103 for a one cycle traversal on a hypothetical 1910 mm diameter track is given in column 4 for comparison.

Materials 103 and 200 exhibit the best abrasion resistance of the test materials according to both estimates (i.e. from the mechanical properties and from the

TABLE 16

The reciprocal of the breaking work of the test materials and values relative to that for filled UHMWPE (103) compared to the relative wear (to material 103) calculated for one cycle on a 1910 mm diameter path on an abrasive paper disc (from Thorp [6])

Material code no.	Breaking work reciprocal, $R \times 10^2$ (m^2 MN^{-1})	R / R_{103}	$\Delta v / \Delta v_{103}$
103	0.8	1.0	1.0
200	1.3(2)	1.6	1.0
202	1.3(0)	1.6	6.0
201 [a]	2.7	3.4	19.8 [a]
102 (A/B)	4.0	5.0	1.9, 1.3
100 (A/B)	6.1	7.6	1.6, 0.4
101 [b]	13.8	17.3	3.0

[a] Faulty material with dispersed cavities.
[b] Contains PTFE and PE.

calculated wear for $N = 1$, $d = 1910$ mm, $t = 1$ min). The faulty (cavity-filled) nature of material 201 is evident from the exceptionally high relative wear, although the mechanical properties predict an intermediate ranking order. The nylon based materials wear to a lesser extent and material 202 to a greater extent than predicted by their bulk properties.

Abrasive wear in single traversal or non-transfer film conditions is met in practice by chute and hopper liners used in moving abrasive materials, by components in agricultural and earth-moving machinery, by wear strips and chain guides, and by impellers in pumps handling abrasive fluid wastes. Materials recommended by manufacturers for these applications include silica-filled UHMWPE (103) and polyurethane-based elastomers 200, 201 and 202, although the polyurethane elastomer grade with the best recommended abrasion resistance was not tested.

Manufacturers' test data, using a thrust washer configuration immersed in abrasive fluid (SAE 30 oil containing 10% alumina powder), shows the abrasion resistance of material 200 to be far superior to 202 which in turn is more abrasion-resistant than grade 201. Further, in slurry (sand/water) tests, the performance of an unidentified grade (presumably either the most abrasion-resistant grade or material 200) was stated to be about on a par with UHMWPE and the best grade of nylon. The estimated relative one-cycle wear ratios shown in Table 15 indeed places material 200 on a par with the polyethylene, ranking equal in second/third positions, with nylon 6 (100B) at top and nylon 6 (102B) a close fourth.

The abrasive wear resistances predicted from the typical tensile test properties of the bulk polymers (see Table 16) does not bear this out since materials 200 and 202 are placed (equal) second to UHMWPE with the best of the nylons only ranking fifth. Differences in the manufacturing processes and heat treatments can alter the physical properties of polymers. Manufacturers give typical values (or ranges) of properties to cover these expected variations. Although such differences may contribute, in part, to the discrepancies between values listed in columns 3 and 4 of Table 16, they are unlikely to give differences of the magnitude observed.

Although the tensile strength (uts) and elasticity (elongation) are important factors relating to the abrasion resistance of a polymer, surface phenomena, such as molecular orientation, the presence of surface films, and changes in the surface structure incurred during manufacture (as in extrusion) or on sliding against an abrasive counterface, are also involved. Thus materials containing PTFE (such as 101) become covered with a surface film of PTFE on sliding which reduces friction, frictional heating, and hence surface deterioration. The surface molecular structure of an extruded nylon rod will differ and have a different abrasion resistance from that of a cast nylon component. The amount of crystallisation occurring at the surface of a cast component may differ from that occurring in the bulk owing to differences in temperature gradient during cooling or because of the effect of special mould coatings. Accordingly, the surface structures of both extruded and cast nylon 6 can be expected to exhibit greater abrasion resistance than that predicted by bulk material properties.

Mechanical properties of the bulk polymeric material can only give an approxi-

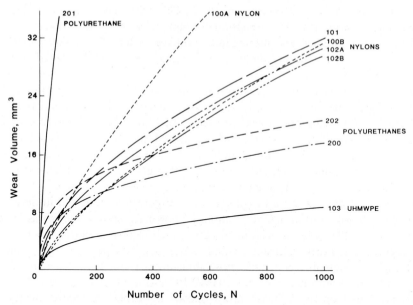

Fig. 15. The wear volume of the materials [calculated using eqn. (14)] against the number of cycles (N) traversed by the test pins on a 99 mm diameter wear path on SiC-coated abrasive paper for N extended to 1000 cycles. (From Thorp [6].)

mate indication of surface wear resistance, therefore, in cases in which the structure of the surface layers is similar to that in the bulk material.

5.2.2.3. Polymer wear under transfer film conditions. The experimental average wear volume per pin for each test material after traversing N cycles on a circular path of specified diameter on an abrasive paper disc is given by eqn. (14) which is valid to at least 300 cycles. Figure 15 shows for each material the plot of the (calculated) wear volume for an intermediate path diameter of 99 mm against the number of cycles, with N extended to 1000 cycles. Extrapolation to a high value of N clarifies and extends the trends observed within the experimental ($N = 15$–280) range.

Material 103 clearly has outstanding overall abrasion resistance, particularly at high N (i.e. under transfer film conditions) whereas the faulty polyurethane (201) shows least wear resistance. With increase in N, the nylons are surpassed by polyurethanes 200 and 202. Further extrapolation to 10^3, 10^4 and 10^6 cycles is illustrated in Table 17, which lists the calculated wear volumes, the wear relative to the most abrasion-resistant material (103) and the ranking order of the materials for each value of N.

Clearly, the UHMWPE (103) is the most abrasion-resistant material under continued cycling (i.e. transfer film) conditions, followed by the polyurethane-based elastomers 200 and 202. The relative wear of these latter polyurethanes (i.e. $\Delta v_{200}/\Delta v_{103}$ and $\Delta v_{202}/\Delta v_{103}$) decreases with increase in the number of cycles (see

TABLE 17

Wear volumes per pin [calculated from eqn. (14)], the relative wear to that given by material 103 (filled UHMWPE) and the ranking order of the abrasion resistance for (a hypothetical) 10^3, 10^4 and 10^6 cycles on a 99 mm diameter abrasive paper wear path (from Thorp [6])

| Material code no. | Number of cycles, N | | | | | | | | |
| | 10^3 | | | 10^4 | | | 10^6 | | |
	Δv (mm^3)	$\Delta v/\Delta v_{103}$	Ranking order	Δv (mm^3)	$\Delta v/\Delta v_{103}$	Ranking order	Δv (mm^3)	$\Delta v/\Delta v_{103}$	Ranking order
103	9.1	1.0	1	21.7	1.0	1	124	1.0	1
200	15.9	1.7	2	33.5	1.5	2	148	1.2	2/3
202	19.8	2.2	3	38.4	1.8	3	144	1.2	2/3
101	30.7	3.4	4/5	106.3	4.9	4	1272	10.3	4
102A	30.9	3.4	4/5	119.1	5.5	5	1770	14.3	5
102B	33.1	3.7	7	159.5	7.4	6/7	3696	29.8	6
100B	31.4	3.5	6	161.8	7.4	6/7	4295	34.6	7
100A	51.2	5.7	8	290.0	13.4	8	9295	75.0	8
201	163.4	18.0	9	638.6	29.4	9	9756	78.7	9

Table 17), whereas that of the other test materials increases. This is in agreement with literature evidence [18] which suggests that transfer films from polymers with high elongations (e.g. 102, 200 and 202 with respective break elongations of 334, 219 and 240%) are beneficial and reduce the wear rate. Conversely, the transfer from less ductile polymers (e.g. 100, 101, 102, with respective elongations of 20, 10 and 30%) is detrimental and increases wear.

Material 101, a nylon 6/6 containing PTFE and PE fillers, ranks fourth place in Table 17, having better abrasion resistance than the unfilled nylons, especially at the higher values of N. The PTFE/PE transfer films deposited by this material clearly reduce friction (see Table 6) and wear under conditions of repeated sliding on an abrasive counterface. According to Table 17, however, these beneficial effects decrease with increase in N.

The relative abrasion resistance of the unfilled nylon 6 materials tested is poor, decreasing with increase in the number of cycles. The loose wear debris (lumpy transfer) associated with these nylons is clearly detrimental.

The rupture work of the bulk base polymer indicates that material 201 should have had moderate abrasion resistance had it been of good quality. However, the presence of numerous cavities virtually caused the material to disintegrate under the abrasive test conditions.

6. Conclusions

6.1. Wear resistance

In reality, the wear of these materials will lie somewhere between that observed on steel gauze and abrasive paper, depending on the predominant wear mode (i.e.

fatigue or micro-cutting and plastic deformation) and whether transfer films play a part.

Unfilled nylons 6 (100A and 102B) were found to possess excellent fatigue wear resistances against steel gauze and appeared even to rank well on a hypothetical fresh abrasive paper track of 1910 mm diameter. Yet on repeated abrasion, with accumulation of wear debris, the abrasive wear resistance of the nylon group ranked lowest of all the test materials bar the faulty (cavity-filled) polyurethane (201). Nylon 6 can thus be expected to wear well against rough steel only in single traversal or non-transfer film conditions. This is born out in practice with successful use of nylon as wear guides and strips and as wire rope sheave bushings.

The filled nylon 6/6 (101) is less fatigue-resistant than unfilled nylon 6, ranks even worse according to predicted performance against fresh abrasive paper but improves under transfer film conditions. This material would thus be more competitive for use as dry bearing material (plain bearings and bushings) against a smooth steel shaft.

Filled UHMW polyethylene (103) gave outstanding overall results in the abrasion tests on abrasive paper and also performed reasonably well (but not as well as nylon 6) against steel gauze; a good wear-resistant material in most conditions. Applications recommended by manufacturers include use as chute and hopper liners, guides and star wheels in bottling plant, wear plates and (unfilled) with great success in hip joint prostheses.

Polyurethane 200 ranked a close follower to filled UHMWPE (103) in the abrasive paper tests yet gave by far the worst performance against steel gauze. This is clearly a material with high resistance to plastic deformation and cutting wear but with low resistance to frictional-type (fatigue) wear. These results endorse the manufacturer's recommendations for use of this material as marine stern tube bearings and lower sleeve bearings in vertical sewage pumps, both situations requiring high abrasion resistance yet avoiding frictional (fatigue) wear because of the presence of a fluid. Polyurethane 202 has fairly similar abrasive (cutting) wear properties to polyurethane 200 but shows a somewhat better performance on steel gauze. Little comment can be made on the third polyurethane sample (201) owing to its faulty condition. However, at very low pressures (i.e. 1 N m^{-2}) which do not cause the cavities to collapse, the material ranks well [second to nylon 6 (102B)] against steel gauze. The bulk mechanical properties further predict a reasonable wear resistance against a rough counterface in single traversal conditions. Both findings would endorse one of its recommended uses as rope sheave bushings.

In conclusion, manufacturers' recommendations regarding applications for the test materials are mostly endorsed by the wear tests. The results illustrate the importance of analysing the type of wear likely to occur and whether conditions are conducive or not to the formation of a stable polymer (or filler) transfer film on the metal counterface. Good transfer film capability is wasted in single traversal conditions.

6.2. Friction properties

Friction tests do not confirm some manufacturers' claims of inherent low-friction properties except in the case of UHMWPE. The results show to the contrary that most of the test materials have high coefficients of friction against steel, indicating their limited use as dry bearing materials except at low sliding speeds. Lubrication reduces the coefficient of friction against steel up to three-fold or more. In fact, according to the results, two polyurethane-based elastomers (200 and 201) could not be recommended for use as dry bearing materials. This is endorsed to some extent by the manufacturer who promotes material 202, with frictional properties on a par with the nylons, for dry bearing applications.

6.3. Silica-filled (UHMW) polyethylene

Of all the materials tested, silica-filled UHMWPE (103) has the lowest coefficients of friction against unlubricated steel at moderate to heavy loads and within the selected sliding speed range (0.01–1.0 m s^{-1}), the best abrasion resistance against cutting and plastic deformation, and at high pressures good fatigue wear resistance. Further, the material has good transfer film capability, the polyethylene film smeared on the steel counterface acting beneficially to reduce friction and wear.

In terms of bulk mechanical properties, material 103 has the highest (calculated) rupture work of the test materials, thereby inferring high wear resistance, as confirmed by experiment.

The silica-filled (UHMW) polyethylene can clearly be recommended as an excellent low-friction, wear-resistant dry bearing material against steel under the load/speed/temperature conditions used in the friction and wear tests.

List of symbols

a	pressure exponent
b	sliding speed exponent
c	sliding time exponent
d	wear path diameter (mm)
k	torque calibration constant (N m mm^{-1})
k	wear proportionality constant [eqn. (6)]
p	nominal contact pressure (N m^{-2})
r	wear path radius (m)
s	linear mass wear rate (mg min^{-1})
t	sliding time (min)
Δv	wear volume (mm^3)
v	sliding speed (m s^{-1})
x	recorder deflection (mm)
A	pin nominal contact area (m^2)
C	proportionality constant (mm^{-1})
D	wear proportionality constant (mg min^{-1})

G	proportionality constant (mg min^{-1} mm^{-1})
G^1	proportionality constant (mm^3 min^{-1} mm^{-1})
K	specific wear rate at unit pressure (mm^3 N^{-1} m^{-1})
H	wear proportionality constant
L_0	normal force at zero applied force (N)
L_a	normal applied force (N)
N	number of cycles traversed by test pins during test duration
R	reciprocal of the rupture work (m^2 MN^{-1})
R_μ	percentage reduction in μ_i (%)
ΔW	average weight loss per pin (mg)
V	sliding speed [eqn. (4)] (m min^{-1})
α	pressure exponent
$\alpha(T)$	Thorp's value of α
$\alpha(R\&F)$	value of α from Ratner & Farberova
β	wear path diameter exponent
γ	wear path diameter exponent
μ	coefficient of friction
μ_i	initial coefficient of friction
μ_f	final coefficient of friction
ν_{sp}	specific wear rate (mm^3 m^{-2} m^{-1})
ρ	material density (mg mm^{-3})

PTFE	polytetrafluorethylene
UHMWPE	ultra-high molecular weight polyethylene
SiC	silicon carbide

References

1 M.J. Neale (Ed.), The Tribology Handbook, Butterworths, London, 1973, p. F3.
2 J.C. Anderson, Tribol. Int., 15 (1982) 255.
3 J.K. Lancaster, Tribology, 5 (1972) 249.
4 Data Item No. 76029, ESDU (1976).
5 Polymer Materials for Bearing Surfaces Selection and Performance Guide, National Centre of Tribology, Risley, Warrington, Gt. Britain, 1983.
6 J.M. Thorp, Tribol. Int., 15 (1982) 59.
7 J.M. Thorp, Tribol. Int., 15 (1982) 69.
8 J.M. Thorp, in National Conference Publication No. 80/12, Conference on Lubrication, Friction and Wear in Engineering 1980, Melbourne, December 1–5, 1980, Institution of Engineers, Barton, A.C.T., Australia, 1980, pp. 68–72.
9 J.M. Thorp, Tribol. Int., 14 (1981) 121.
10 J.M. Thorp, PEL Rep. No. 676, 1980.
11 R.P. Steijn, in D.A. Rigney and W.A. Glaeser (Eds.), Source Book on Wear Control Technology, American Society for Metals, Metals Park, OH, 1978, p. 371.
12 C.J. Smithells (Ed.), Metals Reference Book, Butterworths, London, 5th edn., 1976, p. 1281.
13 Sh.M. Bilik and V.I. Donskikh, Russ. Eng. J., L1, 11 (1972) 60.
14 F.P. Bowden and D. Tabor, The Friction and Lubrication of Solids, Part II, Clarendon Press, Oxford, 1964, Chap. 13, p. 214.

15 S.B. Ratner, in D.I. James (Ed.), Abrasion of Rubber, MacLaren, London, 1967, pp. 23–35.
16 G.S. Klitenik and S.B. Ratner, in D.I. James (Ed.), Abrasion of Rubber, MacLaren, London, 1967, pp. 64–73.
17 S.B. Ratner and I.I. Farberova, in D.I. James (Ed.), Abrasion of Rubber, MacLaren, London, 1967, pp. 297–312.
18 J.K. Lancaster, Wear, 14 (1969) 223.
19 D.C. Evans and J.K. Lancaster, in D. Scott (Ed.), Treatise on Materials Science and Technology, Vol. 13, Academic Press, 1979, pp. 86–139 (see p. 109).
20 Y. Uchiyama and K. Tanaka, Wear, 58 (1980) 223.
21 J.H. Chen and C.R. Ursell, SAE Pap. 790904 (1979).
22 M.M. Khruschov, Wear, 28 (1974) 69.
23 S.K. Rhee, Wear, 29 (1974) 391.
24 B.J. Briscoe, Tribol. Int., 14 (1981) 231.

Chapter 5

Effects of Various Fillers on the Friction and Wear of PTFE-Based Composites

KYUICHIRO TANAKA

Faculty of Engineering, Kanazawa University, Kanazawa 920 (Japan)

Contents

138

Abstract

In this chapter, the fundamental behavior on the transfer, friction and wear of unfilled PTFE are first briefly surveyed and discussed on the basis of the molecular and morphological structures of PTFE. Secondly, the fundamental behavior on the friction, wear and abrasiveness of PTFE incorporating various fillers are surveyed in detail, mainly on the basis of the work which has been carried out in the author's laboratory, and the effects of various fillers on the friction and wear of PTFE-based composites are described. Thirdly, the effect of fillers are discussed on the basis of the properties of unfilled PTFE and microscopic examinations carried out on the worn surfaces of composites and also on the counterface. Thus, the mechanism of the wear-reducing action of fillers in PTFE are explained. Furthermore, the effects of water lubrication on the friction and wear of PTFE incorporating various fillers are presented and discussed.

1. Introduction

Polytetrafluoroethylene (PTFE) exhibits a very low coefficient of friction and retains useful mechanical properties at temperatures from -260 to $+260°C$ for continuous use. The crystalline melting point is $327°C$, much higher than that of most other semicrystalline polymers. Furthermore, PTFE is nearly inert chemically and does not absorb water, leading to excellent dimensional stability. These characteristics of PTFE are very useful in the matrix polymer of polymer-based composites which are used in sliding applications. On the other hand, PTFE is subjected to marked cold flow under stress and reveals the highest wear among the semicrystalline polymers. However, these disadvantages are very much improved by incorporating a filler into the PTFE. A noticeable characteristic of PTFE is that the increase of wear resistance when a filler is incorporated is much greater than that in any other semicrystalline polymer. Owing to the characteristics mentioned above, PTFE is a very important matrix polymer in the polymer-based composites for sliding applications. The unique tribological properties of PTFE are due to its peculiar molecular and morphological structures.

There are many kinds of PTFE-based composite for sliding applications because various fillers are incorporated into PTFE and one or more materials can be used simultaneously. In addition, many properties of polymer composites can generally be altered by changing the fabrication techniques. In this chapter, therefore, the fundamental effects of various fillers on the friction and wear of filled PTFE are described and discussed in order to understand the fundamental tribological properties of PTFE-based composites.

2. Molecular and morphological characteristics of PTFE

The repeat unit in the PTFE molecule has 13 or 15 chemical repeat units ($-CF_2-$ group) depending on the temperature, and the molecule has no branches and is not

Fig. 1. Electron micrograph of a fractured PTFE surface.

bulky, which results in a most smooth molecular profile. It has been pointed out that the smooth molecular profile leads to low friction and easy formation of a thin film transferred on the countersurface in the sliding of PTFE [1]. The molecular weight of PTFE used in a molded product is generally of the order of 10^6 or 10^7 and is much higher than that of any other linear polymer. However, the molecular weight is very much smaller in the PTFE powders used as the filler in metal or polymer composites. It has been shown that the single crystal of PTFE consists of hexagonal lamellae 20–50 nm thick and that the molecules are normal to the lamellae and therefore folded within them [2].

It is well known that many semicrystalline polymers have the morphological structure called spherulite. However, the morphology of PTFE has not yet been completely understood. It is known that PTFE forms extended chain crystals similar to polyethylene grown under elevated pressure and that the lamellae spray apart in spherulitic fashion [3]. On the other hand, it has been pointed out that PTFE has a unique morphological structure, referred to as a banded structure [4]. Figure 1 shows an electron micrograph of a fracture surface of PTFE used as the specimen in the author's work. The banded structure is clearly seen and the width of bands, i.e. crystalline lamellae length, seems to be approximately 0.3 μm and relatively small, indicating that the lamellae are not composed of the completely extended chain crystals because the length of 0.3 μm corresponds to the extended chain molecule of PTFE having a molecular weight much lower than that of normal PTFE. Speerschneider and Li [5] have reported that the bands are composed of crystalline slices about 20 nm thick, oriented normally to the length of the band and that

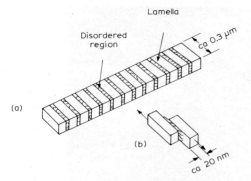

Fig. 2. Schematic representations of the banded structure of PTFE and the slipping of lamellae.

slipping between the individual slices can easily occur. This model for PTFE morphology was applied to explain the transfer characteristics by Makinson and Tabor [6] and also the high wear of PTFE by Tanaka et al. [7]. At present, however, it is more reasonable to consider that the crystalline slices in the model should be replaced by the rod-like lamellae as shown in Fig. 2. On the basis of the striations in the bands seen in Fig. 1, the lamellae thickness seems to be about 20 nm.

3. Tribological characteristics of PTFE

3.1. Transfer

It is well known that polymer transfer to the countersurface occurs during the sliding of polymers and that the transferred material plays an important role in their wear characteristics. When PTFE is slid on a very smooth glass or steel surface at a very low speed, it produces a very thin transferred film. Pooley and Tabor [1] found that the friction and transfer properties of PTFE and high density polyethylene (HDPE) were different from those of other polymers having bulky side groups in the molecular chain. They observed the transfer of an extremely thin PTFE film, about 2.5 nm thick, and suggested that the low friction and light transfer of PTFE and HDPE during sliding were essentially due to their smooth molecular profiles. Tanaka and Miyata [8] also observed an extremely thin PTFE transfer film as shown in Fig. 3(a) on a glass surface at a sliding speed of 0.18 mm s^{-1}. However, they observed that the transfer properties of HDPE were different from those of PTFE and were similar to those of other polymers having a spherulite structure. The transfer of polymers other than PTFE contrasts with the film-like transfer of PTFE, being generally as small lumps or short streaks as shown in Fig. 3(b). Tanaka and Miyata suggested that the HDPE molecule, having a smooth profile similar to PTFE, also possesses an excellent ability for film formation, provided that easy drawing of the molecules from the spherulite structure is possible during sliding.

Tanaka and Yamada [9,10] studied transfer in multiple passes, i.e. the wear process, of semicrystalline polymers on the surface of a smooth steel disk (ca. 0.02

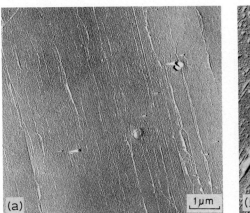 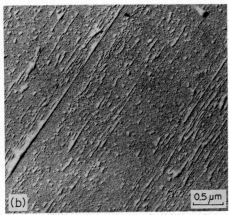

Fig. 3. Electron micrographs of the transferred materials after the first traverse of (a) PTFE and (b) polypropylene.

μm c.l.a. roughness) at various sliding speeds. In this work, the thickness of the polymer layer transferred to the steel surface was measured using electrical capacitance and it was named the reduced thickness. There was a great variation in the reduced thickness obtained at different positions over the frictional track on the steel disk. The average reduced thickness is plotted and the range of the thicknesses obtained is also shown by the bars in Fig. 4. With PTFE and HDPE, the reduced thickness at 10000 revolutions is slightly dependent upon the speed. The increase of the reduced thickness with the number of disk revolutions is relatively small at higher speeds after 1000 revolutions. The reduced thickness after 1000 revolutions at 0.01 m s^{-1} is generally smaller than that at the higher speeds. Figure 5 shows electron micrographs of frictional tracks at speeds of 0.01 and 1 m s^{-1} after 1000 revolutions. It is seen that the thickness of the transferred layer at 1 m s^{-1} is much greater than that at 0.01 m s^{-1} and this is consistent with the result of reduced thickness measurements. It was observed that, in the case of PTFE, many lumps were generally scattered over the transferred layer.

3.2. Friction

Figure 6 shows the relationship between the coefficient of friction of PTFE and the number of traverses on the same track in various environments [8]. The results were obtained in experiments in which a PTFE cylinder, 4 mm in diameter and 3 mm long, was slid over a glass plate in a direction perpendicular to the cylindrical axis at 0.018 cm s^{-1} under a load of about 3 N. The coefficient of static friction, μ_s, decreases rapidly with successive traverses up to about the third traverse, but takes a constant value after the initial decrease. However, the coefficient of kinetic friction, μ_k, is independent of the number of traverses. The static friction is generally two or more times greater than the kinetic friction and both decrease with increase in humidity. It was found that the static friction of PTFE was very sensitive to the

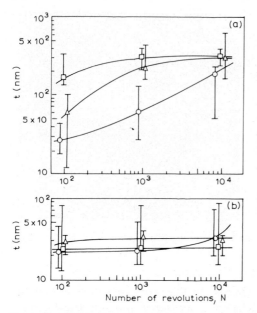

Fig. 4. Variation in the reduced thickness, t, of the transferred polymer layer at various sliding speeds with the number of disk revolutions, N. ○, 0.01 m s^{-1}; □, 0.1 m s^{-1}; △, 1 m s^{-1}. (a) PTFE; (b) HDPE.

direction of pre-rubbing of the frictional surfaces [1,8]. The static friction in sliding parallel to the pre-rubbing direction was much lower than that in sliding perpendicular to it. This must also be due to the smooth profile of the PTFE molecule.

From the rubbing experiments of PTFE at various speeds and at various temperatures on a glass disk, the variations of the coefficient of friction and wear rate with sliding speed have been obtained [7]. The result is shown in Fig. 7, where the temperature of the glass disk is taken as a parameter. The coefficient of friction becomes smaller with decreasing sliding speed and decreases with increasing temper-

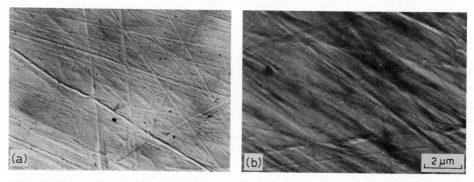

Fig. 5. Electron micrographs of frictional tracks rubbed on PTFE after 1000 disk revolutions at (a) 0.1 m s^{-1} and (b) 1 m s^{-1}.

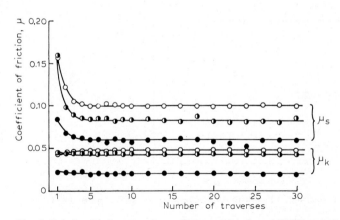

Fig. 6. Relationship between the coefficient of friction of PTFE and the number of traverses in various environments. $\bigcirc$, Vacuum (10^{-7} torr); $\circledcirc$, air at room conditions (56% R.H.); $\bullet$, humid air (78% R.H.).

ature. These are important characteristics of PTFE. The effect of speed on the friction of PTFE at high temperatures was studied using a PTFE pin rubbing against stainless steel [11]. The result is shown in Fig. 8. This shows that the increase of friction with speed occurs even at temperatures above the melting point.

3.3. Wear

In Fig. 7, it is noted that the wear rate versus speed curves have wear rate peaks and these shift to higher speeds as the temperature is increased. A master curve was derived from the experimental results in Fig. 7 by translating the curve obtained at a temperature T horizontally by an amount log a_T to fit the reference curve at temperature T_0. The temperature dependence of log a_T could not be expressed by

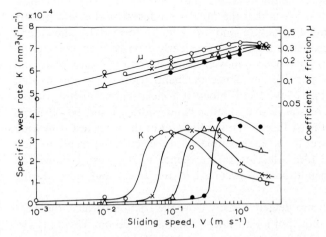

Fig. 7. Variations in specific wear rate, K, and coefficient of friction, μ, with sliding speed for PTFE sliding against a glass surface at $\bigcirc$, 23°C; $\times$, 50°C; $\triangle$, 70°C; $\bullet$, 100°C.

144

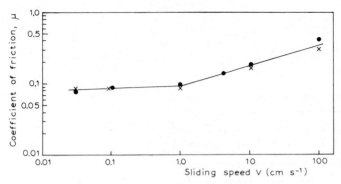

Fig. 8. Variation of coefficient of friction of PTFE with sliding speed at ×, 280°C and ●, 330°C.

the William–Landel–Ferry equation, but by the Arrhenius form. However, a value of about 7 kcal mol^{-1} for the activation energy was deduced by using the temperature dependence of log a_T. Furthermore, the master curve could also be derived in the case of the coefficient of friction seen in Fig. 7. The activation energy derived from Fig. 7 may be assumed to be the activation energy for the slippage between crystalline lamellae in bands, and the small value of 7 kcal mol^{-1} suggests that the destruction of the banded structure of PTFE occurs easily without any melting of the sliding surface as the result of the easy slipping of crystalline lamellae.

Electron microscopic examinations of the frictional surface of PTFE rubbed against a glass disk surface at various speeds and at various disk temperatures were made in detail [7]. Figure 9 is an example of an electron micrograph of the worn surface. In the micrograph, the banded structure appears very clearly and the long films and fibers are laid down on the bands. On the basis of the shadowing angle used in the preparation of the replica, the thickness of the films is estimated at about 30 nm. This value was similar to the thickness of very thin films adhering to the glass disk countersurface. Figure 9 suggests that the fiber is produced by the deformation and destruction of the banded structure and the film is formed by the lateral connection of fibers. The area where the bands are clearly exposed must correspond to the places where the films have been removed by transfer to the glass plate and have not been rubbed after the transfer. Since the band width is similar to that observed on the fracture surface of the PTFE block, it is clear that no melting occurs at the friction surface. The reason is that smaller band widths would be observed because of rapid cooling of the surface at the termination of the wear test if melting occurs at the surface. Figure 10 is an example of an electron micrograph and diffraction pattern of PTFE film adhering to the replica of the worn surface and indicates that molecular orientation has occurred in the fibers and films during sliding.

Since the film of about 30 nm detaches easily from the friction surface, the wear proceeds discretely by the removal of a unit thickness of about 30 nm. Owing to the mechanism of wear mentioned above, the wear rate of PTFE must have a very high value in comparison with the other semicrystalline polymers having a spherulitic

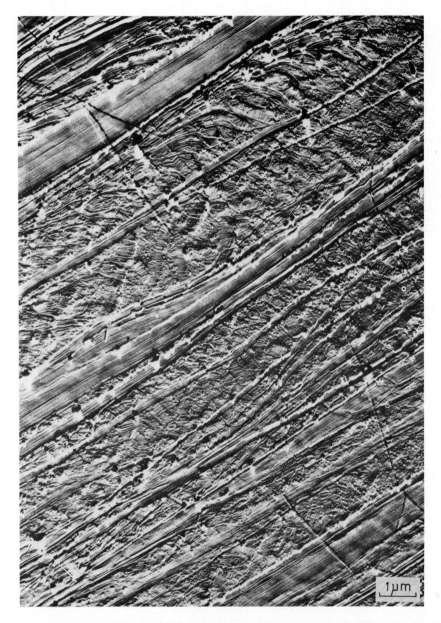

Fig. 9. Electron micrograph of the worn surface of PTFE (load, 1.5 kg; sliding speed, 0.3 m s^{-1}; in vacuum).

structure, because the destruction of spherulite does not occur easily whereas melting of a thin surface layer does.

PTFE exhibits little flow at temperatures above the melting point, which allows measurement of the wear rate over the high temperature range. Therefore, the effects

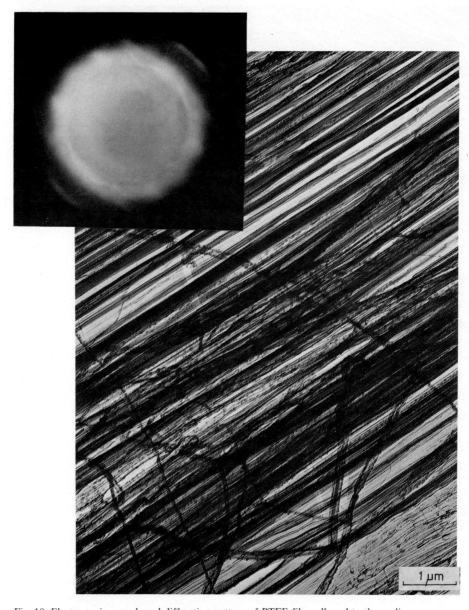

Fig. 10. Electron micrograph and diffraction pattern of PTFE film adhered to the replica.

of temperature and sliding speed on the friction and wear of PTFE were studied in the rubbing on a stainless steel disk over the temperature range 280–380°C [11]. Friction showed little variation with temperature whereas an abrupt increase in wear rate occurred at the melting point. The variation of specific wear rate with tempera-

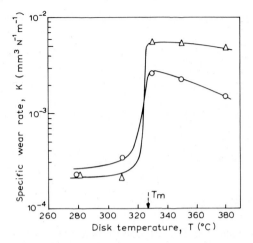

Fig. 11. Effect of temperature on the specific wear rate of PTFE at high temperatures and sliding speeds of $\bigcirc$, 0.01 m s^{-1} and $\triangle$, 0.1 m s^{-1}.

ture is shown in Fig. 11. However, the wear rate does not increase with temperature in the range above the melting point and this is much different from other semicrystalline polymers. It has been pointed out that the surface melting due to frictional heating occurs easily at an engineering speed in the case of typical semicrystalline polymers and much high wear rate appears when the sliding speed exceeds a certain critical value which depends upon the type of polymer [12].

4. Friction and wear of glass and carbon fiber-filled PTFE

4.1. Friction

Lancaster showed that glass fiber (GF)-filled PTFE exhibits higher friction than carbon fiber (CF)-filled PTFE [13]. Tanaka et al. [14,15] studied friction and wear of 3 mm diam. GF (20 wt.%)- and CF (20 wt.%)-filled PTFE pins sliding against smooth steel and glass disk surfaces at various speeds under a load of 5 N. Figure 12 shows the variation of the coefficient of friction with sliding speed for the specimens. In their work, a relatively cheap kind of carbon fiber was used. It was produced from a special pitch material, little graphitized, and had a tensile strength of about 1000 MPa and a tensile modulus of about 40 000 MPa. The mean fiber length of the GF and CF filling was about 0.1 mm. The friction of fiber-filled PTFE is little dependent upon speed. It should be noted that the friction of carbon fiber-filled PTFE is generally greater than that of glass fiber-filled PTFE. However, the difference between the coefficients of friction is relatively small.

4.2. Wear

The relation between specific wear rate and sliding speed for fiber-filled PTFE specimens sliding on steel and glass disk is shown in Fig. 13. It should be noted that

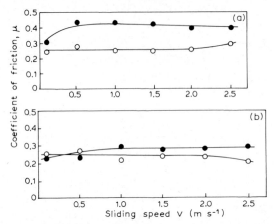

Fig. 12. Variation of the coefficient of friction with sliding speed for GF- and CF-filled PTFE rubbed against (a) steel and (b) glass disks. O, GF–PTFE; ●, CF–PTFE.

the wear-reducing effect of carbon fibers is rather similar to that of glass fibers. However, the wear rates of filled PTFE on glass are generally greater than those on steel and this must be due mainly to the higher temperature generated by frictional heat at the real area of contact between the filled PTFE and glass surface because glass has much lower heat conductivity than steel. It is also noted that the wear rate of GF- and CF-filled PTFE is little dependent upon the sliding speed in the case of the steel disk.

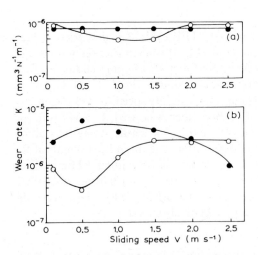

Fig. 13. Variation of wear rate with sliding speed for GF- and CF-filled PTFE rubbed against (a) steel and (b) glass disks. O, GF–PTFE; ●, CF–PTFE.

4.3. Abrasiveness of fiber-filled PTFE

Lancaster studied the abrasiveness of various polymer-based composites using a bronze sphere rubbed against the composite disks [16]. According to his data, the wear rates of bronze were 0.2, 8 and 62 (10^{-6} mm^3 N^{-1} m^{-1}) in the cases of Type I CF (25 wt.%)-, Type II CF (25 wt.%)- and GF (30 wt.%)-filled PTFE specimens, respectively. This indicates that GF-filled PTFE has a much high abrasiveness than CF-filled PTFE and the abrasiveness of CF depends on its type. However, it was observed that the abrasiveness of GF- and CF-filled PTFE was very small in the experiments from which the data shown in Figs. 12 and 13 were obtained. Typical profiles of frictional tracks on glass and steel disks are shown in Fig. 14. With filled PTFE on glass, the abrasiveness of the GF-filled PTFE decreased and that of the CF-filled PTFE increased with decreasing sliding speed. A relatively small number

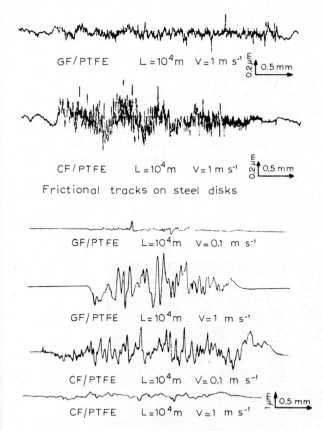

Fig. 14. Profiles of frictional tracks on steel and glass disks after rubbing against GF- and CF-filled PTFE. (L = sliding distance; load = 50 N.)

of deep scratches was generally produced in the tracks on glass, while a fair number of scratches was produced in the tracks on steel. Furthermore, considerable amounts of transferred material must have existed on the tracks rubbed against filled PTFE. The profiles shown in Fig. 14 indicate that the degree of abrasiveness of the carbon fibers used in this work is similar to that of the glass fibers. On the other hand, it was shown that the fiber-filled polyacetal exhibited much high abrasiveness than the fiber-filled PTFE.

4.4. Microscopic examinations on the wear process

Optical and electron microscopic examinations have been made on the worn surfaces of PTFE composites and frictional tracks after wear tests under various conditions [14,15]. The optical microscopic examinations indicated that the proportion of the fibrous fillers deposited on the worn surfaces of composites was very much greater than that seen on composite surfaces cut by means of a razor. This suggests that the proportion of fibrous fillers in the frictional surface layer increases during the initial higher wear rate stage of the wear process until it reaches a greater value at the stationary lower wear rate stage. From the scanning electron microscopic (SEM) study of a fracture surface obtained by impacting GF-filled PTFE at liquid nitrogen temperature, it was suggested that fibrous fillers can move slightly in the PTFE matrix. This is one reason why the fiber-rich surface layer is produced in the PTFE composites incorporating short fibrous fillers during rubbing.

Scanning electron micrographs of the frictional surfaces of composites are illustrated in Fig. 15. Figure 15(a) suggests that there are fibrous films of PTFE on the frictional surface. Figure 15(b) suggests that the fibrous filler such as GF and CF supports preferentially the contact load between the PTFE composites and the countersurface. It was observed that relatively short fibers were produced by the breaking of fibers during rubbing. Figure 16 shows an example of a scanning electron micrograph of fibers existing on the frictional track side. It is seen that the

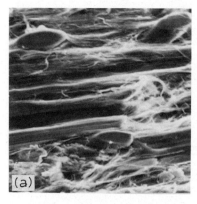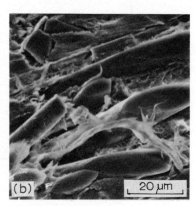

Fig. 15. SEM micrographs of worn surfaces of fiber-filled PTFE specimens rubbed under a load of 50 N against a glass disk. (a) GF–PTFE, 1 m s^{-1}; (b) CF–PTFE, 2.5 m s^{-1}.

151

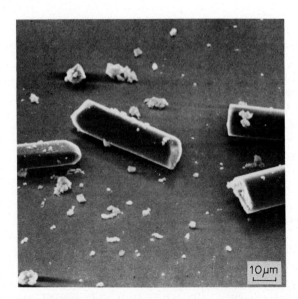

Fig. 16. SEM micrograph of the broken pieces of CF on the frictional track side (glass disk; load, 50 N; sliding speed, 2 m s^{-1}).

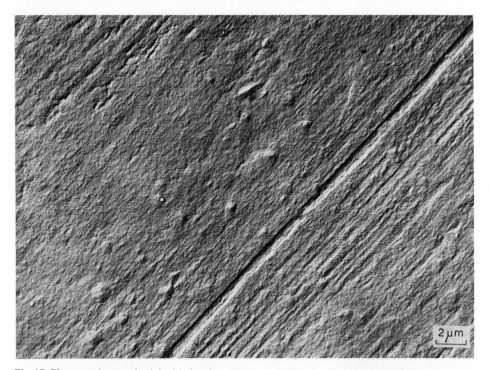

Fig. 17. Electron micrograph of the frictional track on a steel disk after rubbing against GF–PTFE at 1 m s^{-1} under 50 N load.

fibers are relatively short and are partially worn on their cylindrical surfaces. This suggests that the breaking of fibers occurs at the frictional surface and the fiber axis of the broken pieces becomes nearly parallel to the frictional surface, leading to wear of the cylindrical surface, and thus they are removed from the frictional surface. The broken pieces of fiber on the composite surface may play a role in the wear-reducing action of fibers incorporated in PTFE.

Figure 17 shows the existence of coherent transferred film on the frictional track. The film seems to include many very fine particles which may be considered as filler wear debris. The transferred films of PTFE-based composites seems to be much thicker than those of unfilled PTFE.

5. Friction and wear of PTFE incorporating various fillers

Tanaka and Kawakami [17] studied the friction and wear behaviour characteristic of the various types of fillers in PTFE-based composites and determined the factors that influenced the wear-reducing actions of the fillers. In their work, GF-filled PTFE was also used as the reference specimen. The experiments were carried out over a speed range of 0.1–2.5 m s^{-1} under loads of 10 and 50 N at room temperature. The flat ends of 3 mm diam. PTFE-based composite pins were rubbed against mild steel disk surfaces (ca. 0.03 μm c.l.a. roughness). The PTFE-based composites used as the specimens are listed in Table 1. Although the fillers incorporated in the ZrO$_2$–PTFE and TiO$_2$–PTFE specimens were particles that had passed through a 350 mesh sieve, the TiO$_2$ particles were agglomerates composed of

TABLE 1
PTFE-based composite specimens

Specimen	Filler material	Filler content (wt.%)	Shape and size of filler
GF–PTFE	Glass	25	Fiber; mean diameter, 7 μm; mean length, about 100 μm
Bronze–PTFE	Bronze	40	Particle; several microns in size
ZrO$_2$–PTFE	ZrO$_2$	40	Particle; passed through 350 mesh sieve
TiO$_2$–PTFE	TiO$_2$	20	Particle; passed through 350 mesh sieve but is an agglomeration of fine particles of less than 0.3 μm in size
MoS$_2$–PTFE	MoS$_2$	20	Powder; a few microns in size
Graphite–PTFE	Graphite	15	Powder; a few microns in size
Turcite B (commercial product)	Unknown but contains Cu and Sn atoms	Unknown	Mainly particle; about 30 μm in size; matrix is a fluorocarbon

Fig. 18. SEM micrographs of the fractured surfaces of (a) TiO$_2$–PTFE and (b) ZrO$_2$–PTFE.

very fine particles less than 0.3 μm in size. The ZrO$_2$ particles ranged from several microns to about 50 μm in size. Figure 18(a) and (b) show SEM micrographs of the fracture surfaces obtained by impacting TiO$_2$–PTFE and ZrO$_2$–PTFE, respectively, at liquid nitrogen temperature. It is seen that very fine TiO$_2$ particles are incorporated in the TiO$_2$–PTFE specimen and there are large particles of ZrO$_2$ in the ZrO$_2$–PTFE specimen. The polymer-based composite Turcite B, a commercial bearing material, was also used as a specimen because its matrix polymer material seemed to be PTFE. According to X-ray analysis of Turcite B, it seems that the material of the filler is bronze.

5.1. Variation in the coefficient of friction and wear depth with sliding distance

Figure 19 shows some typical variations in the coefficient of friction and wear depth with sliding distance. In most experimental runs, the steady wear stage, which had a lower wear rate, followed the initial transient stage of higher wear rate. However, an initial transient stage of lower wear rate, as seen in Fig. 19(c), was observed only with TiO_2–PTFE sliding under low load at lower speeds. Only with ZrO_2–PTFE was the initial friction considerably lower than the stationary value of the friction, as seen in Fig. 19(b). In contrast, GF–PTFE generally exhibited a

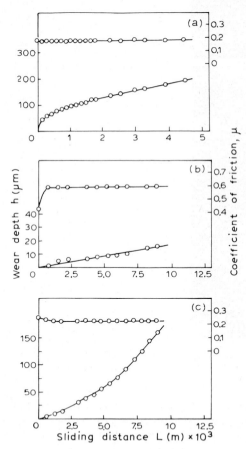

Fig. 19. Typical variation in the coefficient of friction and wear depth with sliding distance. (a) Graphite–PTFE (50 N; 0.5 m s^{-1}); (b) ZrO_2–PTFE (50 N, 0.1 m s^{-1}); (c) TiO_2–PTFE (10 N, 0.5 m s^{-1}).

higher initial friction than the stationary values. In other cases, friction was weakly dependent upon sliding distance. The coefficient of friction and the wear rate were obtained during the steady stages of friction and wear, respectively.

5.2. Friction

Figure 20(a) and (b) show the relationships between the coefficient of friction and the sliding speed for various specimens under loads of 10 and 50 N, respectively. With the exception of ZrO_2–PTFE, the coefficient of friction is slightly dependent upon the speed. Most specimens had a coefficient of friction in the range 0.2–0.3, which is similar to the coefficient of friction of unfilled PTFE. The coefficient of friction for many specimens tended to decrease as the load increased, but for TiO_2–PTFE, MoS_2–PTFE and graphite–PTFE specimens, the load had little effect on the coefficient of friction. ZrO_2–PTFE exhibits considerably higher friction than other specimens. Turcite B exhibits friction properties that are very similar to those of bronze–PTFE.

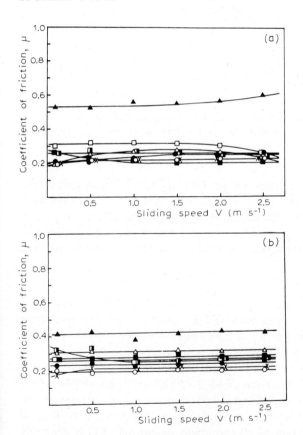

Fig. 20. Relationships between the coefficient of friction and the sliding speed for loads of (a) 10 N and (b) 50 N. △, TiO_2–PTFE; ▲, ZrO_2–PTFE; □, GF–PTFE; ■, bronze–PTFE; ○, graphite–PTFE; ●, MoS_2–PTFE; ×, unfilled PTFE; ◨, Turcite.

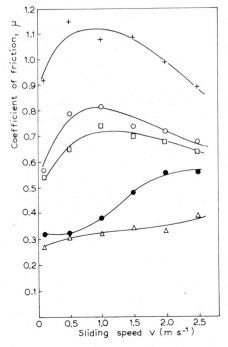

Fig. 21. Variation of the coefficient of friction with sliding speed for various filled polyacetals rubbed against a steel disk under a load of 50 N. +, Unfilled PAC; ○, GF–PAC; ●, CF–PAC; △, PTFE–PAC; □, glass beads–PAC.

In contrast to PTFE-based composites, the coefficients of friction of polyacetal-based composites are dependent very much upon the material used as the filler and, in some cases, very dependent upon the sliding speed, as seen in Fig. 21 [18]. Such characteristics of the friction of polyacetal-based composites seem to be common to the filled semicrystalline polymers that flow easily at temperatures above the melting point.

5.3. Wear

Variations in the specific wear rates with sliding speed for various specimens under loads of 10 and 50 N are shown in Fig. 22(a) and (b), respectively. In contrast to the coefficient of friction, the specific wear rate is strongly controlled by the type of filler. The specific wear rate of PTFE-based composites is generally weakly dependent upon the sliding speed. The wear of MoS_2–PTFE and TiO_2–PTFE under high load varies substantially with speed and is considerably higher. Bronze–PTFE shows the lowest wear. The specific wear rate of Turcite B is similar to that of bronze–PTFE. The specific wear rate of all specimens except GF–PTFE generally increases with load, indicating that the worn volume of PTFE-based composites generally does not increase in direct proportion to load.

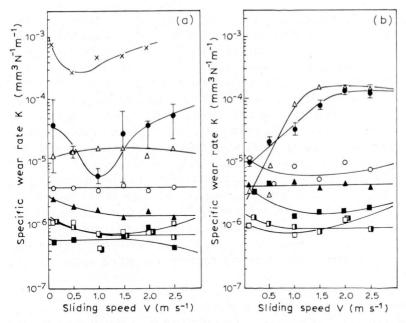

Fig. 22. Variation in specific wear rate with sliding speed for loads of (a) 10 N and (b) 50 N. ×, Unfilled PTFE; △, TiO$_2$–PTFE; ▲, ZrO$_2$–PTFE; □, GF–PTFE; ■, bronze–PTFE; ○, graphite–PTFE; ●, MoS$_2$–PTFE; ▨, Turcite.

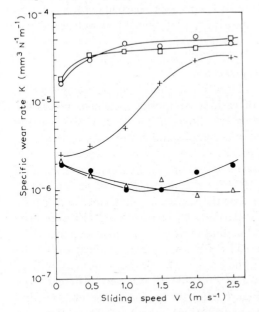

Fig. 23. Variation of specific wear rate with sliding speed for various filled polyacetals rubbed against a steel disk under a load of 50 N. +, Unfilled PAC; ○, GF–PAC; ●, CF–PAC; △, PTFE–PAC; □, glass beads–PAC.

The variations in the specific wear rate with sliding speed of polyacetal-based composites containing various fillers are shown in Fig. 23 [18]. It should be noted that the effect of fillers on wear reduction in polyacetal differs markedly from that in PTFE. Carbon fiber-filled polyacetal displays a very much lower wear rate at higher speeds than the unfilled polyacetal. In contrast, the polyacetal composites incorporating glass fiber and beads exhibit much high wear than unfilled polyacetal at lower speeds. It is noted also that PTFE powder reduces very much the wear rate of polyacetal. According to Tanaka and Uchiyama [12], melting of the matrix polymer, i.e. polyacetal, on the surface layer of polyacetal-based composites must occur during rubbing, because of frictional heating, even at a speed of 0.1 m s^{-1}. Therefore, there will be many direct contacts between the filler in the composites and the countersurface, i.e. steel, during rubbing and this must be the reason why glass fiber- and bead-filled polyacetal composites exhibit high wear. It seems that the effect of various fillers on the wear characteristics of polyacetal-based composites, which was mentioned above, appears in the case of other filled polymers in which the matrix polymer can flow easily at temperatures above the melting point.

5.4. Abrasiveness of the composites

Measurements of the profiles of frictional tracks obtained under various sliding conditions indicated that the abrasiveness of the PTFE-based composites incorporating various fillers sliding against steel was generally very low, even with hard fillers, while the composites incorporating glass fiber, bronze or ZrO_2 particles showed greater abrasiveness than the composites filled with TiO_2 particles, MoS_2 or graphite. However, it was found that the abrasiveness of fiber-filled polyacetal was much greater than that of fiber-filled PTFE [15].

5.5. Microscopic examinations on the wear process

Scanning electron microscopy (SEM) and transmission electron microscopy (TEM) examinations of the frictional surfaces of various composite specimens reveal important information about the action of various fillers in the wear of PTFE-based composites. Figure 24 shows typical SEM micrographs and X-ray maps corresponding to each micrograph for the worn surfaces of some composite specimens. The X-ray maps were taken using X-rays characteristic of one of the elements of the filler material. With GF–PTFE and bronze–PTFE, a comparison between the SEM micrograph and the X-ray map indicates that the surface of glass fiber and bronze on the frictional surfaces of composites is partially covered with a PTFE film. The surface of the glsss fiber is very smooth while that of bronze has many scratches and parts of the scratches are covered with a PTFE film. X-ray maps indicated that the PTFE film on the frictional surface of the composites contains iron as well as elements from the filler. The worn surface of ZrO_2–PTFE was very different from that of GF–PTFE and bronze–PTFE. Figure 24(c) shows that, in general, no film was observed on the ZrO_2 particles on the worn surface of ZrO_2–PTFE and that the worn surface of the ZrO_2 particles was very smooth and free from scratches. By contrast, no TiO_2 particle was generally observed on the worn surface of TiO_2–PTFE

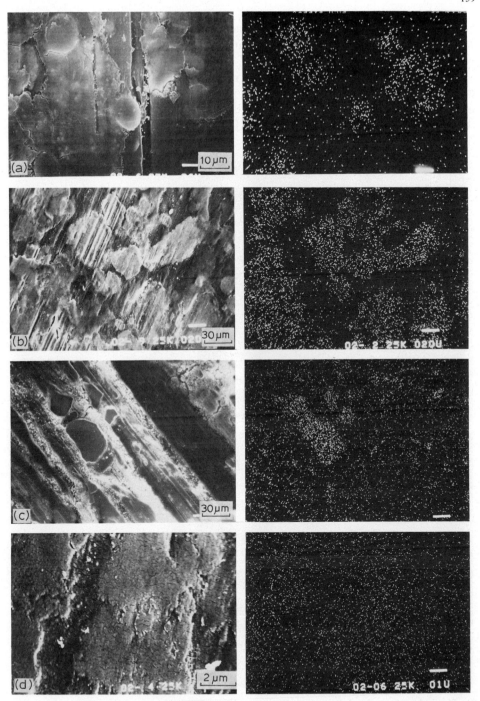

Fig. 24. SEM micrographs (left-hand photographs) of the worn surfaces of PTFE-based composites and X-ray maps (right-hand photographs) corresponding to the areas of the micrographs (load, 10 N; sliding speed, 0.1 m s^{-1}): (a) GF–PTFE, silicon X-ray map; (b) bronze–PTFE, copper X-ray map; (c) ZrO$_2$–PTFE, zirconium X-ray map; (d) TiO$_2$–PTFE, titanium X-ray map.

by means of SEM and TEM of the surface replica. However, X-ray examination
with the Ti $K\alpha$ line indicated that many fine TiO_2 particles were distributed on the
worn surface as seen in Fig. 24(d).

The worn surface of MoS_2–PTFE is very similar to that of graphite–PTFE under
mild sliding conditions. However, there was a difference under severe sliding
conditions. With MoS_2–PTFE, large flake-like pieces of material are formed at the
frictional surface; this is shown in Fig. 25. The graphite filler at the frictional surface

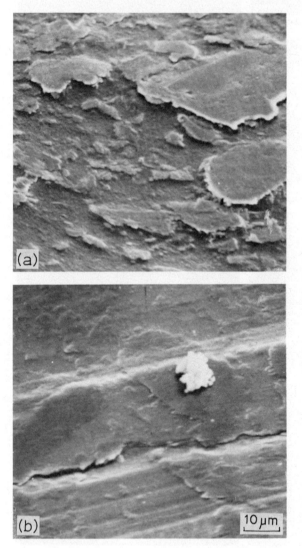

Fig. 25. SEM micrographs of the worn surfaces of (a) MoS_2–PTFE and (b) graphite–PTFE (load, 50 N;
sliding speed, 1.5 m s^{-1}).

seems to be embedded in the PTFE matrix. The flake-like material, consisting of MoS_2–PTFE, may be easily detached from the frictional surface and this may increase the wear rate of MoS_2–PTFE under severe sliding conditions.

Electron micrographs of replicas taken from the worn surfaces of TiO_2–PTFE

Fig. 26. Electron micrographs of the worn surfaces of (a) MoS_2–PTFE (10 N; 0.1 m s^{-1}) and (b) TiO_2–PTFE (10 N; 1.5 m s^{-1}).

and MoS_2–PTFE are shown in Fig. 26. The worn surfaces of these composites are similar to that of unfilled PTFE. There are many long fibers that are produced by large-scale destruction of the banded structure of PTFE. No such long fibers were

Fig. 27. Electron micrographs of the wear fragments that adhered to the replica of the frictional tracks in (a) MoS_2–PTFE (50 N; 0.5 m s^{-1}) and (b) TiO_2–PTFE (10 N; 0.5 m s^{-1}).

observed on the worn surfaces of composites incorporating glass fibers, bronze and ZrO_2 particles.

Electron micrographs of wear fragments that adhered to replicas of the disk surfaces rubbed against TiO_2–PTFE and MoS_2–PTFE are shown in Fig. 27. The dark spots in the fragment seen in Fig. 27(b) must be TiO_2 particles and the electron diffraction pattern taken from the fragment in Fig. 27(a) indicated that the fragment included MoS_2 powder. This indicates that small fillers such as TiO_2 particles and MoS_2 powders are removed with the PTFE film transferred to the counterface.

5.6. Friction of ion sputter-etched surfaces of PTFE-based composites

To examine the coefficient of friction between the filler materials and the steel counterface, the flat end surfaces of various speciment pins were etched by ion sputtering in vacuum and the etched surfaces were rubbed against the steel disk at 0.1 m s^{-1} under a load of 10 N. The filler material at the frictional surface of the pin preferentially rubs against the steel disk because the sputtering rate of PTFE, which is an organic material, is greater than that of any filler material. The coefficients of friction for the etched specimens are shown in Fig. 28. Although the friction for etched TiO_2–PTFE is relatively low, this may be due to the fact that the TiO_2 particles were too small to show the true friction. The etched specimens were also rubbed at 0.1 m s^{-1} under a load of 10 N against a steel disk which had a thin PTFE film transferred to it by pre-rubbing against unfilled PTFE. The measured coefficient of friction provides information about the friction-reducing effect of a PTFE film at the interface between the filler material and the steel disk. The results of this experiment together with results from the friction of unetched specimens rubbed against the disk without a pre-transferred PTFE film (Fig. 20) are also shown in Fig. 28. The friction between the etched specimens and the steel disk with a PTFE film is approximately the same with every composite and the friction of unetched specimens rubbed against the disk without a PTFE film is generally less than the friction between the etched specimens and the disk without a PTFE film.

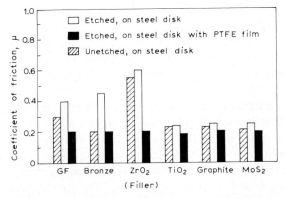

Fig. 28. Results of friction experiments using PTFE-based composites with ion sputter-etched frictional surfaces.

6. Roles of various fillers incorporated in PTFE

6.1. Main mechanism of wear reduction by fillers incorporated in PTFE

As seen in Fig. 22(a), there is a very remarkable reduction of wear by the fillers in PTFE. Lancaster [13] studied the effect of carbon fiber reinforcement on the wear of various polymers and pointed out that the fibers exposed during rubbing preferentially support part of the applied load and reduce the wear rate of the reinforced polymers. This load-supporting action of fibers may be enhanced by the fact that the fiber-rich surface layer has been produced during the initial rubbing stage of fiber-filled PTFE. Briscoe et al. [19] considered that the wear-reducing action of a filler in a high density polyethylene-based composite was due to the formation of a strongly adhering film of polymer transferred to the counterface. This mechanism may also be applicable to the wear of PTFE-based composites because a thicker film transfers to a counterface rubbed by PTFE-based composites compared with the case of unfilled PTFE.

However, there is other important mechanism which contributes to wear-reduction by a filler in PTFE. As described earlier, unfilled PTFE shows much high wear as a result of the destruction of the banded structure due to easy slippage between the crystalline lamellae in the bands. In contrast to unfilled PTFE, electron microscopic examinations of the worn surfaces of composites does not show large-scale PTFE film. This indicates that large-scale destruction of the banded structure of PTFE does not occur during the wear process of PTFE-based composites. Considering that the wear of PTFE itself decreases considerably when it is filled with any filler, it is therefore reasonable to consider that any filler incorporated in PTFE prevents the occurrence of large-scale destruction of the banded structure. It has been shown that wear reduction by fillers in PTFE is very much greater than that in other polymers. The banded structure characteristic of PTFE is very different from the structures of other polymers. Therefore, the most important mechanism of wear-reduction by the filler must be a reduction of the large-scale destruction of the banded structure.

6.2. Load-supporting action of the filler

The load-supporting action of fibers proposed by Lancaster [13] for the wear-reducing action of the fibers incorporated in polymers is reasonable and may be explained theoretically by modifying the analysis for the effect of discontinuous fiber on strengthening the composite [20].

For simplifity, consider the fiber parallel to the loading direction and contacting the counterface at the end [Fig. 29(a)]. Normally, the matrix polymer has a lower modulus than the fiber and thus the difference in the axial strains between the fiber and the matrix will produce a tangential stress τ and a corresponding compressive loading of the fiber [Fig. 29(b)]. Thus, the contact load is transferred from the matrix to the fiber at the cylindrical surface as well as at the end surface. The compressive stress, σ_z, in the fiber builds up approximately as shown in Fig. 29(c). The maximum stress, σ_f, in the fiber is attained at the lower end of the fiber which is the contact

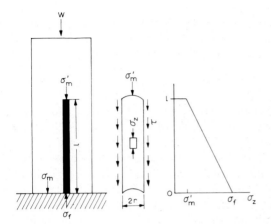

Fig. 29. Theoretical analysis of the load-supporting action of fiber-filled polymer. (a) Contact model of the polymer-based composite incorporating fiber; (b) build-up of compressive stress σ_z in a fiber due to a tangential surface stress τ; (c) variation in compressive stress σ_z with position in a fiber contacted at the counterface.

face of the fiber. The stress σ_f is the contact stress of the fiber and can be expressed as

$$\sigma_f = \frac{2\pi r l \tau}{\pi r^2} + \sigma'_m$$

$$= \frac{2l\tau}{r} + \sigma'_m \tag{1}$$

where σ'_m is the compressive stress of the matrix polymer in the bulk composite pressed on the smooth counterface under the load W, r is the radius of the fiber and τ is the tangential stress applied by the matrix and will be approximately the shear strength of the matrix polymer. According to King and Tabor [21], the ratio of shear strength to yield pressure for thermoplastic polymers is considered to be in the range 0.3–0.7. Therefore, the tangential stress τ is generally considered not to be substantially smaller than the compressive stress σ'_m of the matrix, since the compressive stress σ'_m of the matrix polymer is generally smaller than the yield pressure of the matrix polymer. The stress σ_m of the matrix polymer contacting the counterface may be somewhat different from the bulk matrix stress σ'_m. However, eqn. (1) indicates that the fiber preferentially supports the applied load.

When fibers are used as fillers, the load-supporting action is most effective because l/r is large. When the filler is a particle such as bronze or ZrO_2, the value of l/r may be approximately unity and thus the particle also preferentially supports the load. When the filler is a lamellar solid lubricant such as MoS_2 or graphite, there is a tendency to produce a flake which has a flat surface approximately parallel to the counterface and the value of l/r will be extremely small, resulting in no preferential load-supporting action.

Although, in the above discussion, it was assumed that the fiber was parallel to the loading direction, similar conclusions may be derived by considering the resolved stress parallel to the loading direction even for a fiber inclined to the counterface.

6.3. Effect of the material and the shape and size of the filler on the friction of PTFE-based composites

As seen in Fig. 20, the coefficient of friction for various PTFE-based composites is weakly dependent upon the incorporated filler. A thin PTFE film generally exists at the interface between the filler and the disk surface and the existence of this PTFE film leads to the similar friction values that are obtained between the composites filled with various fillers and the steel disk. Small particles or powder fillers are generally embedded in the PTFE matrix at the frictional surfaces of composites (Fig. 26). Consequently, the friction of various PTFE-based composites is generally weakly dependent upon the type of filler. However, it was observed that there was no PTFE film on the surface of large particles of ZrO_2 filler and that the ZrO_2–PTFE specimen exhibited a friction similar to that of the etched ZrO_2–PTFE specimen rubbing against the steel disk. This suggests that a PTFE-based composite incorporating filler with too large a size does not show a low friction similar to unfilled PTFE. It is reasonable to consider that PTFE-based composites generally exhibit a low friction similar to unfilled PTFE and their friction is almost independent of the material and the shape and size of the filler provided that the filler size is not too large. This characteristic of PTFE-based composites is essentially due to the excellent film-forming ability of PTFE.

6.4. Effect of the material and the shape and size of the filler on the wear of PTFE-based composites

The fact that TiO_2–PTFE exhibits much greater wear than composites incorporating ZrO_2 and bronze particles indicates that the wear-reducing action of fillers of very small size is poor. The wear of MoS_2–PTFE was also generally much greater than that of composites incorporating other fillers except for TiO_2 particles. Although the wear of graphite–PTFE is generally much less than that of MoS_2–PTFE and TiO_2–PTFE, CF–PTFE exhibits much lower wear than graphite–PTFE. Electron microscopic examinations of the worn surfaces of composites incorporating TiO_2, MoS_2 and graphite indicated similar surface features to those of unfilled PTFE and these fillers were embedded in the PTFE matrix on the frictional surfaces. The film that transferred to the counterface included many particles of TiO_2 or MoS_2, as seen in Fig. 27. These facts support strongly the conclusion that the wear-reducing action of fillers of very small size is generally poor in PTFE-based composites. This is due to the fact that very small fillers on the frictional surfaces cannot prevent large-scale destruction of the banded structure of the PTFE matrix and thus very small fillers are easily removed from the frictional surface together with the PTFE film transferred to the counterface. However, graphite may adhere to PTFE more strongly than MoS_2 and graphite–PTFE may exhibit considerably less wear than MoS_2–PTFE.

In contrast to very small fillers, fillers such as ZrO_2, bronze, GF and CF greatly reduce the wear of PTFE. These fillers are firmly embedded in the PTFE matrix on the frictional surface and preferentially support the load, and the filler surface exposed at the frictional surface rubs against the counterface. Consequently, the composites incorporating these fillers show lower wear. However, it must be noted that the wear-reducing action of ZrO_2, which has a very high hardness, is less than that of bronze, GF anf CF. This must be due to the fact that many large particles of ZrO_2 at the frictional surface are rubbed directly against the counterface because there is not much PTFE film at the sliding interface of large particles of ZrO_2. The lack of a PTFE film at the interface may be due to the difficulty with which the film is transferred over a large area of the interface. A high temperature may be generated at the sliding interface of a ZrO_2 particle as the result of the high friction of ZrO_2, as seen in Fig. 28. This may also contribute to the difficulty that a PTFE film at the sliding interface of ZrO_2 has in existing.

In contrast, many parts of the sliding interfaces for GF, CF and bronze fillers are covered with a PTFE film which reduces the inherent wear of glass, carbon and bronze, resulting in effective wear reduction for the composites incorporating these fillers. Since the mechanical properties of glass are very different from those of bronze, the fact that GF–PTFE shows a similar wear rate to bronze–PTFE indicates that the filler material is of a little importance in the wear-reducing action of PTFE-based composites.

7. Effect of water lubrication on the friction and wear of PTFE-based composites

Lancaster has studied the friction and wear properties of various CF-filled polymers sliding against metals in water, aqueous solutions and organic fluids [22,23] and found that the wear of CF-reinforced polymers as well as unfilled polymers under water lubrication was generally greater than that under dry conditions. He attributed the higher wear rate in water to the fact that the counterface could not be modified by transfer, because he could not observe the transferred film on the counterface rubbed under water lubrication. Tanaka [24] also found that the wear of unfilled semicrystalline polymers increased generally under a boundary lubrication with water. However, he found that the amount of transferred polymer in water lubrication was similar to that in dry conditions and suggested that the increase of wear under water lubrication may be due to a modification of the surface structure of polymers by water rather than the effect of modification of the counterface by polymer transfer. According to Pratt [25], the mica-filled PTFE and the 20% bronze/20% graphite-filled PTFE show the low wear rate in underwater applications.

Yamada and Tanaka [26] studied the friction and wear of various PTFE-based composites under boundary lubrication with water using the 3 mm diam. flat-ended specimen pins sliding against a stainless steel disk at a speed of 0.01 m s^{-1} under a load of 10 N. The PTFE-based composites shown in Table 1 and CF(40%)–PTFE were used as specimens in their work. The friction and wear under water lubrication were compared with those under dry conditions. The results are described below.

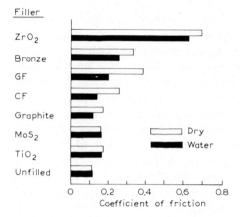

Fig. 30. Effect of water lubrication on the coefficient of friction of various PTFE-based composites (load, 10 N; sliding speed, 0.01 m s^{-1}).

7.1. Friction

Figure 30 shows the coefficient of friction obtained under water lubrication and dry conditions for various specimens. With the exception of the composites incorporating MoS$_2$ and TiO$_2$, the coefficient of friction under water lubrication is somewhat lower than that under dry conditions. It should be noted that ZrO$_2$–PTFE exhibits much higher friction than other specimens under water lubrication as well as dry conditions. This may be due to a greater ploughing component in the friction of ZrO$_2$ particles incorporated in the composite specimen because ZrO$_2$–PTFE abraded the counterface under water lubrication considerably.

7.2. Wear

Figure 31 shows the specific wear rates under water and dry conditions for various specimens. It is seen that the wear of PTFE-based composites increases, in

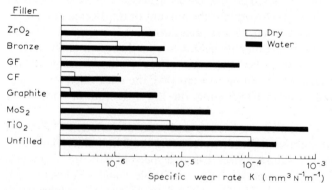

Fig. 31. Effect of water lubrication on the specific wear rate of various PTFE-based composites (load, 10 N; sliding speed, 0.01 m s^{-1}).

general, very much under water lubrication. The wear of unfilled PTFE under water lubrication is about three times as large as that under dry conditions. In contrast to unfilled PTFE, the wear of PTFE-based composites is generally 10–30 times higher than that under dry conditions. It should be noted that the composites incorporating small-size fillers such as MoS_2, graphite powders and TiO_2 particle exhibit an especially marked increase in wear under water lubrication compared with dry conditions. On the other hand, the wear of GF–PTFE is much greater than that of CF–PTFE under water lubrication. In addition, GF–PTFE shows much greater wear than CF–PTFE even under dry conditions at a low sliding speed of 0.01 m s^{-1}.

7.3. Microscopic examinations on the wear process of the PTFE-based composites under water lubrication

Electron microscopic examination of the counterface rubbed by various specimens indicated that the transfer of composites to it under water lubrication was similar to that under dry conditions. Lancaster attributed higher wear of CF-filled polymers under water lubrication to little transfer. However, the higher wear of PTFE-based composites observed in the experiments of Yamada and Tanaka under water lubrication was not due to low transfer. The amount of transfer of polymer-based composites under water lubrication may be influenced by the sliding conditions.

Figure 32 shows TEM micrographs of the worn surfaces of some specimens rubbed under water lubrication. These micrographs suggest that there is a very small amount of PTFE film at the interface between the fillers in PTFE-based composites and the counterface during rubbing of the composites under water lubrication. Figure 33 shows SEM micrographs of the worn surfaces of some specimens rubbed under water lubrication. It is seen that the surface of the fillers incorporated in the PTFE matrix are only slightly covered with PTFE film. Therefore, SEM examination of the worn surfaces also suggests that there exists little PTFE film at the interface between the fillers and the counterface during rubbing of PTFE-based composites under water lubrication. Thus, the existence of little or no PTFE film at the interface between the fillers and the counterface during rubbing must cause higher wear to PTFE-based composites under water lubrication compared with the case of dry conditions.

However, it was found that there is another cause for higher wear of PTFE-based composites under water lubrication. A thin PTFE film adhering weakly on the worn surfaces of the composites rubbed under water lubrication was removed by taking the replica several times from each worn surface by means of the replica technique used in TEM microscopy and then the worn surfaces were examined by SEM. Figure 34 shows SEM micrographs of the worn surfaces of GF- and CF-filled PTFE after the multiple replications. Very dark portions seen in the micrographs correspond to the cavities on the worn surface. The cavity was also seen on the worn surface of bronze–PTFE after the multiple replications. Such cavities on the worn surface must be produced by the detachment of a small piece of filler from the frictional surface of the composite. The small piece seems to be a very short fiber, a broken piece of

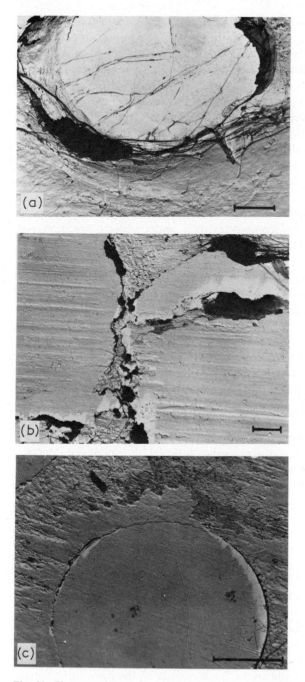

Fig. 32. Electron micrographs of the worn surfaces of PTFE-based composites rubbed under water lubrication. (a) GF–PTFE; (b) bronze–PTFE; (c) CF–PTFE. The bar in the photographs represents 4 μm.

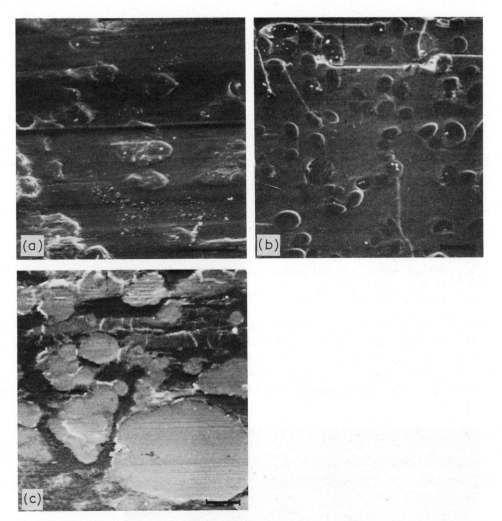

Fig. 33. SEM micrographs of the worn surfaces of PTFE-based composites rubbed under water lubrication. (a) GF–PTFE; (b) CF–PTFE; (c) bronze–PTFE. The bar in the photographs represents 25 μm.

fiber or a fragment of the filled particle produced in the frictional surface layer during rubbing of the composite. As seen in Fig. 34, there were many cavities on the worn surface of the GF–PTFE, but only a few on those of the CF- or bronze-filled PTFE. The existence of cavities indicates that the separation of fillers embedded in the PTFE matrix in the composite surface layer occurs somewhat easily under water lubrication and this may be due to the permeation of water molecules to the interface between the filler in the surface layer of composites and the PTFE matrix. The GF–PTFE shows much higher wear than the CF- or bronze-filled PTFE and

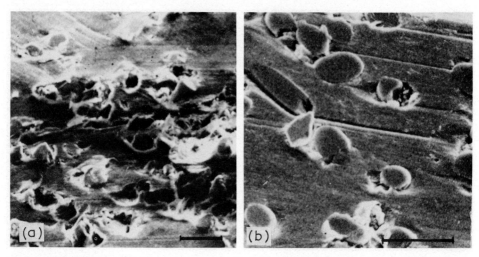

Fig. 34. SEM micrographs of the worn surfaces of fiber-filled PTFE after removal of the surface film. (a) GF–PTFE; (b) CF–PTFE.

this is consistent with the fact that there were many more cavities in the case of GF–PTFE than other specimens. With GF–PTFE specimen only, it was also found that many large lumps adhered to the counterface. Figure 35 shows the lumps adhering to the counterface rubbed against GF–PTFE and the lump seems to be produced by the accumulation of wear debris from the composite. This may be due

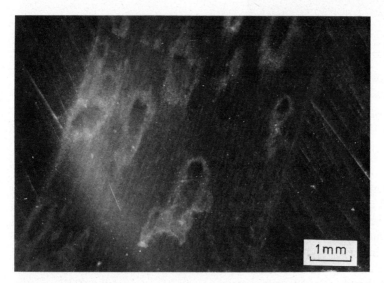

Fig. 35. Optical micrograph of the lumps adhering on the frictional track of stainless steel disk rubbed against GF–PTFE under water lubrication.

to the fact that glass is a very hydrophilic material. In addition, friction of GF–PTFE showed great fluctuation during rubbing because of the adhering lumps.

Although very small-size fillers such as MoS_2, graphite powders and TiO_2 particles are easily removed from the frictional surface together with the PTFE film transferred to the counterface under water lubrication as well as dry conditions, it is noted that graphite–PTFE shows a much lower wear rate than TiO_2–PTFE and MoS_2–PTFE. The permeation of water molecules to the interface between the filler and the PTFE matrix in the composite surface layer may also play an important role in the wear behavior of the composites incorporating very small-sized fillers.

8. Conclusions

In this chapter, the fundamental behavior on the transfer, friction and wear of unfilled PTFE were first briefly surveyed and discussed on the basis of the molecular and morphological structures of PTFE. Secondly, the fundamental behavior on the friction and wear of PTFE incorporating various fillers were considered, mainly on the basis of the work carried out in the author's laboratory and were discussed on the basis of the results which had been obtained with unfilled PTFE. Although there are many problems on the tribological properties of the PTFE-based composites which need further investigation, this chapter will be useful for understanding and predicting the tribological behavior of the PTFE-based composites used in sliding applications.

List of symbols

μ_s	coefficient of static friction
μ_k	coefficient of kinetic friction
T	temperature
T_0	reference temperature
a_T	horizontal shift factor determined by reduction of the wear rate vs. speed curves at temperature T to that at any reference temperature T_0
τ	tangential stress at the interface between the fiber and the matrix
σ_z	compressive stress in the fiber
σ_f	maximum stress in the fiber or contact stress of the fiber
l	fiber length
τ	radius of the fiber
σ_m'	compressive stress of the matrix polymer
W	load
V	sliding speed
L	sliding distance
N	number of the disk revolution
h	wear depth

174

K	specific wear rate
t	thickness of the transfer film
PTFE	polytetrafluoroethylene
HDPE	high density polyethylene
GF	glass fiber
CF	carbon fiber
SEM	scanning electron microscope
TEM	transmission electron microscope

References

1 C.M. Pooley and D. Tabor, Proc. R. Soc. London Ser. A, 329 (1972) 251.
2 P.H. Geil, Polymer Single Crystals, Interscience, New York, 1963, p. 183.
3 B. Wunderlich, Macromolecular Physics, Vol. 1, Academic Press, New York, 1973, p. 332.
4 C.W. Bunn, A.J. Cobbold and R.P. Palmer, J. Polym. Sci., 38 (1958) 363.
5 C.J. Speerschneider and C.H. Li, J. Appl. Phys., 33 (1962) 1871.
6 K.R. Makinson and D. Tabor, Proc. R. Soc. London Ser. A, 281 (1964) 49.
7 K. Tanaka, Y. Uchiyama and S. Toyooka, Wear, 23 (1973) 153.
8 K. Tanaka and T. Miyata, Wear, 41 (1977) 383.
9 K. Tanaka and Y. Yamada, in K.C. Ludema (Ed.), Wear of Materials, 1983, ASME, New York, 1983, pp. 617.
10 K. Tanaka, Wear, 75 (1982) 183.
11 K. Tanaka and S. Ueda, Wear, 34 (1976) 323.
12 K. Tanaka and Y. Uchiyama, in L.H. Lee (Ed.), Advances in Polymer Friction and Wear, Vol. 5B, Plenum Press, New York, 1974, pp. 499–530.
13 J.K. Lancaster, J. Phys. D, 1 (1968) 549.
14 K. Tanaka, Y. Uchiyama, S. Ueda and T. Shimizu, in T. Sakurai (Ed.), Proc. JSLE–ASLE Int. Lubrication Conf., Tokyo, June 9–11, 1975, Elsevier, Amsterdam, 1976, pp. 110–118.
15 K. Tanaka, J. Lubr., 99 (1977) 408.
16 J.K. Lancaster, in A.D. Jenkins (Ed.), Polymer Science, Vol. 2, North-Holland, Amsterdam, 1972, Chap. 11, p. 1035.
17 K. Tanaka and S. Kawakami, Wear, 79 (1982) 221.
18 K. Tanaka, S. Ueda and T. Sasaki, Preprint of the JSLE Annual Meeting (Nagoya), 1977, p. 117 (in Japanese). Part of data has been published in ref. 15.
19 B.J. Briscoe, A.K. Pogosian and D. Tabor, Wear, 27 (1974) 19.
20 C.R. Barrett, W.D. Nix and A.S. Tetelman, The Principles of Engineering Materials, Prentice-Hall, Englewood Cliffs, NJ, 1973, p. 321.
21 R.F. King and D. Tabor, Proc. Phys. Soc. London Sect. B, 66 (1953) 728.
22 J.K. Lancaster, Wear, 20 (1972) 315.
23 J.K. Lancaster, Wear, 20 (1972) 335.
24 K. Tanaka, J. Lubr. Technol., 102 (1980) 526.
25 G.C. Pratt, in M.O.W. Richardson (Ed.), Polymer Engineering Composites, Applied Science Publishers, London, 1977, Chap. 5, p. 254.
26 Y. Yamada and K. Tanaka, Junkatsu, 29 (1984) 209 (in Japanese).

Chapter 6

Friction and Wear of Metal Matrix–Graphite Fiber Composites

ZWY ELIEZER

Department of Mechanical Engineering, Materials Science and Engineering Program, The University of Texas, Austin, TX 78712 (U.S.A.)

Contents

Abstract

The friction and wear behavior of metal matrix graphite fiber composites is dependent on manufacturing, operational, and environmental parameters. In particular, the type of fiber employed seems to have a significant influence. The graphite fibers are bent and broken at the subsurface during sliding. It is found that composites containing low modulus fibers exhibit, in general, lower wear rates than the high modulus composites. It is suggested that this behavior may be due, in part, to the larger amount of work needed to bend and break the low modulus compared with the high modulus fibers. The friction and wear behavior of various composites is also significantly influenced by the fiber orientation with respect to the sliding interface. Based on experiments conducted on aluminum matrix composites at relatively low sliding speed, a new wear equation is proposed. Some potential tribological applications of metal matrix–graphite fiber composites are discussed.

1. Introduction

In recent years, the tribological characteristics of metal matrix–graphite composites have been extensively investigated. The excellent mechanical properties, combined with the in-situ lubricating ability of these graphite-containing composites make them potential candidates for such applications as advanced high speed–high load bearing and seal components, high speed–high current electrical brushes, etc.

The friction and wear properties of chopped graphite fiber–metal matrix composites were studied by Giltrow and Lancaster [1] as early as 1968. Although the wear rates exhibited by these composites when sliding against tool steel were lower than those of the matrix materials (cobalt, nickel, copper, silver, lead), no significant improvement was obtained in friction. Moreover, their high temperature behavior was unsatisfactory, and their general mechanical properties were degraded compared with the bare matrix materials. These drawbacks were eliminated, to a large extent, by improvements in the fabrication process and by the employment of continuous graphite fibers instead of the chopped fibers.

The systematic tribological investigation of the unidirectionally oriented graphite fiber–metal matrix composites started in the early 1970s at the Aerospace Corporation (El Segundo, CA) and has been continued at The University of Texas at Austin and at The International Harvester Company (Hinsdale, IL). This article is an attempt to review and discuss the friction and wear behavior of various types of graphite fiber–metal matrix composites and to give some examples of potential industrial applications.

2. Materials

The composite samples used for tribological characterization were produced by liquid–metal infiltration of a number of graphite yarns which include those from rayon, PAN (polyacrylonitrile), and pitch. The composites were produced by applying a thin deposit of titanium and boron through chemical vapor deposition prior to immersion into the liquid metal [2]. The resulting product was a wire consisting of about 25–45% fiber in a metal matrix. In the final step (consolidation), the wires were cleaned and pressed in vacuum at various temperatures and pressures, depending on the particular matrix material.

Characteristic properties of the graphite fibers, of the matrix materials, and of the

TABLE 1

Some characteristic properties of the graphite fibers used in the composite samples

Precursor	Commercial designation	Tensile strength (GPa)	Young's modulus (GPa)
Rayon	T50	2.40	413
Rayon	T75	2.47	578
PAN	T300	2.63	241
PAN	Fortafil 4R	2.41	218
PAN	Hercules A	3.50	243
PAN	HM 3000	2.43	365
PAN	Modmor I	2.05	317
PAN	Celion 6000	2.76	234
Pitch	VSA 11	1.2	380

composite samples are given in Tables 1–3. The wire qualities designated in Table 3 as A, B, and C refer to the amount of infiltration and surface appearance as follows: class A has substantial (at least 95%) infiltration within the wire bundles and a continuous metal coating on the surface; class B has incomplete infiltration within

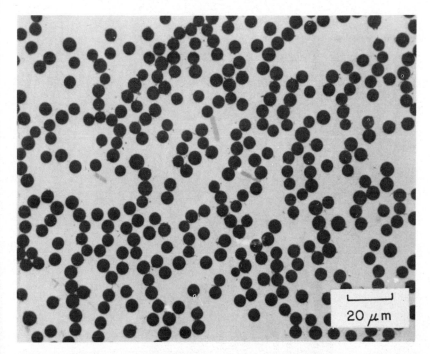

Fig. 1. Micrograph of a typical graphite fiber–metal matrix composite.

TABLE 2
Matrix compositions

Matrix material	Al	Mg	Si	Cr	Cu	Fe	Sn	Ag	P	Ti	Sb	Pb	Mn	Remarks
Al 201	Bal	0.3			4.7			0.7		0.3			0.3	
Al 6061	Bal	1	0.6	0.25	0.3									
Al A 413	Bal	0.1	12		0.6	1.3								
Al 1100	99 +													Commercially pure Al
Alloy SSC 155		0.11			Bal			0.04	0.05					
Alloy 954	11				Bal	4								
Babbitt 1					4.5		91				4.5			
Babbitt 4					3		75				12	10		

TABLE 3

Some characteristics of the composite materials used for tribological evaluation

Composite designation	Fiber type	Matrix material [a]	Wire quality [b]
C1	T50	Al 201	
C2	T300	Al 201	
C3	T50	Al 6061	
C4	T75	Al 6061	
C5	Hercules A	Al 6061	
C6	Fortafil 4R	Al 6061	
C7	T75	Al A413	
C8	Fortafil 4R	Al A413	
C9	T50	Cu	
C10	T50	Tin	
C11	T300	Cu–1% Sn	A
C12	T300	Cu–1% Sn	B
C13	T300	Cu–1% Sn	C
C14	T300	Cu–10% Sn	B
C15	T50	Cu–10% Sn	B
C16	HM 3000	Cu–1% Sn	A
C17	HM 3000	Cu–3% Sn	A
C18	HM 3000	Cu–4% Sn	A
C19	HM 3000	Cu–4% Sn	B
C20	HM 3000	Cu–8% Sn	A
C21	Celion 6000	Alloy SSC 155	A
C22	HM 3000	Al 1100	
C23	T50	Cu–6% Sn	
C24	HM 3000	Cu–6% Sn	
C25	Modmor I	Cu–6% Sn	
C26	T300	Alloy SSC 155	
C27	T300	Cu–6% Sn	
C28	T300	Alloy 954	
C29	T300	Ag–20% Cu	
C30	T300	Ag–28% Cu	
C31	Celion 6000	Alloy 954	
C32	Celion 6000	Ag–28% Cu	
C33	Celion 6000	Ag–26% Cu–2% Ni	
C34	VSA 11	Alloy SSC 155	
C35	VSA 11	Ag–28% Cu	
C36	T50	Babbitt 1	
C37	T300	Babbitt 1	
C38	T50	Babbitt 4	
C39	T300	Babbitt 4	

[a] Matrix compositions are given in Table 2.
[b] For definition of wire quality, see text.

the fiber bundles but a continuous metal coating on the surface; class C has incomplete infiltration within the fiber bundles and a discontinuous metal coating on the surface. A typical composite micrograph is shown in Fig. 1.

3. Friction and wear behavior

3.1. General remarks

The mechanical properties of the graphite fiber composites are known to depend on the properties of the fibers, of the matrix, of the interface between fiber and matrix, and on the fiber volume fraction. Obviously, the tribological properties are expected to depend, at least to some extent, on the same variables. In addition, the tribological properties are expected to depend on the process variables (speed, load, environment), on the nature of the counterface, and on the relative orientation of the unidirectional fibers with respect to the counterface and to the sliding direction. Experiments were conducted to determine the effect of most of the above parameters on the friction and wear values. The results of such experiments are discussed below.

3.2. Fiber type

In polymeric matrices, the type of graphite fiber reinforcement has a significant effect on the tribological properties. High strength graphite fiber–polymer matrix composites were reported to be more wear resistant [3] and more sensitive to the counterface material [4] than high modulus fiber composites. The effect of type of fiber on the friction and wear properties of metal matrix composites has been reported in a number of papers. Amateau et al. [5] studied this effect on various aluminum alloy composites. These composites contained about 30 vol.% fiber, had a strength within 80% of that predicted by the rule of mixtures, and exhibited excellent fatigue properties. The experiments were conducted in the unlubricated condition on a pin-on-ring machine at room temperature in a normal laboratory environment. The pins were made of the composite materials identified as C1–C8 in Table 3, while the ring counterface (35 mm in diameter) was made of AISI 4620 steel heat treated to a hardness of about 60 R_c. The sliding speed was about 0.7 m s^{-1}, while the normal load varied between 6 and 30 N. Under these experimental conditions, it was found that the type of fiber had a significant influence on the wear rate and friction coefficient. The composites containing high modulus fibers exhibited lower friction and higher wear rates than those containing lower modulus fibers. By and large, the friction and wear values for the composites were lower than those of the bare matrix for both types of fiber.

In a subsequent study, the friction coefficients and the wear rates of a high modulus (HM 3000) fiber and a high strength (T300) fiber in a Cu–1% Sn matrix were compared. Both the friction and the wear rate were higher for the high modulus fiber composite.

Strong indications that the type of fiber might be one of the most significant factors in the wear and friction behavior of metal matrix composites were also obtained in an investigation involving various types of bronze and silver matrices [6]; again, the composites with high modulus fibers appear to have wear rates that are noticeably larger than those with high strength fibers (Fig. 2). The difference in friction behavior between the high strength and high modulus fiber composites is also evident (Fig. 3). The high strength fibers are reported to have about 30% higher

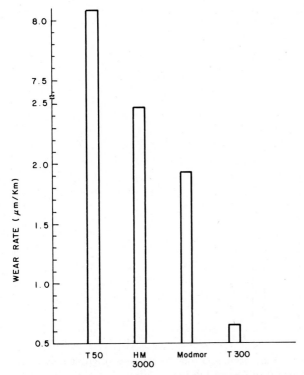

Fig. 2. Wear rates for Cu–6% Sn matrix reinforced with various types of graphite fiber. The elastic modulus is highest for T50 and lowest for T300 (see Table 1).

tensile strength and about 50% higher fracture strain than high modulus fibers [7]. Recalling that graphite fibers in a deformable matrix bend during sliding, as shown in Fig. 4, it can be understood that high strength (lower modulus) fibers can bend to a larger extent than the high modulus fibers without breaking. Consequently, the work of fracture to generate a wear particle is greater in a composite containing high strength fibers [8]. Moreover, the high modulus fibers are characterized by a larger graphite grain size and a higher degree of graphitization [2]. The lower work of fracture combined with the larger graphite grain size, satisfactorily explains the higher wear rates of composites containing high modulus fibers.

The effect of type of fiber on the friction coefficient, however, is inconsistent for the various types of matrix discussed above. In the case of Al-alloy matrices, the coefficient of friction is lower for the high modulus fibers than for the high strength fibers. The reverse is true for the composites containing Cu-alloy and Ag-alloy matrices. This apparent inconsistency can be partially explained by taking into account the nature of the surface films formed at the sliding interfaces. This aspect will be further considered in Sect. 4.

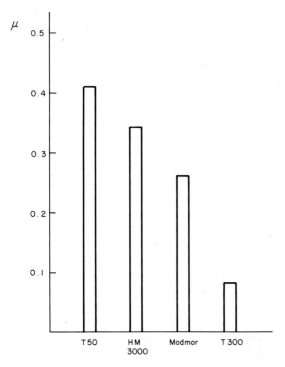

Fig. 3. Friction coefficients for Cu–6% Sn matrix reinforced with various types of graphite fiber. The elastic modulus is highest for T50 and lowest for T300 (see Table 1).

3.3. Matrix materials

The effect of matrix material has been studied in several systems including bronze matrices with various tin content (samples C16, C17, C18, C20, and C24), two more complex copper base alloys (samples C26 and C28), silver–copper alloys with various copper content (samples C29, C30, C32 and C33), and two babbitt alloys (samples C36–C39).

A very good correlation was obtained between the friction coefficient and the tin content of the matrix, as well as between wear rate and tin content for the bronze matrix composites [2]. The results are exhibited in Figs. 5 and 6. The increase in tin content is equivalent to an increase in matrix hardness [9]. Thus, both the friction and the wear decrease with increasing matrix hardness. This inverse dependency on hardness suggests that, under the described experimental conditions, adhesion might be the main wear mechanism [10].

However, a delamination model [11] seems to explain equally well some of the important features of the wear process observed in this family of composites, i.e. the sheet-like shape of the debris particles, the wear volume dependence on the fiber volume fraction, and the higher wear observed for high modulus composites [7]. It will be shown later (Sect. 4) that, in actuality, the wear process in composites is

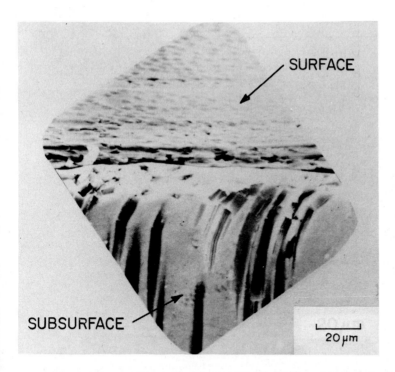

Fig. 4. Scanning electron micrograph of the surface and subsurface of a composite pin at steady state. The bending and breaking of the graphite fibers at the subsurface can be clearly seen.

extremely complex and cannot be explained by any single wear theory. Auger analysis of the sliding surfaces combined with SEM observations in sections normal to the sliding surface will lead to the tentative conclusion that the friction coefficient is determined mainly by the properties of the graphite-containing surface film, while the wear behavior is determined mainly by the properties of the subsurface.

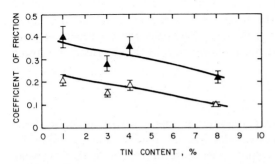

Fig. 5. Coefficient of friction as a function of tin content of the matrix. HM 3000 fibers; class A wires. ▲, 60 m s^{-1}; △, 120 m s^{-1}.

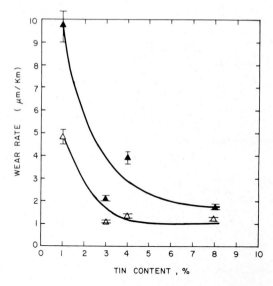

Fig. 6. Wear rate as a function of tin content of the matrix. HM 3000 fibers; class A wires. ▲, 60 m s^{-1}; △, 120 m s^{-1}.

With the other composite systems mentioned in this section, although some correlation between the tribological behavior of the composites and of the matrix is occasionally observed [6], no definite interdependence between the wear of the composites and the mechanical properties of the matrix can be revealed.

3.4. Fiber–matrix interface

Several samples were characterized with respect to the amount of infiltration of liquid metal between the fibers and the wire surface. They were classified in three classes (A, B and C) as described in Sect. 2. Apparently, the wires with higher infiltration (A) exhibit somewhat better wear behavior than wires B or C, although the friction coefficient shows a slight increase (Fig. 7). The lower wear resistance of class B or C samples is most probably related to the existence of debonded fibers/matrix interfaces in these materials. Such specimens will eventually provide large graphite fiber debris, leading to better lubrication at the interface and lower friction coefficients.

3.5. Fiber volume fraction

A controlled variation of the fiber fraction in metal matrix graphite fiber composites is difficult to achieve experimentally. The few observations that could be made on composites of varying fiber volume fraction seem to indicate that the wear rate decreases when the fiber volume fraction increases. An attempt to explain this effect was made by Amateau [7]. He argued that a higher fiber volume fraction leads

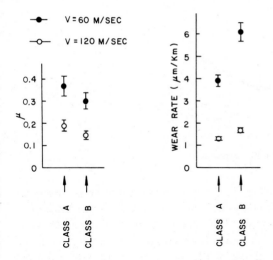

Fig. 7. Influence of wire quality on friction (left) and wear (right). HM 3000/Cu–4% Sn. Composite C18 contains class A wires. Composite C19 contains class B wires. ●, sliding speed 60 m s^{-1}; ○, sliding speed 120 m s^{-1}.

to smaller matrix platelets and, presumably, to better pinning of these platelets to the wear surface, thus reducing the wear rate.

3.6. Sliding speed

Typical variations of friction coefficient and wear rate with sliding speed are shown in Figs. 8 and 9 for some bronze matrix composites. The minimum values observed at about 120 m s^{-1} may result from a competitive effect between oxidation and matrix softening; thus, at sliding speeds lower than 120 m s^{-1}, the beneficial effect of surface oxidation exceeds the detrimental effect of matrix softening while the reverse is true at higher sliding speeds. Similar trends were observed by Hirst and Lancaster [12] with brass pins on steel at heavy loads.

It is interesting to note that, if the friction coefficient and the wear rates are

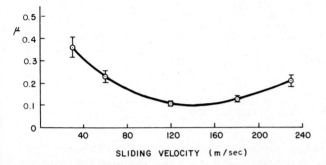

Fig. 8. Coefficient of friction as a function of sliding speed. C20 composite.

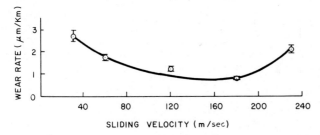

Fig. 9. Wear rate as a function of sliding speed. C20 composite.

plotted against interface temperature (Fig. 10), a monotonic increase is observed with a tendency of saturation for the friction coefficient. A similar dependence of μ on temperature was also discussed by Steijn [13] in relation to the frictional behavior of some metal–plastic systems. This behavior tends to support our previous suggestion that, at high sliding speed, the softening of the matrix is the rate-determining factor.

3.7. Load

Very few experimental results are available that describe the variation of μ and wear rate with applied normal load. From what is available [14], it seems that the wear rate increases when the normal load increases.

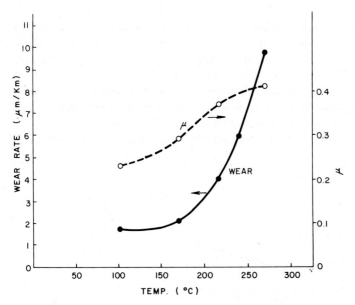

Fig. 10. The interrelation between friction, wear, and brush temperature. HM 3000 composites; class A wires; different amounts of tin in the matrix. Sliding speed 60 m s^{-1}. The temperatures were measured by copper–constantan thermocouples mounted 5 mm below the wear surface of the brush.

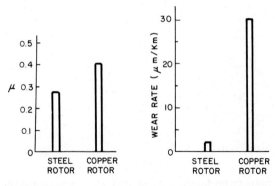

Fig. 11. Influence of rotor material on friction coefficient (left) and wear rate (right). Composite C24.

3.8. Counterface

A copper–tin composite (specimen C24) was run against steel and copper-plated rotors. Typical results are plotted in Fig. 11. Both the coefficient of friction and the wear rate are higher for the copper-plated rotor. These results are in accord with the higher compatibility between the copper rotor and the copper–tin matrix than between the steel rotor and the copper–tin matrix as discussed by Rabinowicz [15] in connection with the effect of surface energy on friction and wear.

3.9. Environment

A few experiments were conducted with composite C24 sliding against a steel counterface in several environments (air, H_2, CO_2). The results are depicted in Fig. 12. It can be seen from these results that the lower oxidation rates obtained in the less active environments (e.g. H_2, CO_2) are beneficial. Films formed while testing in these environments effectively prevent metal-to-metal contact between the slider and rotor, thus avoiding high friction and wear.

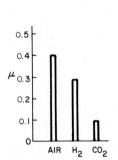 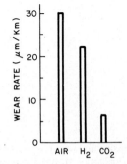

Fig. 12. Influence of test environment on friction coefficient (left) and wear rate (right). Composite C24.

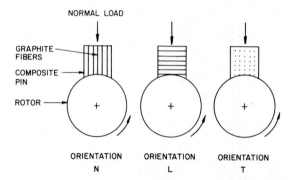

Fig. 13. Definitions of fiber orientations. N, fibers perpendicular to sliding surface; L, fibers parallel to both sliding surface and sliding direction; T, fibers parallel to the sliding surface but perpendicular to the sliding direction.

3.10. Fiber orientation

The tribological properties of graphite fiber polymer matrix composites show a relatively strong dependency on fiber orientation [16]. It is expected that a similar dependency will be observed in metal matrix composites.

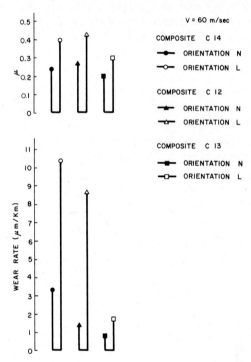

Fig. 14. The dependence of the coefficient of friction and wear rate on the graphite fiber orientation with respect to rotor surface and sliding direction. Orientation N, fibers perpendicular to the sliding surface; orientation L, fibers parallel to both sliding surface and sliding direction.

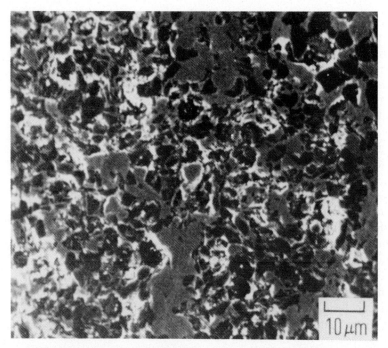

Fig. 15. Scanning electron micrograph of composite tested with fibers perpendicular to the sliding surface (orientation N).

Fig. 16. Scanning electron micrograph of composite tested with fibers parallel to the sliding surface but perpendicular to the sliding direction (orientation T).

The results presented until now in this paper referred mainly to the behavior of composites with fibers oriented perpendicular to the counterface (orientation N). Experiments were also performed employing the other two fiber orientations; namely, fibers parallel to the sliding surface and oriented in the sliding direction (orientation L), and fiber parallel to the sliding surface but oriented normal to the sliding direction (orientation T). A schematic representation of these three configurations is given in Fig. 13. The effect of fiber orientation is different for different families of metal matrix composites. For example, the friction and wear of graphite fiber–aluminum alloy matrix composites are very little dependent on fiber orientation [17]. On the other hand, the wear behavior of copper alloy composites is quite dependent on fiber orientation. Some typical results for this latter family are shown in Fig. 14. Evidently, orientation N gives a much lower wear rate than orientation L. In general, orientation T gives values close to those given by orientation L. Some explanation for this effect can be obtained by studying the nature, size, shape and distribution of the wear debris particles. In the case of orientation N (Fig. 15), the wear debris particles are small (less than about 10 μm) and are dependent, to a large extent, on the average distance between fibers and on the degree of bending that a fiber can withstand without breaking. In the T orientation (Fig. 16), however, large segments of fiber can be plucked out from the matrix. In both cases, the debris contains graphite, matrix, and counterface elements.

4. Wear mechanisms

In order to determine the wear mechanism(s) prevalent in composite materials, experiments were conducted on several types of composite, employing two different types of friction machine [18,19]. Simultaneously, experiments were conducted on two commercially pure metals. A comparison between the results of the two series of experiments, supplemented by scanning electron microscopy and Auger electron spectroscopy observations, revealed the main characteristics of the wear mechanisms in composites.

Pins made of the Al-alloy composites C3 and C22 were run against a grey cast iron counterface on a brake-type friction machine (Fig. 17). The fibers were oriented normal to the counterface.

On each test, the shaft was rotated at the speed of the electric motor and disconnected by disengaging the clutch when the angular velocity, w_0, reached the desired value. At that moment, the specimen was pressed against the counterface. The resulting frictional torque caused the counterface to decelerate and to stop after a certain time. This motion was recorded on a strip chart recorder (Fig. 18). The slope of the angular velocity vs. time curve found from Fig. 18 is the angular deceleration. The frictional torque was calculated by multiplying the rotational inertia, I, through the angular deceleration, α

$$T = I\alpha \tag{1}$$

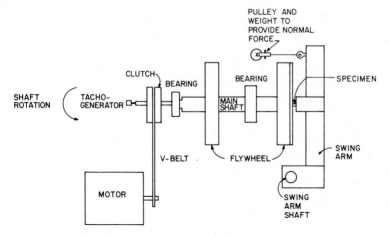

Fig. 17. Plan view of the brake-type friction machine.

The average friction coefficient, μ, during a braking cycle was calculated from

$$T = rL\mu \tag{2}$$

where r is the radius of the circular friction track and L the normal load.

From eqns. (1) and (2)

$$\alpha = r\mu\frac{L}{I} \tag{3}$$

The time t needed to bring the flywheel from its initial velocity, w_0, to rest is given by

$$t = \frac{w_0}{\alpha} \tag{4}$$

The results obtained with two commercially pure metals (copper and aluminium)

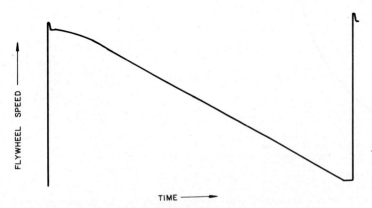

Fig. 18. A typical record of flywheel deceleration during a braking cycle.

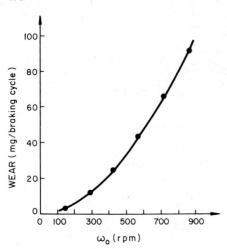

Fig. 19. Wear braking cycle as a function of initial flywheel velocity, w_0, for commercially pure copper.

and with the two composites mentioned above (C3 and C22) are presented in Figs. 19–22.

The wear–speed dependence depicted in Figs. 19 and 20 can be described by an equation of the form

$$W = k_1 w_0^2 \tag{5}$$

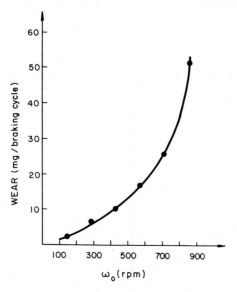

Fig. 20. Wear per braking cycle as a function of initial flywheel velocity, w_0, for commercially pure aluminum.

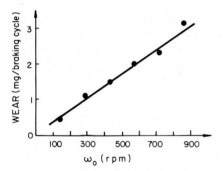

Fig. 21. Wear per braking cycle as a function of initial flywheel velocity, w_0, for the C3 composite.

The sliding distance during a braking cycle is

$$x = \frac{r\alpha t^2}{2} \tag{6}$$

Substituting α and t from eqns. (3) and (4) into eqn. (6) gives

$$w_0^2 = \frac{2\mu L x}{I}$$

Equation (5) then becomes

$$W = k_1 \frac{2\mu L x}{I} \tag{7}$$

or

$$W = k_1' L x \tag{8}$$

where $k_1' = 2 k_1 \mu / I$.

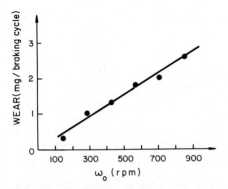

Fig. 22. Wear per braking cycle as a function of initial flywheel velocity, w_0, for the C22 composite.

Equation (8) shows that, in the case of commercially pure metals (Al, Cu), the wear volume is proportional to the product of load L and sliding distance x, in accord with both the adhesion [10] and the delamination [11] theories of wear. This is not the case, however, with the composite sample. From Figs. 21 and 22, the wear–speed dependence can be described as

$$W = k_2 w_0 \tag{9}$$

From eqns. (3) and (4)

$$w_0 = \frac{r\mu Lt}{I} \tag{10}$$

Substituting eqn. (10) into eqn. (9)

$$W = k_2' Lt \tag{11}$$

where $k_2' = k_2 r\mu/I$. (μ is approximately constant during a braking cycle, according to Fig. 18.) Equation (11) shows that, in the case of the two composites studied, the wear volume is proportional to the product of load L and sliding time t. This relation cannot be explained by the existing wear theories, suggesting that the wear mechanism in composite materials might be fundamentally different from the wear mechanism in pure metals.

Similar experiments were conducted on a pin-on-disc type friction machine [19]. The pins were made of composite C1 and bare Al 201 matrix, while the counterface

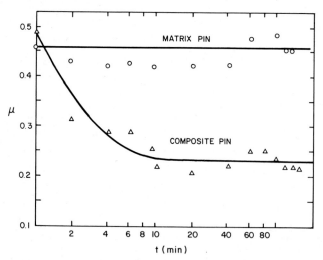

Fig. 23. The variation of coefficient of friction with sliding time. Pin-on-disk friction machine; load, 4.46 N; velocity, 0.37 m s^{-1}; matrix pin, Al 201; composite pin, C1.

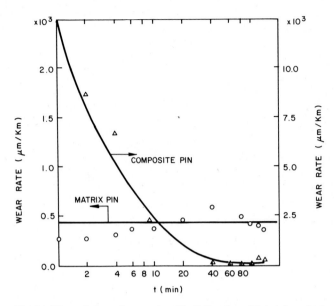

Fig. 24. The variation of wear rate with sliding time. Materials and experimental conditions as in Fig. 23.

was a disc of commercially pure iron. Figures 23 and 24 exhibit the variation with sliding time at constant load and speed, for both the friction coefficient and the wear rate. For the Al 201 pin, the friction coefficient and the wear rate are essentially

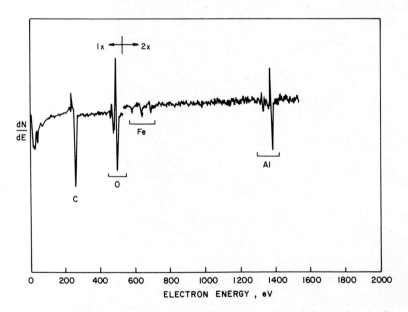

Fig. 25. Auger spectrum of the iron disc surface after 1 min of sliding against the C1 composite pin.

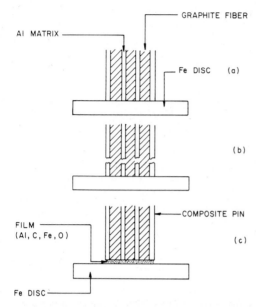

Fig. 26. A simplified model for the wear of composites during initial, transient, and steady state sliding. (a) Initial configuration. (b) The aluminum matrix is pulled from between the fibers by adhesion forces and transferred to the counterface. (c) Unable to support the applied tangential load, the graphite fibers break and become part of the interface film.

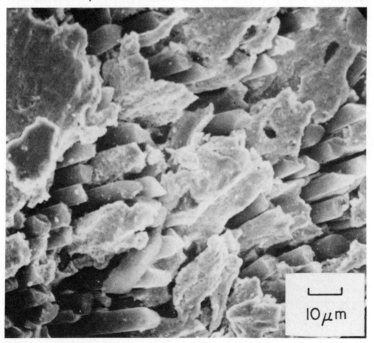

Fig. 27. Scanning electron micrograph of a region at the trailing edge of the composite specimen. The bare graphite fibers are clearly seen.

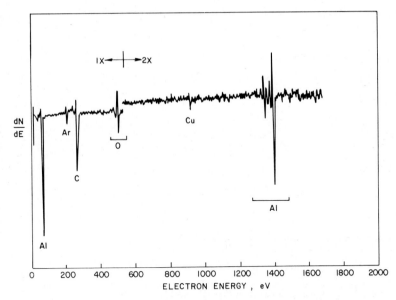

Fig. 28. Auger spectrum of the same specimen as in Fig. 25 after argon ion sputtering for 30 min.

independent of sliding time. The composite pin, however, shows high initial values, a long transient period and low steady state values. The high initial values are due to the high degree of adhesion between matrix and counterface, as suggested by the Auger spectrum presented in Fig. 25. It can be seen that, besides the iron peaks, carbon, oxygen and aluminum peaks are also present. Since the adhesion between graphite and iron is poor, the presence of graphite at the interface suggests the sequence of events schematically depicted in Fig. 26. Evidence of matrix being pulled out from between the fibers is presented in Fig. 27. The graphite-containing film satisfactorily explains the relatively low values achieved by the friction coefficient and the wear rate at steady state. Aluminum was probably transferred to the iron disc as metallic particles, which subsequently underwent oxidation. From the decrease of the oxygen peak during sputtering (Figs. 28 and 29), the thickness of this transient oxide layer was estimated to be about 250 Å. A similar mechanism of transfer and subsequent oxidation was also observed in other sliding systems [20,21]. It is interesting to note that iron does not transfer to the aluminum matrix in the initial stages of sliding (Fig. 30).

At steady state, the composite pin surface contained aluminum, graphite and oxygen, while the disc was covered by a film of aluminum, graphite, oxygen and iron.

The effect of sliding speed is shown in Figs. 31 and 32. The friction coefficient is independent of speed for both the matrix and the composite pins. The wear rate, however, is independent of velocity for the Al 201 pin, but decreases with increasing velocity for the composite pin. It can be shown [19] that such behavior can again be

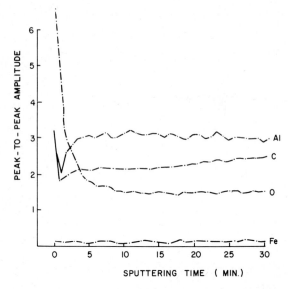

Fig. 29. Sputtering profile for the same specimen as in Figs. 25 and 28.

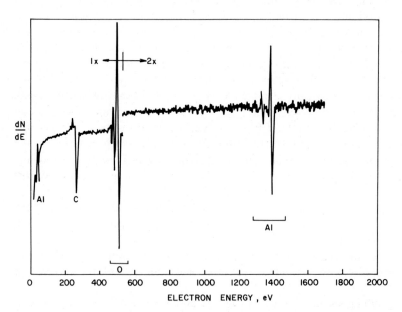

Fig. 30. Auger spectrum of the composite C1 pin surface after 1 min sliding at a velocity of 0.35 m s^{-1} against an iron disc. Notice the absence of any iron peak; no iron was transferred from the disc to the pin during the initial sliding period.

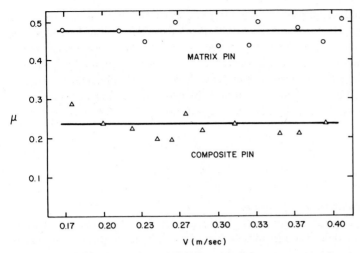

Fig. 31. The variation of coefficient of friction with sliding speed at a load of 4.46 N. The matrix pin is Al 201; the composite pin is C1.

described by eqn. (11), showing that the wear volume is proportional to the product of load and time. This variation of the wear rate with sliding speed can be due to possible changes in the properties of the interface film or of the subsurface when the sliding speed is varied. If it is assumed that the value of the friction coefficient is

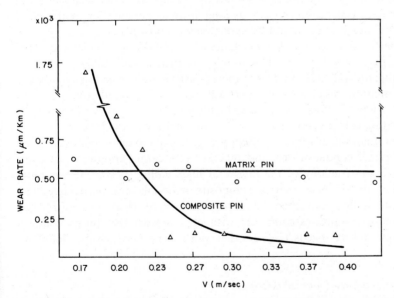

Fig. 32. The variation of wear rate with sliding speed. Materials and experimental conditions same as in Fig. 31.

mainly dictated by the properties of the interface film, the insensitivity of μ to velocity changes (Fig. 31) indicates that the film properties are independent of velocity, at least in the range of sliding speeds employed in these experiments. Consequently, the wear rate dependence on velocity cannot be explained in terms of changes in the interface film properties. The deformation of the subsurface is clearly shown in Fig. 4; the fibers are bent and broken beneath the sliding interface. It is suggested that the wear rate depends on the extent of this velocity-dependent subsurface damage.

5. Potential applications

5.1. Brushes for homopolar motor–generators

Homopolar motor–generators are inertial storage devices which can store large amounts of energy and deliver them in relatively low voltage, high current pulses [22]. These machines are extremely competitive with other means of energy storage (capacitors and batteries) in terms of cost, safety, and efficiency. Homopolar motor–generators operate on the principle of the Faraday disc, i.e. a conducting rotor turning in a magnetic field. In order to store a sizable amount of energy, the rotor must be large and the angular velocity must be high. As a consequence, the brushes are subjected to very high rubbing speeds which approach the speed of sound in air. In addition, efficiency considerations put a limit on the maximum allowable mechanical and electrical losses. This means that the coefficient of friction should be kept under a certain value, as should the voltage drop between rotor and brushes. Economic considerations require as long a life as possible for the brushes; in other words, the brush wear should be kept as low as possible.

Brushes currently employed for such machines are, by and large, made of sintered copper–graphite materials (Fig. 33). This family of brushes was studied by a large number of investigators [23–27]. Binders in these brushes range from almost all lead, through various mixtures of lead and tin, to brushes containing mainly tin.

Experimental evidence [25] shows that, at high sliding speeds (above 220 m s^{-1}), the frictional heating is large enough to result in the melting of the binder.

In order to maintain the advantages of copper–graphite mixtures (relatively good electrical and thermal conductivities, and inherent lubricating abilities), but overcome their deficiencies (melting of the binder at high sliding speeds), the employment of graphite fiber–copper matrix composite materials has been suggested. Experiments done under controlled laboratory conditions tend to confirm the superior behavior of composites compared with their powder metallurgy counterparts at very high sliding speeds [26]. Further work is necessary for a definitive assessment.

5.2. Graphite fiber-reinforced journal bearings

A journal bearing material should have good embeddability, conformability, compatibility, corrosion resistance, thermal conductivity, yield strength, and fatigue

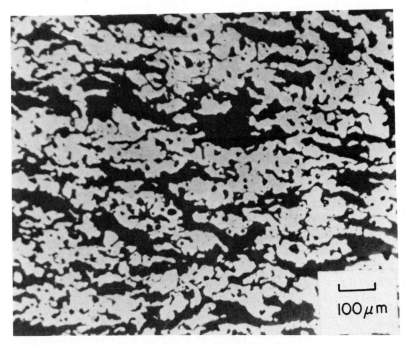

Fig. 33. Optical micrograph of a typical copper–graphite sintered brush.

resistance [28,29]. Self-lubrication would be desirable as a margin of safety in case of a loss of conventional lubrication. Babbitt is the preferred material for most lubricated journal bearings. The term babbitt is used to include tin–lead alloys ranging from 90% tin and no lead to 80% lead and less than 5% tin. Antimony is added to increase hardness. Copper and other elements are usually added in small amounts. The babbitts are relatively soft and have excellent conformability and embeddability. They have almost no corrosion problems and their compatibility with lubricated steel is excellent. These alloys would be ideal bearing materials except for their low strength. If babbitt could be strengthened significantly without degrading its other characteristics, a much improved bearing material would result. Many attempts have been made to strengthen babbitt by further alloying. However, reinforcement with strong graphite fibers appears to be most promising, not only because of strength enhancement, but also because of the expected self-lubrication characteristics.

Several types of graphite fiber-reinforced babbitt were characterized with respect to their tribological properties in both dry and lubricated conditions [30–32]. For comparison purposes, samples made of the plain babbitt matrix were also employed.

It has been found that, at room temperature under no external lubrication, the composite material exhibited a much lower friction coefficient than the pure babbitt specimen; the same was true at high temperature ($132 \pm 1°C$) under boundary lubricated conditions. The wear rate of the composite was about one order of magnitude lower than that of pure babbitt.

6. Conclusions

This article reviewed the main tribological characteristics of graphite fibers–metal matrix composites and indicated the status of some practical applications. The full potential of the metal matrix composites as tribological components has yet to be realized.

List of symbols

I	rotational inertia
L	normal load
T	frictional torque
k_1, k_2, k_1', k_2'	proportionality constants
r	radius of circular friction track
t	time
w_0	initial angular velocity of flywheel
x	distance slid during braking cycle
α	angular deceleration
μ	coefficient of friction

References

1 J.P. Giltrow and J.K. Lancaster, Wear, 12 (1968) 91.
2 Z. Eliezer, C.H. Ramage, H.G. Rylander, R.H. Flowers and M.F. Amateau, Wear, 49 (1978) 119.
3 D. Scott, J. Blackwell, P.J. McCullagh and G.H. Mills, Wear, 15 (1970) 257.
4 J.P. Giltrow and J.K. Lancaster, Wear, 16 (1970) 359.
5 M.F. Amateau, W.W. French and D.M. Goddard, Proc. Int. Conf. Composite Mater., 1975, Vol. 2, Metallurgical Society of AIME, New York, 1976, pp. 623–643.
6 M.F. Amateau, R.H. Flowers and Z. Eliezer, Wear, 54 (1979) 175.
7 M.F. Amateau, in J.E. Hack and M.F. Amateau (Eds.), Mechanical Behavior of Metal–Matrix Composites, American Institute of Mining, Metallurgical and Petroleum Engineers, New York, 1983, pp. 213–225.
8 Z. Eliezer, V.D. Khanna and M.F. Amateau, Wear, 53 (1979) 387.
9 Metals Handbook, Vol. 1, American Society for Metals, Metals Park, OH, 1961, p. 1027.
10 D. Tabor, in J.M. Blakely (Ed.), Surface Physics of Materials, Vol. II, Academic Press, New York, 1975, p. 513.
11 N.P. Suh, Wear, 25 (1973) 111.
12 W. Hirst and J.K. Lancaster, Proc. R. Soc. London Ser. A, 259 (1960) 228.
13 R.P. Steijn, Met. Eng. Q., 7(2) (1967) 9.
14 Z. Eliezer, V.D. Khanna and M.F. Amateau, Wear, 51 (1978) 169.
15 E. Rabinowicz, Friction and Wear of Materials, Wiley, New York, 1965, p. 30.
16 T. Tsukizoe and N. Ohmae, J. Lubr. Technol., Oct. (1977) 401.
17 M.F. Amateau, W.W. French and D.M. Goddard, Aerosp. Rep. ATR-75 (9450)-3, May 1975.
18 Z. Eliezer, C.J. Schulz and H.E. Mecredy, Wear, 52 (1979) 133.
19 K.J. Pearsall, Z. Eliezer and M.F. Amateau, Wear, 63 (1980) 121.
20 M. Kerridge, Proc. R. Soc. London Ser. B, 68 (1955) 400.

21 Z. Eliezer, C.J. Schulz and J.W. Barlow, Wear, 46 (1978) 397.
22 R.A. Marshall, Wear, 37 (1976) 233.
23 I.R. McNab, Proc. Holm Conf. Electr. Contacts, 1977, Illinois Institute of Technology, Chicago, IL, pp. 105–144.
24 J.M. Casstevens, H.G. Rylander and Z. Eliezer, Wear, 48 (1978) 121.
25 J.M. Casstevens, H.G. Rylander and Z. Eliezer, Wear, 49 (1978) 169.
26 J.M. Casstevens, H.G. Rylander and Z. Eliezer, Wear, 50 (1978) 371.
27 M. Brennan, Z. Eliezer, W.F. Weldon, H.G. Rylander and H.H. Woodson, Proc. 9th Int. Conf. Electr. Contact Phenom., Illinois Institute of Technology, Chicago, IL, Sept. 11–15, 1978, pp. 577–582.
28 Metals Handbook, Vol. 1, American Society for Metals, Metals Park, OH, 1961, pp. 843–863.
29 A.W.J. deGee, Wear, 36 (1976) 33.
30 B.W. Boyle, M.Sc. Thesis, The University of Texas at Austin, 1981.
31 T.C. Chou, M.Sc. Thesis, The University of Texas at Austin, 1982.
32 T.C. Chou and Z. Eliezer, Wear, 82 (1982) 93.

Chapter 7

Friction and Wear Performance of Unidirectionally Oriented Glass, Carbon, Aramid and Stainless Steel Fiber-Reinforced Plastics

TADASU TSUKIZOE * and NOBUO OHMAE

Faculty of Engineering, Osaka University, Osaka 565 (Japan)

Contents

Abstract

Friction and wear of unidirectionally oriented fiber-reinforced plastics (FRP) against mild steel have been investigated. The law of mixtures in the calculation of the friction coefficient of FRP is deduced, and the validity of this law is discussed by a comparison of computed values with experimental data. The wear performances of seven different kinds of FRP are summarized. A model for wear of FRP is proposed

---

* Present address: Iron and Steel Technical College, Nishikoya, Amagasaki, Hyogo 661, Japan.

stating that the wear proceeds by wear-thinning of the fiber reinforcements, subsequent breakdown of the fibers and by peeling off of the fibers from the matrix. Based on this model, the wear equation for FRP is also proposed.

Seizure of FRP as well as friction and wear of hybrid FRP, SiC–FRP and mica dispersion-reinforced polyester resin are also described. The finite element method was applied to stress analysis of FRP under friction processes. The principal stress distributions at the FRP surface and around the fiber reinforcement were computer-simulated.

1. Introduction

The tribological behavior of fiber-reinforced plastics (FRP) has not been fully understood, despite the recently increased demand for FRP as a high-specific-strength material. When we consider the wide application of FRP in many machine components, a study of their friction and wear becomes of technological and industrial importance. Pioneering research by Lancaster and Giltrow has shown a number of significant factors affecting the friction and wear of FRP [1–8]. Their experimental results using carbon fiber-reinforced plastics (CFRP), for instance, have revealed that the wear performance of CFRP under lubricated conditions is influenced by the types of carbon fiber and matrix material, the counterface and its surface roughness, the experimental temperature and also by the formation of transferred film. Tsukizoe and Ohmae have studied the friction and wear of unidirectionally oriented FRP in contact with carbon steel and have discussed the influences of the volume fraction of the fibers and the kinds of fiber, as well as tribological anisotropy with respect to sliding direction [9–19].

This chapter deals with friction and wear properties between unidirectionally oriented FRP and carbon steel. A wide variety of FRP were prepared for experiments; the fiber reinforcements used were high-tensile-strength carbon fiber, high-modulus carbon fiber, E-glass fiber, aramid fiber (Kevlar-49) and stainless steel fiber, while epoxy resin, polyester resin and PTFE were employed as matrix material.

The law of mixtures in the calculation of the friction coefficient of FRP is deduced, and the validity of this law is discussed by a comparison of computed values with experimental data.

The wear performances of seven different kinds of FRP were summarized. A model is proposed stating that the wear of FRP proceeds by wear-thinning of the fiber reinforcements, subsequent breakdown of the fibers and by peeling off of the fibers from the matrix. On the basis of this model, the estimation of specific wear rates of FRP is also proposed stating that the Young's modulus, interlaminar shear strength and friction coefficient are the influential factors on wear performances.

Some other topics related to the tribology of FRP are described; seizure of FRP, friction and wear of hybrid FRP, SiC fiber-reinforced epoxy resin and mica dispersion-reinforced polyester resin, the stress distribution near the surface of FRP during friction using the finite element method are studied. The microscopic structure of the aligned fibril, which may provide low friction, is investigated using the field ion microscopy.

2. Experimental

2.1. Material

The unidirectionally oriented fiber-reinforced plastics were prepared by the leaky-mold method, the procedure of which was described in detail elsewhere [14]. With this method, a wide variety of volume fraction of fibers can be chosen and a uniform dispersion of the fibers in the matrix is possible.

Table 1 shows the seven kinds of FRP mainly used in friction and wear experiments. The high-tensile-strength carbon fibers (type II) and the high-modulus carbon fibers (type I) were pyrolyzed at 1800 and 2500°C, respectively. The stainless steel fibers had a compositions of 12–15% Ni, 16–18% Cr, 2–3% Mo and Fe. Typical

TABLE 1

Constitution of the composites tested

No.	Symbol of FRP	Fiber reinforcements	Resin matrixes
1	HS-CFRP	High-strength carbon fiber	Epoxy resin or polyester resin
2	HM-CFRP	High-modulus carbon fiber	Epoxy resin
3	NT-CFRP	High-strength carbon fiber (no surface treatment)	Epoxy resin
4	GFRP	E-glass fiber	Epoxy resin or polyester resin
5	SFRP	Stainless steel fiber	Epoxy resin or polyester resin
6	AFRP	Aramid fiber (Kevlar-49)	Epoxy resin
7	CFRTP	High-strength carbon fiber	PTFE

TABLE 2

Mechanical properties of fiber reinforcements and matrixes

Fiber reinforcement and matrix	Diameter of fiber (μm)	Tensile strength (GPa)	Young's modulus (GPa)	Density (g cm^{-3})	Fibers per strand
High-strength carbon fiber	6.8	2.55	235	1.8	3000
High-modulus carbon fiber	8.0	2.20	358	1.95	3000
E-glass fiber	9.4	2.16	61	2.5	2400
Stainless steel fiber	8.0	1.67	186	7.9	5000
Aramid fiber	12.3	2.75	130	1.5	3070
Epoxy resin		0.069	0.33	1.9	
Polyester resin		0.069	3.24	1.2	
PTFE		0.025	0.32	2.1	

TABLE 3
Compounding ingredients and molding conditions

	Resin	Hardening agent	Cure accelerator	Release agent	Molding conditions	
					Hardening	After-curing
Epoxy resin	Epikote 828, 100 wt. %	MCD, 90 wt. %	BDMA, 1 wt. %	Silicon grease	150°C, 3 h	150°C, 3 h
Polyester resin	Polymal 6305, 100 wt. %	DDM, 0.7 wt. %	P106, 0.3 wt. %	Silicon grease	Room temp., 2 h	120°C, 2 h

material properties of the fiber reinforcements and the resin matrixes are shown in Table 2 and the fabrication parameters of FRP are tabulated in Table 3.

The mechanical properties of several kinds of unidirectionally oriented FRP are listed in Tables 4–6.

2.2. Testing procedures

The friction-testing apparatus of the pin-on-disk type and the wear-testing apparatus of the flat-on-flat type are described elsewhere [19]. Carbon steel used for a counter-surface in friction and wear tests had a carbon content of 0.20–0.25%. Friction and wear experiments were conducted unlubricated in air (temperature 20–25°C, relative humidity 55–60%).

TABLE 4
Mechanical properties of FRP (epoxy composites)

	Volume fraction (V_f, %)	Tensile strength (GPa)	Young's modulus (GPa)	Bending strength (GPa)	Modulus of rigidity (GPa)	Barcol hardness	Interlaminar shear strength (MPa)
HS-CFRP	42	1.22	102	1.39	110	65	76
	52	1.32	125	1.47	117	66	74
	59	1.69	140	1.72	128	72	73
	65	1.82	148	1.78	141	75	78
HM-CFRP	65	1.12	240	1.14	208	58	70
NT-CFRP	65	1.71	171	1.48	105	70	53
GFRP	60	1.11	33	1.07	37	72	
	68	1.33	41	1.18	40	76	
	76	1.53	50	1.32	47	77	78
SFRP	56	0.59	60	0.73	58	68	
	62	0.72	74	0.75	66	73	
	69	0.87	84	0.96	82	76	
	75	0.89	88	1.11	97	82	56
AFRP	40	0.73	36	0.37	27	50	67
	50	0.96	41	0.41	29	51	62
	60	1.08	52	0.48	34	50	71
	70	1.39	62	0.52	36	57	56

TABLE 5
Mechanical properties of FRP (polyester composites)

	Volume fraction (V_f, %)	Tensile strength (GPa)	Young's modulus (GPa)	Bending strength (GPa)	Modulus of rigidity (GPa)	Barcol hardness	Interlaminar shear strength (MPa)
HS-CFRP	42	1.15	103	0.97	88	65	
	50	1.48	134	1.13	99	68	
	57	1.62	139	1.22	104	69	
	65	1.68	147	1.29	116	70	61
GFRP	52	0.96	44	0.80	35	64	
	58	1.11	52	0.86	40	64	
	70	1.38	53	1.06	43	68	41
SFRP	48	0.64	87	0.68	55	67	
	54	0.70	92	0.76	59	72	
	70	0.88	116	1.05	77	82	51
	76	1.18		1.16	83	85	
AFRP	70						30

TABLE 6
Mechanical properties of FRP (PTFE composites)

	Volume fraction (V_f, %)	Young's modulus (GPa)	Bending strength (GPa)	Modulus of rigidity (GPa)	Shore hardness	Interlaminar shear strength (MPa)
CFRTP	42	93	0.13	8.6	72	
	67	109	0.14	9.0	65	9
AFRTP	42	32	0.12	14.0	68	
	67	42	0.12	18.6	64	
SFRTP	42	47			70	
	67	54			68	

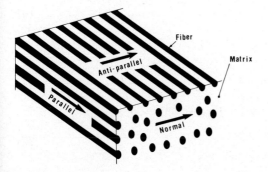

Fig. 1. FRP block indicating sliding directions.

Three different sliding directions are to be identified in order to clarify the tribological anisotropy of unidirectionally oriented FRP with respect to sliding direction. Figure 1 illustrates an FRP block indicating three sliding directions, i.e. parallel, anti-parallel and normal directions. The wear volume of FRP was calculated from the change in surface profiles measured with a Talysurf. The wear of carbon steel could be neglected under the experimental conditions in Sect. 4.2 when compared with that of FRP. However, at high applied pressures and sliding velocities, especially when a carbon steel was slid against the SiC fiber-reinforced plastics, the wear of carbon steel became large. Some experimental results of the wear of carbon steel are shown in Sect. 6.2.

3. Law of mixtures for calculating the friction coefficient

When a counter-surface slides against an FRP surface, both applied load, W, and tangential force, F, are supported by fibers and matrix, so that the friction coefficient μ can be given by

$$\mu = \frac{F}{W} = \frac{F_f + F_m}{W_f + W_m} \tag{1}$$

where suffixes f and m denote fiber and matrix, respectively. Then we can assume

$$A_f = V_f A, \qquad A_m = V_m A = (1 - V_f) A \tag{2}$$

where A is the nominal area of contact, and V_f and V_m are the volume fractions of the fibers and matrix, respectively. When a peeling off of the fibers from the matrix, under the action of shear deformation, does not occur, we can assume that the shear strain, γ_{A_f}, is equal to γ_{A_m}.

$$\gamma_{A_f} = \gamma_{A_m} \tag{3}$$

If G_{A_f} and G_{A_m}, the moduli of rigidity of material underneath the contacting surface, are equal, then the shear stress, τ, becomes constant, i.e.

$$\tau_f = \frac{F_f}{A_f} = \tau_m = \frac{F_m}{A_m} \tag{4}$$

From eqns. (2) and (4), we obtain

$$F_f = V_f F, \qquad F_m = V_m F \tag{5}$$

By taking into account the relation $W = W_f + W_m$, we have

$$\frac{F}{\mu} = \frac{F_f}{\mu_f} + \frac{F_m}{\mu_m} \tag{6}$$

and, finally

$$\frac{1}{\mu} = V_f \frac{1}{\mu_f} + V_m \frac{1}{\mu_m} \tag{7}$$

From this equation, we are able to calculate the friction coefficient, μ, of FRP when the friction coefficients of the fibers, μ_f, and of the matrix, μ_m, are given.

4. Results

4.1. Friction

The measurement of the friction coefficient was performed when the top of the cone indenter became flattened (approximately 0.1 mm diam.) and the fluctuation of friction force became small, typically after a sliding distance of about 10 mm [13].

The relationships between the friction coefficient and the volume fraction of the fibers are shown in Figs. 2–4. The thick solid, dotted and chain lines in these figures show the calculated friction coefficient obtained from eqn. (7). As there is good agreement between the theoretical and the experimental results, the assumption of eqn. (3) might be reasonable for the friction of FRP at light applied forces and low sliding velocities, in which case fracture of FRP at the sliding surface does not take place. From the results in Figs. 2–4, it is clear that carbon fiber is the best reinforcement as far as the friction of FRP is concerned. The friction anisotropy depending on the fiber orientation relative to the sliding direction cannot be recognized in these figures.

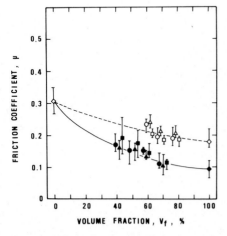

Fig. 2. Influence of the volume fraction of fibers on the friction coefficient of epoxy composites. SFRP: $\bigcirc$, parallel; $\triangle$, anti-parallel; $\square$, normal. CFRP: $\bullet$, parallel; $\blacktriangle$, anti-parallel; $\blacksquare$, normal. $v = 25 \ \mu\text{m s}^{-1}$; $W = 1.52$ N.

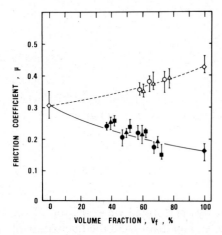

Fig. 3. Influence of the volume fraction of fibers on the friction coefficient of epoxy composites. GFRP: O, parallel; △, anti-parallel. AFRP: ●, parallel; ▲, anti-parallel; ■, normal. $v = 25 \ \mu \mathrm{m \ s^{-1}}$; $W = 1.52$ N.

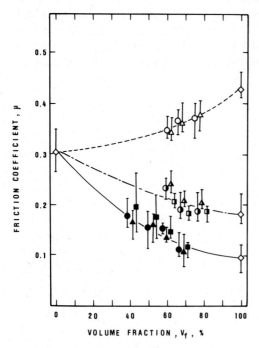

Fig. 4. Influence of the volume fraction of fibers on the friction coefficient of polyester composites. GFRP: O, parallel; △, anti-parallel. SFRP: ◑, parallel; ▲, anti-parallel; ◨, normal. HS-CFRP: ●, parallel; ▲, anti-parallel; ■, normal. $v = 25 \ \mu \mathrm{m \ s^{-1}}$; $W = 1.52$ N.

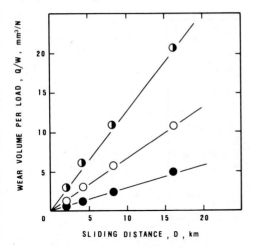

Fig. 5. Wear volume per load of AFRP versus sliding distance. Volume fraction, $V_f = 70\%$; applied pressure, $p = 1.5$ N mm^{-2}; sliding velocity, $v = 0.83$ m s^{-1}. ○, Parallel; ◑, anti-parallel; ●, normal.

4.2. Wear

The results of our wear tests have shown that wear volume increases linearly with sliding distance; no running-in period of wear has been apparent. Typical wear curves of AFRP and CFRTP are shown in Figs. 5 and 6. It is clear that CFRTP has better wear resistance than AFRP. This may suggest that the self-lubricating abilities of both carbon fibers and PTFE mitigate wear damage.

The influences of applied pressure, p, and sliding velocity, v, are shown in Figs. 7

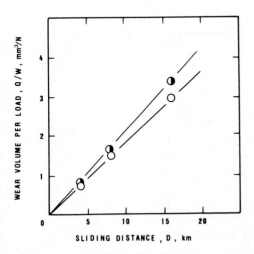

Fig. 6. Wear volume per load of CFRTP versus sliding distance. Volume fraction, $V_f = 67\%$; applied pressure, $p = 1.5$ N mm^{-2}; sliding velocity, $v = 0.83$ m s^{-1}. ○, Parallel; ◑, anti-parallel.

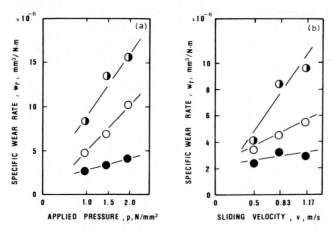

Fig. 7. Influences of (a) applied pressure and (b) sliding velocity on the specific wear rate of AFRP (V_f = 70%). ○, Parallel; ◑, anti-parallel; ●, normal. (a) $v = 0.83$ m s^{-1}; $D = 8$ km. (b) $p = 1.0$ N mm^{-2}; $D = 8$ km.

and 8. Since wear increases linearly with sliding distance, it is possible to characterize the wear in terms of a specific wear rate, w_r, which has units of mm^3 N^{-1} m^{-1}. The value of w_r increases with p up to 2.0 N mm^{-2} and with v up to 1.17 m s^{-1}. This may be due partially to a deterioration in adhesive bonding between the fiber reinforcement and the matrix, resulting from the greater generation of friction heat at high pressures and velocities. The temperature of the carbon steel specimen measured at a spot 1 mm beneath the friction surface reaches approximately 70°C for the friction experiments of AFRP at 2.0 N mm^{-2} applied pressure and 0.83 m

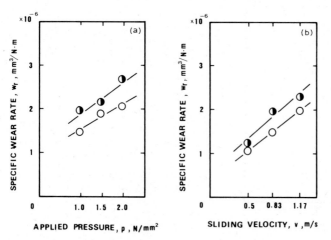

Fig. 8. Influences of (a) applied pressure and (b) sliding velocity on the specific wear rate of CFRTP (V_f = 67%). ○, Parallel; ◑, anti-parallel. (a) $v = 0.83$ m s^{-1}; $D = 8$ km. (b) $p = 1.0$ N mm^{-2}; $D = 8$ km.

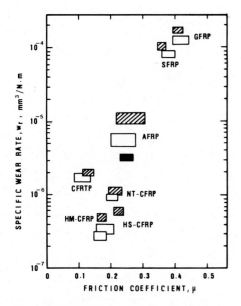

Fig. 9. Relationship between specific wear rate and friction coefficient. □, Parallel; ▨, anti-parallel; ■, normal; $p = 1.5$ N mm^{-2}; $v = 0.83$ m s^{-1}; $D = 16$ km.

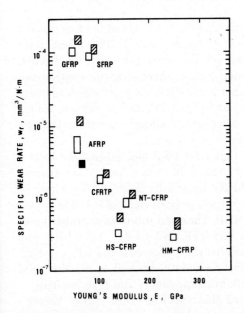

Fig. 10. Relationship between specific wear rate and Young's modulus. □, Parallel; ▨, anti-parallel; ■, normal; $p = 1.5$ N mm^{-2}; $v = 0.83$ m s^{-1}; $D = 16$ km.

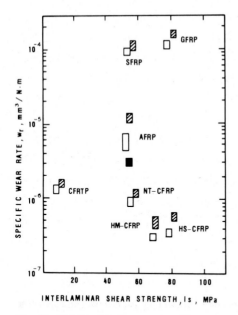

Fig. 11. Relationship between specific wear rate and interlaminar shear strength. □, Parallel; ▨, anti-parallel; ■, normal; $p = 1.5$ N mm^{-2}; $v = 0.83$ m s^{-1}; $D = 16$ km.

s^{-1} sliding velocity. Thus, the generation of friction heat affects wear behavior of FRP considerably.

Figures 9–11 summarize the wear performance of seven kinds of unidirectionally oriented FRP, the volume fraction of which is approximately 70%. These three figures show the relationships between the specific wear rate and the friction coefficient, the Young's modulus and the interlaminar shear strength, respectively. As for the tribological anisotropy, it seems that every FRP has good wear resistance in parallel sliding but poor wear resistance in anti-parallel sliding. In normal sliding, every FRP except AFRP suffered "seizure" after several kilometers sliding distance, so these data are not given in the three figures.

From the results in Fig. 9, it is evident that HS-CFRP and HM-CFRP have a small specific wear rate of the order of 10^{-7} mm^3 N^{-1} m^{-1} and a low friction coefficient of 0.2. In contrast, GFRP and SFRP show a large specific wear rate of the order of 10^{-4} mm^3 N^{-1} m^{-1} and a high friction coefficient of 0.4. The lowest friction coefficient of 0.1 is obtained for CFRTP. The good tribological properties of the carbon-fiber-reinforced plastics group (CFRTP, NT-CFRP, HS-CFRP and HM-CFRP) may be caused by the good mechanical properties of FRP, e.g. high Young's modulus and high interlaminar shear strength, as well as by the good tribological properties of the fibers, e.g. self-lubricating ability and high strength.

From the results in Fig. 10, it is noticed that the FRP with higher Young's modulus always shows a better wear resistance and the coefficient of correlation between them is calculated as high as -0.74. When we look at the results of the

CFRP group in Fig. 11, it will be noticed that the interlaminar shear strength also has a close relation with the specific wear rate, the coefficient of correlation being estimated at -0.88. The analysis using systems methodologies has also revealed that the most significant parameters of the wear of FRP are Young's modulus, tensile strength and interlaminar shear strength [16, 18].

4.3. Seizure

It was mentioned in Sect. 4.2 that sliding in the normal direction caused seizure of every FRP except AFRP. Seizure is frequently referred to the stopping of relative motion as the result of interfacial friction. For metallic friction, seizure is related to abrupt increase in friction as well as to rapid increase in surface temperature. However, for FRP, these rapid changes in sliding conditions were not clearly observable. Thus, in this section, the seizure of FRP is defined as a critically severe sliding condition in which the FRP specimen can not continue to serve as a sliding member due either to a great increase in the fluctuation of friction coefficient or to a thermal decomposition of the matrix.

The seizure of an FRP was investigated using the same wear-testing apparatus in the previous section and the sliding was conducted in the normal direction of epoxy composites containing about 70% volume fraction of fibers. Seizure tests started with an initial experimental condition of 2.0 N mm^{-2} applied pressure and 0.5 m s^{-1} sliding velocity. A stepwise increase in sliding speed was then performed up to 16 km sliding distance. After applying the maximum sliding velocity of 1.6 m s^{-1} with this testing apparatus, normal pressure was raised stepwise. The experimental procedures of stepwise increments of sliding velocity and applied pressure are tabulated in Table 7.

Figure 12 shows the fluctuation of friction coefficients for GFRP, AFRP, HS- and HM-CFRP. In all the seizure tests, GFRP and HS-CFRP suffered seizure at around 30 km sliding distance. The fluctuation in friction coefficient for GFRP

TABLE 7
Experimental details of seizure tests

Sliding distance, D (km)	0 ~ 2	~ 4	~ 6	~ 8	~ 10	~ 12	~ 14	~ 16	~ 18	~ 20
Applied pressure, p (N mm^{-2})	2.0	2.0	2.0	2.0	2.0	2.0	2.0	2.0	2.5	3.0
Sliding velocity, v (m s^{-1})	0.5	0.5	0.7	0.8	1.0	1.2	1.4	1.6	1.6	1.6

Sliding distance, D (km)	~ 22	~ 24	~ 26	~ 28	~ 30	~ 32	~ 34	~ 36	~ 38	~ 40
Applied pressure, p (N mm^{-2})	3.5	4.0	4.5	5.0	5.5	6.0	6.5	7.0	7.5	8.0
Sliding velocity, v (m s^{-1})	1.6	1.6	1.6	1.6	1.6	1.6	1.6	1.6	1.6	1.6

Sliding distance, D (km)	~ 42	~ 44	~ 46	~ 48	~ 50	~ 52	~ 54	~ 56	~ 58	~ 60
Applied pressure, p (N mm^{-2})	8.5	9.0	9.5	10.0	10.5	11.0	11.5	12.0	12.5	13.0
Sliding velocity, v (m s^{-1})	1.6	1.6	1.6	1.6	1.6	1.6	1.6	1.6	1.6	1.6

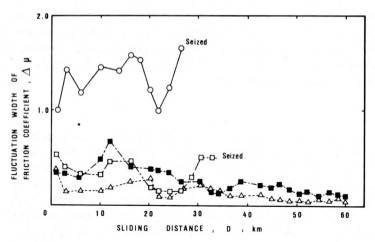

Fig. 12. The range of fluctuation in friction coefficient plotted against distance slid in seizure tests. ○, GFRP; △, AFRP; □, HS-CFRP; ■, HM-CFRP.

shows a remarkably high value at an onset of seizure, while that for HS-CFRP seems relatively low. Chipping of the HS-CFRP specimen occurred when the specimens seized. This may indicate the decomposition of epoxy resin due to high friction heat. Scanning electron microscope observations of the cross-section of seized HS-CFRP revealed non-existence of the matrix between the fiber reinforcements. The measurement of surface temperature was attempted, but it was not possible because of greater wear of FRP as well as greater vibration. For these reasons, the temperature at a spot 4 mm beneath the friction surface was plotted in Fig. 13. It is noticed that the HS-CFRP experienced a surface temperature as high as 200°C when seizure

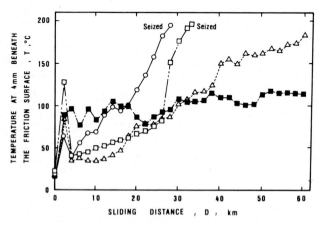

Fig. 13. The temperature measured at a spot 4 mm beneath the friction surface during seizure tests. ○, GFRP; △, AFRP; □, HS-CFRP; ■, HM-CFRP.

occurred. The thermal decomposition temperature of epoxy resin ranged from 150 to 250°C. Since the temperature 4 mm beneath the surface reaches 200°C, the true surface temperature is much higher. Therefore, thermal decomposition of the epoxy resin will inevitably occur. Although a slight phenomenological difference exists in the seizure of FRP, both GFRP and HS-CFRP were unable to serve as specimens for seizure tests after about 30 km sliding distance.

With respect to the practical uses of FRP at high pressures and velocities, the pv value may become of importance. Our experimental results have shown that the maximum pv value of GFRP and HS-CFRP is approximately $7-9 \times 10^6$ N m m^{-2} s^{-1}, while those of AFRP and HM-CFRP are more than 1.5×10^7 N m m^{-2} s^{-1}. Besides the self-lubricating ability of HM-carbon fibers, one of the possible reasons for the good tribological property of HM-CFRP is a transferred lubricating film to steel surfaces as observed by transmission electron microscopy and electron spectroscopy for chemical analysis [10].

5. The wear equation for FRP

From the experimental results and discussions in Sect. 4.2 and the scanning electron microscope observations of worn FRP surfaces, the following model of wear processes can be proposed. The wear of FRP proceeds by
(a) wear-thinning of the fiber reinforcements;
(b) subsequent breakdown of the fibers;
(c) peeling off of the fibers from the matrix.
A schematic drawing of these processes are shown in Fig. 14. Their sequential occurrence governs the wear of FRP. The essential factors affecting these three processes may be
(a) the wear-thinning of the fibers (applied force W and sliding distance D);
(b) the breakdown of the fibers (strain $\mu p/E$ of the FRP caused by friction force, applied force W and sliding distance D);
(c) the peeling off of the fibers from the matrix (interlaminar shear strength I_s of the FRP, strain $\mu p/E$ of the FRP, applied force W and sliding distance D).
Therefore, wear volume, Q, can be given by

$$Q = f(\mu p/E, 1/I_s, W, D) \tag{8}$$

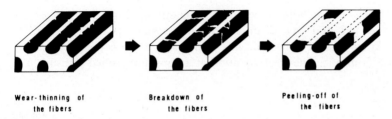

Wear-thinning of   Breakdown of   Peeling-off of
the fibers   the fibers   the fibers

Fig. 14. A model showing the wear processes of FRP.

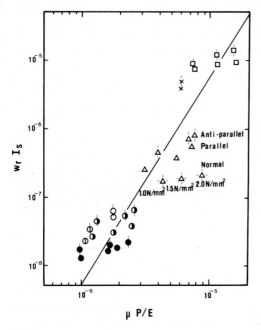

Fig. 15. Relationship between $w_r I_s$ and $\mu p/E$. □, GFRP; ×, SFRP; △, AFRP; ○, NT-CFRP; ◑, HS-CFRP; ◓, HM-CFRP; ●, CFRTP.

As for first-order approximation, we can assume

$$Q = k \frac{\mu p}{E} \frac{1}{I_s} WD \tag{9}$$

where k is a dimensionless constant. Then the specific wear rate, w_r, can be written as

$$w_r = k \frac{\mu p}{E} \frac{1}{I_s} \tag{10}$$

or

$$w_r I_s = k \frac{\mu p}{E} \tag{11}$$

where $w_r I_s$ and $k \mu p/E$ are both dimensionless quantities. Figure 15 shows the relationship between $w_r I_s$ and $k \mu p/E$ under applied pressures, p, of 1, 1.5 and 2 N mm^{-2} for seven kinds of FRP. The solid line in this figure shows the experimental values for $p = 1.5$ N mm^{-2}. The wear equation at a constant applied pressure p is given by

$$w_r I_s = \alpha \left(\frac{\mu p}{E} \right)^{\beta} \tag{12}$$

The values of α and β may be calculated from the experimental line in Fig. 15 and the experimental wear equation of FRP can be written as

$$w_r = 1.40 \times 10^{10} \left(\frac{\mu p}{E} \right)^{3.08} \frac{1}{I_s} \tag{13}$$

When the Young's modulus, E (MPa), and the interlaminar shear strength, I_s (MPa), of the FRP are given and the friction coefficient, μ, between the FRP and a carbon steel is known, we can estimate the specific wear rate w_r (mm^3 N^{-1} mm^{-1}) under the applied pressure $p = 1.5$ N mm^{-2} from the wear equation, eqn. (13).

6. Some topics related to the tribology of FRP

6.1. Friction and wear of hybrid FRP

The unidirectionally oriented FRP reinforced with two different fibers was prepared, in which the epoxy resin was reinforced with carbon fibers and glass fibers, glass fibers and aramid fibers, and aramid fibers and carbon fibers. Table 8 shows the mechanical properties of epoxy composites reinforced with HS-carbon fibers and E-glass fibers as a typical example.

When the FRP is hybrid-reinforced with two fibers, f_1 and f_2, the law of mixtures in the calculation of the friction coefficient is given by

$$\frac{1}{\mu} = V_{f_1} \frac{1}{\mu_{f_1}} + V_{f_2} \frac{1}{\mu_{f_2}} + V_m \frac{1}{\mu_m} \tag{14}$$

where

$$V_{f_1} + V_{f_2} + V_m = 1 \tag{15}$$

Figure 16 shows the friction property of hybrid FRP as function of volume

TABLE 8
Mechanical properties of hybrid FRP (epoxy composites)

Volume fraction (%)			Tensile strength (GPa)	Bending strength (GPa)	Modulus of rigidity (GPa)	Interlaminar shear strength (MPa)
HS-carbon fiber (V_{f_1})	Glass fiber (V_{f_2})	$V_f = V_{f_1} + V_{f_2}$				
70	0	70	1.82	1.78	141	78
43	27	70	1.52	1.62	96	83
35	35	70	1.42	1.60	93	82
22	48	70	1.18	1.35	69	81
15	55	70	1.08	1.23	60	83
0	70	70	1.46	1.18	47	79

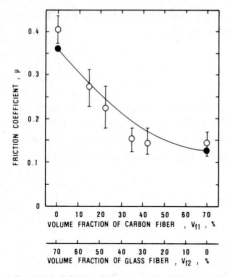

Fig. 16. Influence of the volume fraction of fibers on the friction coefficient of carbon–glass hybrid epoxy composites. $v = 25\ \mu\mathrm{m\ s}^{-1}$; $W = 2.94$ N; parallel.

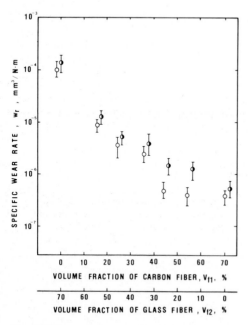

Fig. 17. Wear performances of carbon–glass hybrid epoxy composites. $p = 1.5$ N mm^{-2}; $v = 0.83$ m s^{-1}. ○, Parallel; ◑, anti-parallel.

fraction of the fiber reinforcements. When a medium hybrid FRP ($V_{f_1} = 35\%$, $V_{f_2} = 35\%$) is utilized instead of an HS-CFRP ($V_{f_1} = 70\%$, $V_{f_2} = 0\%$), there is only a slight increase in the friction coefficient.

Wear properties of hybrid FRP are shown in Fig. 17. The specific wear rate of hybrid FRP decreases with an increase in the volume fraction of carbon fibers. Thus, by adding carbon fibers to glass fibers, the improvement in wear property of GFRP is apparent. The specific wear rate of HS-CFRP with 40% volume fraction of fiber reinforcement was found to be 2×10^{-6} mm^3 N^{-1} m^{-1}. The hybrid carbon FRP with the same volume fraction of carbon fibers seems to produce less wear than HS-CFRP. From the experimental results shown in Figs. 16 and 17, for a practical application of FRP where friction and wear become an important problem, the use of hybrid FRP can be recommended in regard to the performance/cost ratio.

6.2. Wear of SiC fiber-reinforced epoxy resin and mica dispersion-reinforced polyester resin

Wear performances of the SiC fiber-reinforced epoxy resin (SiC–FRP) and the mica dispersion-reinforced polyester resin (MDRP) at high loads and velocities were studied under the experimental conditions shown in Table 7. Since the applied pressure and sliding velocity were varied stepwise at each stage of the seizure tests, the specific wear rate had to be calculated by summing individual wear rates, i.e.

$$w_r = \sum_{i=1}^{n} \left\{ \frac{Q_1}{W_1 D_1} + \ldots + \frac{Q_n}{W_n D_n} \right\} \tag{16}$$

Figure 18 represents the relationship between the specific wear rate and the friction coefficient. In this figure, MDRP (8Yn60) 10, for instance, means that the matrix was reinforced with the 60-mesh mica flakes, the volume fraction of which was 10%. There being a slight difference in manufacturing processes of two SiC fibers, SiC–FRP were labelled as (a) and (b). Every MDRP shows an extremely high specific wear rate. It is amazing that the SiC–FRP show no effective wear resistance despite good physical and mechanical properties of the SiC fibers. One of the possible reasons for this experimental result may be due to poor adhesive bonding between the SiC fibers and the epoxy resin. One might assume that the SiC particles act as abrasives once they are detached from the bulk SiC-FRP surface as wear debris. The specific wear rates of HS- and HM-CFRP are higher than those in Fig. 9. This may indicate that, at high loads and velocities, the wear of CFRP is greater. In contrast, the AFRP shows lower specific wear rate than that in Fig. 9

The specific wear rates of carbon steel specimens are plotted in Fig. 19. Because of the fact that the carbon fibers and mica flakes have a self-lubricating ability, the specific wear rates of carbon steel specimens in contact with them are low. The glass fibers and SiC fibers are hard enough to abrade the carbon steel surface. The high specific wear rate of SiC–FRP shown in Fig. 18 may be explained by the abrasive action of fragmented SiC fibers as well as of the steel wear debris.

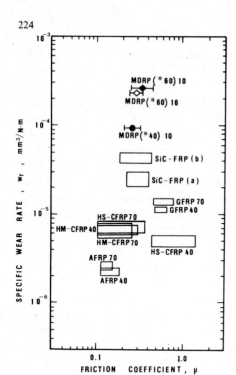

Fig. 18. Specific wear rates and friction coefficients of FRP obtained from sliding in the normal direction.

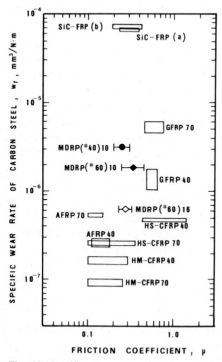

Fig. 19. Specific wear rates of carbon steel plotted against friction coefficients resulting from the experiments shown in Fig. 18.

6.3. Application of the finite element method analysis to fracture of FRP

An attempt was made to analyze the stress distributions at and around the FRP surface under friction processes using the finite element method (FEM). We have recently conducted the FEM analyses on friction and deformation of pure copper and aluminum, and the mechanism of the wear process of pure metals was discussed from the standpoint of fracture mechanics [20]. For an analysis of the stress distribution of an FRP specimen, an elastic program of FEM will be satisfactory [21]. From the observations of worn FRP surfaces, the following three situations are to be considered in regard to the contact of FRP.

(a) both the matrix and the fibers are in contact with a countersurface;

(b) the FRP surface is covered with the matrix, only the matrix being in contact with a counter-surface;

(c) the fibers protrude at the FRP surface owing to wear of the matrix.

Figure 20 shows the principal stress distribution in the case of the situation (a). The computer-simulation was carried out for the HS-CFRP having 40% volume fraction of fibers. The normal pressure and the friction coefficient were set at 1 N mm^{-2} and 0.2, based on the experimental result. The sliding direction is from right to left in this figure. The stress level in the fiber reinforcement is far less than the tensile strength of the HS-carbon fiber while, in the matrix, the stresses over the tensile strength of the epoxy resin are analyzed. At both sides of the lower part of the fibers, the stress levels are found to be high. Since the stress concentration on the periphery of fibers is critical to the debonding of the fibers from the matrix, fracture of FRP is likely to take place in situation (a).

In situation (b) shown in Fig. 21, it is noticed that the stress at the surface is large. Fracture of FRP will initiate at the surface or at the lower right-hand side of the fibers.

100 MPa

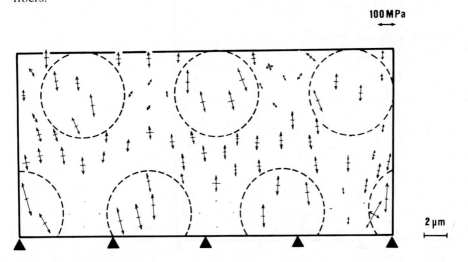

Fig. 20. Principal stress distribution when both the fibers and the matrix are in contact with a counter-surface; situation (a).

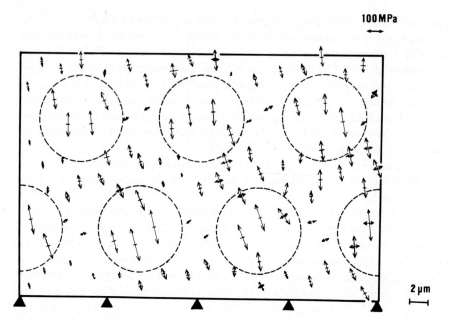

Fig. 21. Principal stress distribution when only the matrix is in contact with a counter-surface; situation (b).

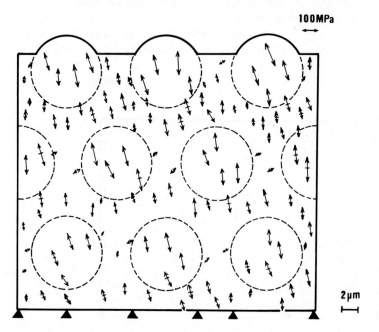

Fig. 22. Principal stress distribution when only the fibers are in contact with a counter-surface; situation (c).

50 MPa

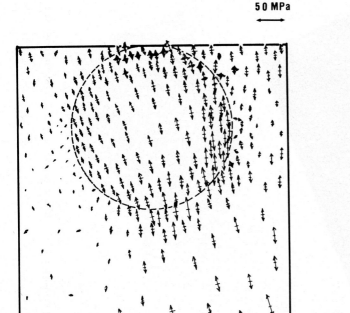

1 µm

Fig. 23. The detail of stress distribution around the one fiber-reinforcement.

Figure 22 shows situation (c), where large stresses within fibers were analyzed. When the protruding fibers make contact with a counter-surface during sliding, wear thinning or breakdown of fibers, diagrammatically shown in Fig. 14, may take place.

A detail of the stress distribution around the fiber reinforcement is shown in Fig. 23. The stresses just above and at the lower right-hand side of the fiber are found to be large. Thus, it may be stated that the fracture of FRP initiates either at the surface or at the lower right-hand side of the fiber. In the latter case, when the stress exceeds the interlaminar shear strength of HS-CFRP, peeling off of the fiber from the matrix will occur. Although the present FEM analyses are not able to argue quantitative wear mechanism of FRP, an analysis of the stress distribution is able to suggest where the fracture is likely to occur.

6.4. The atomic structure of fibril as observed by field ion microscopy

Field ion microscopy (FIM) is increasingly utilized in surface analyses [22,23] since it is able to resolve individual atoms. Almost all of the FIM work on tribology has been carried out using a tungsten field emitter tip, because the imaging procedures using He ions are well-established [24,25]. Adhesion and friction of W in contact with Au, Pt, Ni, W, etc. have been studied in our laboratory and some of the experimental results, together with computer-simulation, were reported elsewhere

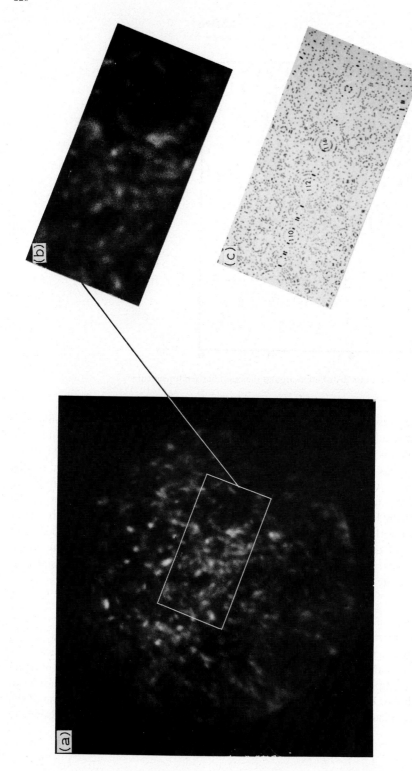

Fig. 24. (a) Field ion micrograph of the HM-carbon fiber repyrolyzed at 3000°C. Ne-ion image at 0.3 Pa. Best image voltage 11.5 kV. (b) The magnified photograph of (a). (c) The computer-simulated FIM pattern.

[26,27]. Quite recently, a field ion micrograph of the HM-carbon fiber was successfully obtained. The good tribological properties of CFRP shown in the preceding sections are obviously correlated with the fibril structure which provides low friction. The FIM of carbon fibers is a critically difficult work, because very little has been known about the electro-polishing of carbon fibers and the selection of image gas to obtain a good FIM photograph.

Figure 24 shows the Ne-ion image of HM-carbon fiber repyrolyzed at 3000°C. The FIM image shown in this figure does not give such a high contrast as that with a W emitter tip. However, a localized concentric circles indicates the agglomeration of fibril and each white spot corresponds to each carbon atom. High orientation of fibril was observed at the periphery of repyrolyzed HM-carbon fibers. The thickness of stacked basal planes was estimated to be 20–50 nm from the FIM photograph and these values showed good agreement with the result of X-ray diffraction analysis. The central region of the repyrolyzed HM-carbon fiber and the peripheral region of an ordinary HM-carbon fiber (pyrolyzed at 2500°C) also showed an orientation of the fibril alignment, while the HS-carbon fiber did not indicate the aligned fibril. Low friction of the HM-CFRP, especially slid in the parallel direction, could be attributed to this aligned fibril around the periphery of the fiber.

7. Concluding remarks

The law of mixtures in the calculation of the friction coefficient of FRP and the wear equation for FRP showed good agreement with the experimental results. Seizure of FRP occurred for HS-CFRP and GFRP in the sliding direction normal to the fiber axis, due either to frictional heat or to chipping of FRP. As for a practical application of FRP to tribology, the use of hybrid FRP instead of genuine CFRP is found to be promising in regard to the performance/cost ratio. The finite element method analysis showed the localized stress concentrations at the FRP surface and around the fiber reinforcement. The above experimental results, together with those obtained by the systems analysis, signified that the Young's modulus, tensile strength and interlaminar shear strength are closely correlated with the wear of FRP.

Acknowledgements

This chapter is based on the work carried out while one of the authors (T.T.) was with Osaka University as professor of precision engineering. The experimental work on the friction and wear of FRP was made by a number of his students and our sincere thanks are due to S. Fukuyama, T. Tanaka, F. Asano, and A. Saeki for carrying out many of the experiments described in this chapter. The authors thank Toray Industries Inc., Nippon Seisen Co. Ltd., Nippon Pillar Packing Co. Ltd., and Nihon Glass Fiober Products Co. Ltd., for offering the fiber reinforcements.

List of symbols

A	nominal area of contact (mm^2)
D	sliding distance (km)
E	Young's modulus (GPa) *
F	tangential force (N)
G	modulus of rigidity (GPa)
I_s	interlaminar shear strength (MPa)
P	applied pressure (N mm^{-2})
Q	wear volume (mm^3)
T	temperature (°C)
V_f	volume fraction of the fiber reinforcements (%)
V_m	volume fraction of the matrix (%)
v	sliding velocity (m s^{-1})
W	applied force (N)
w_r	specific wear rate (mm^3 N^{-1} m^{-1}) **
γ	shear strain
μ	friction coefficient
$\Delta\mu$	fluctuation width of friction coefficient
τ	shear stress (Pa)

FEM	finite element method
FIM	field ion microscopy
MDRP	mica dispersion-reinforced polyester resin
SiC–FRP	SiC fiber-reinforced epoxy resin

References

1 J.K. Lancaster, Proc. Inst. Mech. Eng., 182 (1967/68) 33.
2 J.P. Giltrow and J.K. Lancaster, Proc. Inst. Mech. Eng., 182(3N) (1967/68) 147.
3 J.K. Lancaster, Br. J. Appl. Phys., 1 (1968) 549.
4 J.P. Giltrow and J.K. Lancaster, Wear, 16 (1970) 359.
5 J.K. Lancaster, Wear, 20 (1972) 315.
6 J.K. Lancaster, Wear, 20 (1972) 335.
7 J.P. Giltrow, Composites, March (1973) 55.
8 J.P. Giltrow, ASLE Trans., 16 (1973) 83.
9 N. Ohmae, K. Kobayashi and T. Tsukizoe, Wear, 29 (1974) 345.
10 T. Tsukizoe and N. Ohmae, Tribol. Int., 8 (1975) 171.
11 T. Tsukizoe and N. Ohmae, Ind. Lubr. Tribol., 28 (1976) 19.
12 T. Tsukizoe and N. Ohmae, J. Jpn. Soc. Lubr. Eng., 21 (1976) 330 (in Japanese).
13 T. Tsukizoe and N. Ohmae, Trans. Jpn. Soc. Mech. Eng., 43 (1977) 1115 (in Japanese).

---

* MPa in eqn. (13).
** mm^3 N^{-1} mm^{-1} in eqn. (13).

14 T. Tsukizoe and N. Ohmae, J. Lubr. Technol., 99 (1977) 401.
15 T. Tsukizoe, J. Jpn. Soc. Lubr. Eng., 23 (1978) 11 (in Japanese).
16 M. Yukumoto, T. Tsukizoe and N. Ohmae, J. Jpn. Soc. Lubr. Eng., 23 (1978) 881 (in Japanese).
17 N. Ohmae, M. Yukumoto and T. Tsukizoe, Proc. Eurotribol. 77, Düsseldorf, October 3–5, 1977, Gesellschaft für Tribologie, 1977, Vol. II/III, pp. 57/1–57/4.
18 N. Ohmae, in N.P. Suh and N. Saka (Eds.), Fundamentals of Tribology, MIT Press, Cambridge, MA, 1980, pp. 1183–1196.
19 T. Tsukizoe and N. Ohmae, Fibre Sci. Technol., 18 (1983) 265.
20 E. Yamamoto, M. Kohno, N. Ohmae and T. Tsukizoe, J. Jpn. Soc. Lubr. Eng., 27 (1982) 860 (in Japanese).
21 E. Yamamoto, A. Sakakura, N. Ohmae and T. Tsukizoe, J. Jpn. Soc. Lubr. Eng., 26 (1981) 269 (in Japanese).
22 E.W. Müller and T.T. Tsong, Field Ion Microscopy, Principles and Applications, Elsevier, Amsterdam, 1969, pp. 229–299.
23 R. Wagner, Field-Ion Microscopy in Materials Science, Springer-Verlag, Berlin, 1982, pp. 5–109.
24 R.J. Walko, Surf. Sci., 70 (1978) 302.
25 D.H. Buckley, Surface Effects in Adhesion, Friction, Wear, and Lubrication, Elsevier, Amsterdam, 1981, pp. 252–257.
26 N. Ohmae and T. Tsukizoe, Proc. 4th Int. Conf. Production Eng., Tokyo, August 18–20, 1980, Jpn. Soc. Precision Eng. and Jpn. Soc. Technol. Plasticity, 1980, pp. 989–998.
27 N. Ohmae, J. Jpn. Soc. Lubr. Eng., 28 (1983) 715 (in Japanese).

Chapter 8

Wear of Reinforced Polymers by Different Abrasive Counterparts

KLAUS FRIEDRICH

Polymer und Composites Group, Technical University Hamburg – Harburg, 2100 Hamburg 90 (F.R.G.)

Contents

Abstract

An attempt is made to formulate a framework within which the behavior of polymer composites under wearing conditions can be considered. Experimental results obtained with different types of polymer composite material [short fiber reinforced thermoplastics and continuous fiber (woven fabric) reinforced thermosets] worn under two extremely different conditions (sliding against smooth steel and severe abrasion by different abrasive papers) were used for this approach. The tendencies of wear rate vs. fiber volume fraction typical for the different composites and testing conditions are discussed along with the corresponding microscopic observations of the wear mechanisms. The wear data are correlated with the mechanical properties of the materials and with frictional as well as material factors having an effect on the individual wear mechanisms during the different wear processes. Information from the literature and the conclusions from this study are used to present a hypothetical model of a polymer composite microstructure, which is assumed to have both a high load-carrying capacity and a generally high wear resistance.

1. Introduction

1.1. General aspects

Over the past decades, there has been a remarkable growth in the large-scale industrial application of plastics and fiber-reinforced polymer matrix composites. The features that make them so promising as industrial and engineering materials are their high specific strength (strength/density), high specific stiffness (modulus/density), and the opportunities to tailor material properties through the control of fiber and matrix combination, fiber orientation states and processing technology [1]. In addition, parts made from these materials have the ability to dampen shock and vibration, provide corrosion protection, and operate with little or no maintenance, making them excellent materials for the aircraft and aerospace industry, chemical industry, automotive industry, etc. Applications of polymer composites are concentrated, for example, in mechanical components such as gears, cams, wheels, impellers, brakes, clutches, seals, conveyors, transmission belts, bushes, and bearings. On chemical engineering plants, reinforced plastics are used in the field of piping, duct work, pumps, agitated vessels, separators, and thermocompressors [2–5]. Other important applications are conveyor aids and chute liners, which are used for agricultural, mining, and earth-moving equipment and in industries processing coal, coake, mineral ores, and glass.

In most of these services, the materials are subjecterd to different kinds of mechanical stress and various corrosive environments. Additional problems can arise from friction and wear-loading conditions (Fig. 1) and even if the first two problems are solved, there is still an uncertainty about the third. For certain requests, the coefficient of friction is of the highest importance, but largely it is the mechanical load-carrying capacity and the wear life of the components that determine their acceptability in industrial applications (Table 1). Usually, wear is undesirable not only because it makes necessary frequent inspection and replacement of parts, but it will also lead to deterioration of accuracy (e.g. of machine parts). It can induce vibrations, fatigue, and consequently failure by rupture. Wear can also produce

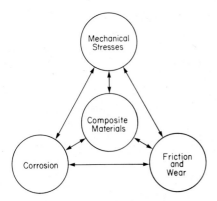

Fig. 1. Possible external loading conditions acting separately or simultaneously on composites in practical applications.

debris, which in turn will cause immobilization of close-fitting precision mechanisms, cause orifice occlusions, and foul electrical contacts [6]. Erosive wear plays an important role in many technical systems, which range from pipes and pumps handling slurries and other corrosive abrasive to aircraft airfoils passing through zones of rain and hail [7].

1.2. Factors influencing wear performance

Though composites are commonly used for such purposes, little has been published on how to select systematically material properties for optimum wear resistance. The major reason is that wear properties are not intrinsic material properties but are very dependent on the system in which the surfaces function [8]. The internal properties of the material tested (microstructure) are only one factor of the tribosystem (Fig. 2) [9]. For polymer composites, this problem is especially delicate because their internal structure often consists of two or more components having extremely

TABLE 1

List of requirements with respect to friction and wear properties of tribologically loaded structural components, which are often made of polymer composites

No.	Structural element	Frictional coefficient (μ)	Wear resistance ($\dot{w}^{-1}$)	Load-carrying capacity (pv)
1	Pipelines		↑	
2	Conveyor belts	↑	↑	
3	Chute liners	↓	↑	
4	Dry bearings	↓	↑	↑
5	Hip replacements	↓	↑	↑
6	Brake coverings	↑	↑	↑

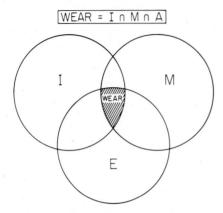

Fig. 2. Wear as a system property [9]. I = internal properties; M = wear mechanisms; E = external conditions.

different properties. Most commonly, hard, high-strength glass or carbon fibers are embedded in a more or less ductile, relatively soft polymer matrix. While hardness, H, elastic modulus, E, and ultimate strength, σ_f, of a composite usually increase with fiber volume fraction, V_f, fracture toughness, K_{Ic}, and/or fracture energy, G_{Ic}, may increase or decrease depending on the particular fiber/matrix combination [10]. The resistance of such a material to the removal of mass from the surface as a result of a mechanical action, i.e. contact pressure, p, and relative motion, v, of a solid counterpart should, on the one hand, be determined by the friction and wear properties of the individual components. On the other hand, it can be expected that geometrical factors and the interaction of these components in the composite play a dominant role. Especially are size, geometry, arrangement of the reinforcement elements, their position relative to the wear surface, and the quality of the fiber/matrix interface very important in this respect.

The other two factors consider the external parameters such as structure and hardness of the counterpart, H_{abr}, sliding velocity, v, apparent contact pressure, p, and the active wear mechanisms during the particular type of wear loading. As a consequence, a certain composite material, which, for example, is optimized for sliding application against smooth steel counterparts (adhesive and mild abrasive wear mechanisms), may prove quite disappointing when used under conditions where hard particles can cause wear of the structural elements mainly by severe abrasive mechanisms (Fig. 3). The relative amount of wear in both cases is dependent on how effectively the material's microstructure can resist the different processes of material separation characterizing the dominant wear mechanisms (Fig. 4).

1.3. Objectives

In this contribution it is shown how some polymer composites of quite different microstructure (varying from injection molded, very short-fiber-reinforced thermoplastics to continuous-fiber-woven fabric/epoxy resin laminates) behave under two

Structure of Tribosystem	Tribological Loading	Type of Wear	Mechanisms	
			Adh.	Abr.
Solid-Solid	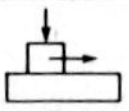	Sliding Wear	××	×
	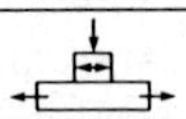	Fretting Fatigue	×	×
Solid-Solid Particles	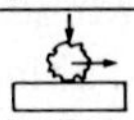	Grooving Wear		×× (2 BODY)
	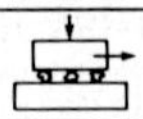	Grain Sliding Wear		×× (3 BODY)

Fig. 3. Schematic of sliding and grooving wear and the corresponding dominant mechanisms of material removal.

extreme cases of abrasive wear loading (i.e. sliding against steel surfaces and abrasion by coarse grinding paper). The results should help to get a feeling for the complicated processes involved in wear of composites under different tribological applications. There is also an attempt to evaluate those factors which contribute strongly to the wear resistance of composites under particular conditions and to derive from these considerations a hypothetical composite microstructure, which is assumed to have very high wear resistance.

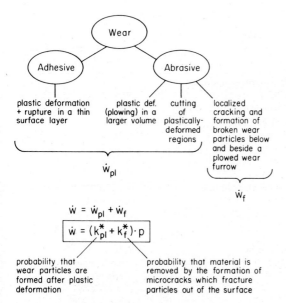

Fig. 4. Schematic subdivision of wear either into contributions of adhesion and abrasion or into contributions of wear due to plastic deformation and fracture mechanisms, respectively.

238

2. Experimental details

2.1. Composite materials

The materials tested can be divided into two major groups: injection molded plaques of short-fiber-reinforced or glass-sphere-filled thermoplastics and thermosetting resin plates reinforced with continuous fibers of different arrangement. A list of the individual composites and their abbreviations are given in Table 2. Details on the particular microstructure of these composite systems and further information on the mechanical properties are reported in previous papers [10,11]. Briefly, the short-fiber-reinforced thermoplastics possess a three-layer fiber orientation distribution for which fibers in the surface layers are arranged as shown in Fig. 5(a). In the central layer ($\approx 1/3$ of plaque thickness, B), the opposite fiber orientation exists. Typical micrographs of the different structure in the glass-sphere-filled systems and in the thermosetting resin composites can be seen on Fig. 5(b)–(f). Schematically, the structure of a unidirectional fiber composite is illustrated, along with possible wear directions, in Fig. 6.

2.2. Wear tests

Sliding wear experiments were performed at different sliding speeds, v, and contact pressures, p (details are given for each diagram), on an apparatus having a 100 Cr 6-ball bearing steel rotor, 60 mm in diameter. The average roughness at the beginning of the test was $R_{tm} = 0.6$ μm. In order to confirm that steady-state wear

TABLE 2
List of polymer composites investigated in this study

Class	No.	Matrix	Reinforcement
Glass sphere and short fiber filled thermoplastics	1	Polyethylene terephthalate (PET)	0–55 wt.% GF 30 wt.% GS
	2	Polyamide 6 (PA 6)	0–70 wt.% GS
	3	Polyamide 6.6 (PA 6.6)	0–50 wt.% GF 20 wt.% CF
	4	Polyethersulfone (PES)	0–30 wt.% GF
	5	Polyphenylene sulfide (PPS)	0–30 wt.% GF 30 wt.% CF
	6	Ethylenetetrafluoroethylene (ETFE)	0–30 wt.% GF 30 wt.% CF
Continuous fiber-reinforced thermosets	7	Epoxy resin (EP)	0–60 vol.% unidirectionally oriented CF 50 vol.% GWF 60 vol.% CWF 70 vol.% AWF

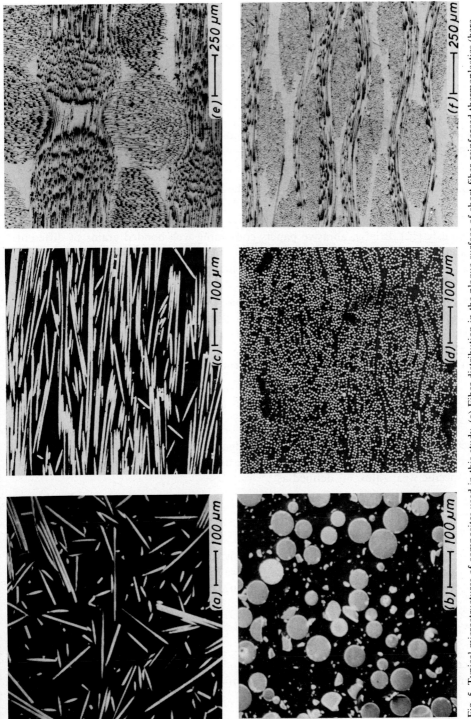

Fig. 5. Typical microstructures of composites used in this study. (a) Fiber distribution in the plaque surface of short fiber-reinforced thermoplastics (here, 30 wt.% GF–PET; $x-y$ plane, cf. Fig. 6); (b) glass sphere-filled PA 6; (c) unidirectionally oriented carbon fiber-reinforced epoxy resin (fiber length > 3 mm) in the $x-y$ plane and (d) $y-z$ plane; (e) $x-y$ plane and (f) $y-z$ plane of a glass fiber woven fabric (GWF)–epoxy matrix composite.

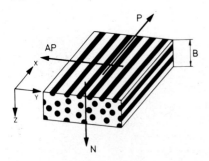

Fig. 6. Schematic structure of a unidirectional fiber composite and possible wear directions. AP = antiparallel, P = parallel, and N = normal to the main fiber arrangement in the specimen.

conditions were reached, measurements of the mass loss, Δm, were carried out after testing times of 1, 5, 10, 60, 120 min.

The set-up of a pin-on-disc type of abrasive wear test has been described in detail previously [12]. Briefly, abrasive papers containing well-bonded grains of different ceramic material (the mean particle size, D, varied between 7 and 100 μm) were fixed on the wear table of the apparatus and moved stepwise with a speed of $v = 300$ mm min^{-1} along the loaded test sample ($p = 2.22$ MPa). After a distance of $L = 500$ mm, the specimens were unloaded, their mass loss measured (to 0.1 mg), and then moved again against a fresh track of abrasive paper (single-path condition). About 4 tracks were made side by side. In all cases, the environmental temperature was $23 \pm 2°$C and the relative humidity $60 \pm 4\%$.

The dimensionless wear rate, $\dot{w}$, was determined as

$$\dot{w} = \frac{\Delta m}{(\rho L A)} \tag{1}$$

where ρ is the density of the material investigated and A the apparent contact area (in most of the cases $A = 3 \times 3$ mm^2). The results are often compared on the basis of the specific wear rates, $\dot{w}_s$, defined as

$$\dot{w}_s = \frac{\dot{w}}{p} \tag{2}$$

$\dot{w}_s$ has the dimension of a (stress)$^{-1}$, but it is more meaningful to use units (mm^3 N^{-1} m^{-1}) whose physical significance is more readily apparent.

Considering the well-known Archard equation [13]

$$\dot{w} = \frac{kp}{H} \tag{3}$$

where k is the probability that a junction between two surfaces leads to the formation of a wear particle (k = wear coefficient), and H is the hardness (flow pressure) of the composite material. Eliminating p on the right-hand side of this equation leads to

$$k^* = \frac{k}{H} \tag{4}$$

k^* is numerically the same as the specific wear rate, $\dot{w}_s$, but when used in the context of sliding wear, it is often referred to in the literature as the "wear factor". The reciprocal value of the wear rate, (i.e. $\dot{w}_s^{-1}$), is considered to be the material's wear resistance.

Studies of the wear mechanisms were performed by the use of scanning electron microscopy (SEM), for which the worn specimen surfaces were prepared by gold coating.

3. Sliding wear against steel counterparts

3.1. Basic correlations and wear test results

3.1.1. Specific wear rate and pv factor

Following eqns. (1)–(4), the specific wear rate, $\dot{w}_s$ (dimensionless wear rate, $\dot{w}$, normalized by the load per apparent contact area, p) is equal to the "wear factor" k^*. The latter depends on the properties of the two materials in contact and the dominant wear mechanisms. According to

$$k^* = \frac{\Delta m}{\rho A t p v} \tag{5a}$$

$$k^* = \frac{\Delta h}{t p v} \tag{5b}$$

$$\dot{w}_t = k^* p v \tag{5c}$$

with $L = vt$, Δh = wear depth, and $\Delta h/t = \dot{w}_t$ as a time-related (depth) wear rate, the wear factor k^* should be a material constant as long as changes in the product of pressure and sliding velocity, pv, are directly proportional to the wear depth, Δh. This is usually the case for moderate pressure, p, and sliding speeds, v, for which the temperature of the materials in contact remains near ambient. The corresponding value of the "wear factor" or "specific wear rate" is designated k_0^* or $\dot{w}_{s0}$ [14]. In many cases, however, higher values of pressure and/or velocity predominate, which cause increases in temperature and/or changes in the dominant wear mechanisms. Under these conditions, the value of k^* is no longer constant but increases with pv. According to eqn. (5c), this results in a progressive increase of the time-related wear rate, $\dot{w}_t$ (Fig. 7). The pv factors can be considered as a performance criterion for bearing materials (load-carrying capacity). They are widely quoted in the literature and may take two forms [15], viz.

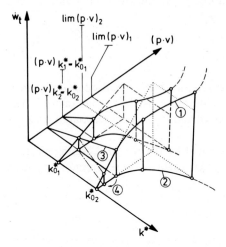

Fig. 7. Interrelationship between time-related wear rate, $\dot{w}_t$, wear factor, k^*, and load-carrying capacity, pv, for two materials of different basic wear factor k_0^* (curve ①). Curve ② illustrates possible changes in k^* with pv above a certain level $(pv)_{k^*=k_0^*}$. Slope ③ indicates, for constant pv, an increase in $\dot{w}_t$ when a material with higher wear factor k_0^* is used. To maintain the same wear rate in a material with higher wear factor, the pv level must be reduced (transition ④).

(a) the pv factor for continuous operation at some arbitrary specific wear rate, e.g. $\dot{w}_t = 25 \ \mu\text{m}/100 \ \text{h} \ (k^* = k_0^*)$ and

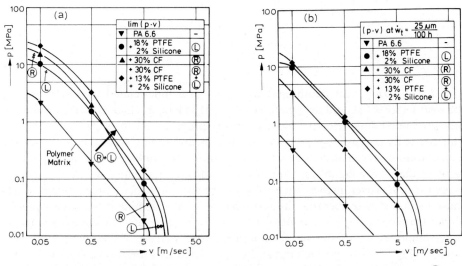

Fig. 8. pv diagrams of unfilled and filled PA 6.6 illustrating the effects of reinforcement, ⓡ, and lubricants, ⓛ, on (a) the limiting pv, and (b) the maximum pv level for constant wear rate $\dot{w}_t$.

TABLE 3

Influence of fibers and lubricants on the basic wear factor, k_0^*, the dynamic frictional coefficient, μ, and limiting pv values of PA 6.6 composites at 3 different sliding velocities

No.	Material	Frictional coefficient, μ (dynamic)	Wear factor k_0^* ($mm^3 \ N^{-1} \ m^{-1}$)	Limiting pv value ($MPa \ m \ s^{-1}$)		
				$v = 0.05 \ m \ s^{-1}$	$v = 0.5 \ m \ s^{-1}$	$v = 5 \ m \ s^{-1}$
1	PA 6.6	0.28	4×10^{-6}	0.105	0.088	0.088
2	PA 6.6 +30% GF	0.31	1.5×10^{-6}	0.438	0.350	0.263
3	PA 6.6 +30% CF	0.20	4×10^{-7}	0.735	0.945	0.280
4	PA 6.6 +18% PTFE + 2% silicone	0.08	1.2×10^{-7}	0.490	1.050	0.420
5	PA 6.6 +30% GF +13% PTFE + 2% silicone	0.14	1.8×10^{-7}	0.595	0.700	0.665
6	PA 6.6 +30% CF +13% PTFE + 2% silicone	0.11	1.2×10^{-7}	1.015	1.505	0.700

(b) the "limiting pv" above which wear increases too rapidly, either as a consequence of thermal effects or of stresses approaching the elastic limit of the material.

The effects of fiber reinforcement and lubricating fillers on the reduction of the basic wear factor k_0^* and the enhancement of the limiting pv value of a thermoplastic matrix is illustrated in Table 3 and Fig. 8(a) and (b), using data taken from the literature [16]. It can be stated that the basic wear factor of a polymer matrix is generally reduced when reinforcing fillers and lubricants are incorporated. With respect to the upper limit of the pv factor, fiber reinforcements are more effective in the range of lower speeds and higher contact pressures (improved mechanical properties), whereas lubricants prove better for higher sliding velocities at lower pressure levels. The latter is probably due to the drastic reduction in frictional coefficient, resulting in a lower amount of frictional heat and thus a lower temperature rise. Best results are obtained for combinations of both, appropriate lubricants and reinforcing filler elements.

3.1.2. Microstructural parameters and wear rate of composites

For two different pv levels, Fig. 9 represents the effect of short-fiber content on the specific wear rate of GF–PA 6.6.

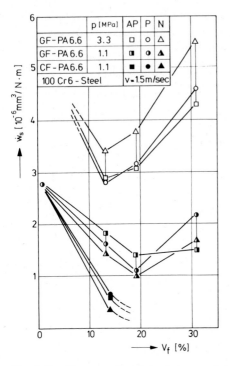

Fig. 9. Specific wear rates of short glass- and carbon fiber-reinforced PA 66.6 against 100 Cr6-steel at $v = 1.5$ m s^{-1} sliding speed and two different pressures, p, as a function of the fiber volume fraction and sliding direction.

For the lower pv level, wear is initially reduced if glass fibers are added, but if the fiber fraction exceeds 20 vol.% (v/o), wear increases again. This fact can be considered to be a result of changes in the dominant wear mechanisms and thus changes in the tendency of wear factor reduction of this particular polymer matrix with increasing amount of fiber reinforcement. On the basis of the same volume fraction, short carbon fibers result in a better reduction of the material's wear factor than do glass fibers. This trend is in accordance with the literature data given in Table 3 as well as with those results on PA 6.6 composites reported by Lancaster several years ago [17].

In the case of a very high pv level, no stable wear rate could be measured for the unreinforced material. Stick–slip behavior was observed and the temperature at the sliding contact was raised to over 120°C, indications that this pv level was much higher than the limiting pv for PA 6.6. Although stable conditions could be achieved with the glass fiber-reinforced PA 6.6 composites, the values of the specific wear rate increase permanently with V_f and they are too high so that this pv range is no longer acceptable for the practical application of these composites, e.g. as bearing materials. The corresponding temperatures measured at the rubbing surfaces of the steel counterparts immediately after stopping the experiment ranged between 70 and 85°C, compared with 35–50°C for the lower pv level.

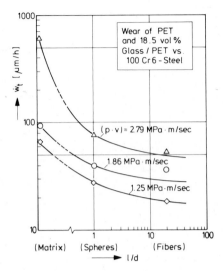

Fig. 10. Effect of glass fillers and their geometry on the time-related wear rate, $\dot{w}_t$, at three different pv levels.

As far as the effect of the sliding direction in Fig. 9 is concerned, no clear trend of the dependence of the wear rate on fiber orientation could be detected. This is probably due to the broad angle distribution of the short fibers relative to the directions applied. Therefore, mean values of a mixture of all three directions are used for further discussions, unless otherwise mentioned.

Figure 10 illustrates the geometrical influence of 18 vol.% glass reinforcement on the time-related wear rate of a PET matrix at three different pv values, which belong to the range between $(pv)_{k^*=k_0^*} < pv < \lim(pv)$, i.e. $k^* = f(pv) > k_0^*$ (cf. Fig. 7). In all these cases, $\dot{w}_t$ and k^* are highest for the neat PET matrix relative to the PET composites, for which fibers with an aspect ratio (length/diameter) of $l/d = 20$ are superior to spheres ($l/d = 1$). The same relationship should hold for the basic wear factor, k_0^*.

As expected from other studies [18–20], the relative reduction in specific wear rate of various composites may also vary as a result of different initial wear rates of the thermoplastic matrices (Fig. 11). In spite of having the highest hardness (Table 4), the extremely brittle PPS matrix exhibits the highest wear rate, but, at the same time, the highest reduction in wear rate is observed when fibers are incorporated in this type of matrix. The higher wear-reducing effect of short carbon fibers relative to glass fibers, already reported for the PA 6.6 matrix, is also observed here for both the brittle PPS and the ductile ETFE material. It must be noted, though, that under the given circumstances, the effective filler volume fraction, V_f, and thus the ratio of exposed area of filler to the apparent contact area ($A_F/A = V_f$), is higher for the carbon fibers. Further, the PET curve of Fig. 11 indicates that the wear-reducing effect by the fillers is obviously limited up to a certain fiber volume fraction above which the wear rate increases again. Similar observations were also made for the PA

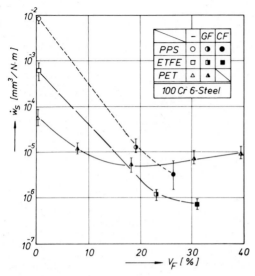

Fig. 11. Specific wear rate $\dot{w}_s$ vs. fiber volume fraction V_f for reinforced thermoplastics against steel ($pv = 3$ MPa m s^{-1}; $p = 0.3$ MPa).

6.6 composites in which this transitional behavior was attributed to changes in the effective wear mechanisms [21].

The data of Fig. 12 generated in a separatet test series, are not quite comparable with those of Fig. 11 because of the different pv conditions applied. Nevertheless, the effect of wear reduction by the reinforcing elements in the thermosetting resin is, in principle, the same as for the thermoplastic matrices. In addition, both systems have in common that, at relatively high levels of fiber loading, wear is mainly determined by the parameters introduced by the fibers (material, geometrical arrangement etc.), as is known from other studies [22]. Neglecting differences in the fiber volume fraction of the EP matrix composites in Fig. 12, the woven fabric structure of aramid fibers is superior to that of glass fibers by more than one order of magnitude. The values of the carbon fiber composites were found between these

TABLE 4
Mechanical properties of the thermoplastic matrices PPS, ETFE, and PET

Polymer	Modulus, E (MPa)	Strength, σ_f (MPa)	Strain at break, ϵ_f (%)	Hardness, H [a] (MPa)	Fracture toughness, K_{IC} (MPa m$^{1/2}$)
PPS	3600	77	3	145	1.2
ETFE	825	27	180	40	7
PET	3200	69	45	114	2.4

[a] Ball indentation hardness after German Standard DIN 53456.

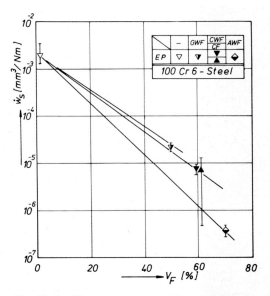

Fig. 12. Specific wear rate $\dot{w}_s$ vs. fiber volume fraction V_f for continuous fiber-reinforced thermosets against steel ($pv = 1.8$ MPa m s^{-1}; $p = 3$ MPa). The scatter bars are the result of averaging the three different sliding directions.

two and, except as a result of minor orientation effects, there was no difference in the average wear rate between the woven fabric and the unidirectional carbon fiber arrangement of the same fiber volume fraction.

3.2. Sliding wear mechanisms

The scanning electron micrographs of Fig. 13 show some of the wear mechanisms involved in the wear processes of unfilled thermoplastics against smooth steel surfaces.

In the case of the very brittle PPS, wear is initiated by the formation of surface cracks parallel to the sliding direction as a result of a high frictional coefficient and a locally high strain rate [Fig. 13(a)]. Subsequently, pieces of material between these cracks are removed from the surface due to adhesive forces to the counterpart and/or a chipping process by the harder asperity tips [Fig. 13(b)]. The very ductile ETFE surfaces, on the other hand, apparently deform plastically before final debris removal by tearing fibril-like pieces out of the deformed surface layer [Fig. 13(c)].

The fact that fibers usually improve the wear resistance of a polymer matrix under these particular wear conditions is probably due to several factors. When the fibers are exposed at the sliding surface, they support part of the applied load. Thus, penetration of the steel asperity tips into the polymer surface is reduced, so that microcutting or microplowing mechanisms are no longer as effective. In addition, since the wear rates of the fiber materials are usually lower than those of the matrices under sliding steel conditions, there should be a reduction in the wear rate

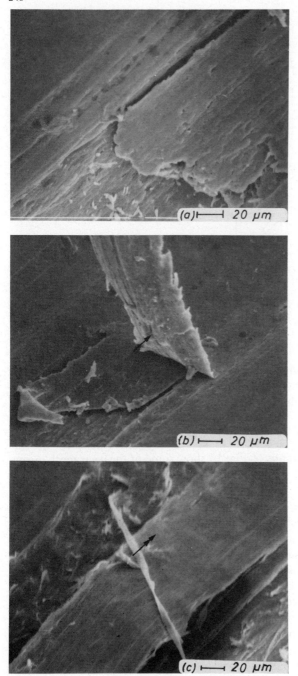

Fig. 13. (a), (b) Surfaces of unfilled PPS and (c) ETFE worn by a steel counterpart. The arrows indicate a high degree of chip and groove formation on the brittle PPS surfaces. The double arrow indicates the sliding direction.

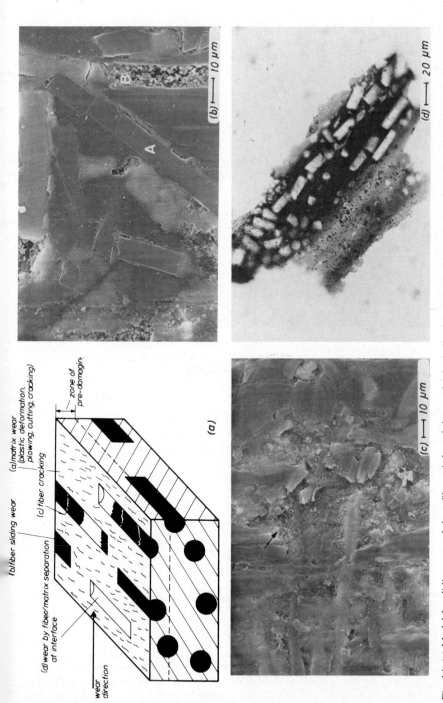

Fig. 14. (a) Model for sliding wear of short fiber-reinforced thermoplastics indicating different wear mechanisms. The corresponding SEM-micrographs of worn composite surfaces show (b) wear-thinned and cracked fibers (A), pulverized matter in former fiber beds (B), and (c) regions where larger material patches have been broken out of the surface (arrow). Photograph (d) is an optical micrograph of a wear particle which contains many broken glass fiber pieces (courtesy H. Voss, Hamburg).

250

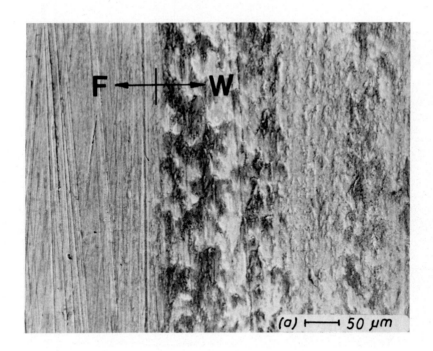

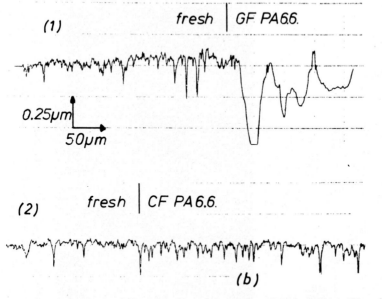

Fig. 15. (a) Appearance of the steel rings in the unworn (fresh) condition (F) and after transfer of third-body layers during the sliding procedure against CWF–EP (W). (b) Profilometer traces of steel rings, perpendicular to the sliding direction, showing the fresh profile on the left and the profile after sliding against (1) GF–PA 6.6 and (2) CF–PA 6.6 over a sliding distance of $L = 3 \times 10^4$ m on the right.

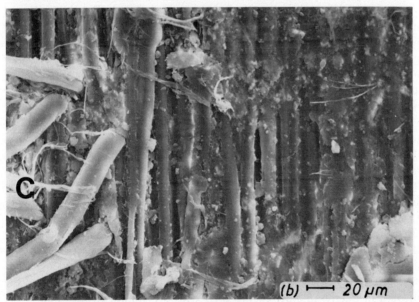

Fig. 16. Typical features on worn surfaces of (a) CF–EP and (b) AWF–EP against steel counterparts. (A) wear-thinned and cracked fibers; (B) regions of pulverized fiber and matrix material; and (C) fibrillized aramid fibers torn out of the woven fabric structure.

of the composites relative to the neat polymers. Examinations of the worn surfaces by scanning electron microscopy (SEM) have shown many of the detailed features characterizing the wear of the fibers and reported by others [23,24]. In particular, wear thinning of fibers, multiple fiber cracking, fiber pulverization and fiber/matrix separation along their interfaces occurred. Schematically, this is illustrated in Fig. 14(a) along with characteristic micrographs of the worn surfaces of GF–PA 6.6 composites with a low [Fig. 14(b)] and a high [Fig. 14(c)] fiber volume fraction. In the contact region of the latter material, characterized by a larger amount of exposed fibers to the counterpart asperities, fiber/matrix separation at interfaces and multiple fiber breakage was associated with the removal of larger material patches from the surfaces. This enhanced contribution of typical abrasive wear features can probably be considered as the reason for the increase in wear rate above a certain fiber volume fraction, as observed for GF–PET and GF–PA 6.6 composites.

In the cases of both Fig. 14(b) and (c), the nature of the typical wear debris consisted of shear-deformed polymer matrix containing small broken fiber elements, pulverized glass particles and wear powder of the metallic counterpart. An example of such a removed composite wear particle is given in Fig. 14(d) [21]. The particles can either be lost from the contact zone immediately after being broken from the composite surface, or remain there for a while as transferred and back-transferred layers [Fig. 15(a)]. In such cases, their polymer component can cushion the counterface asperities and reduce its effective roughness, thus being an important mechanism of wear reduction. But there can also occur an increase in wear when the glass splinter components in these films act as third-body abrasives, leading to enhanced roughening of the steel partner, which may in turn cause an increase in composite wear by preferred abrasive mechanisms. The optimum condition for minimum wear is not easy to determine and is, of course, a function of several factors, in particular the type of polymer in which the fillers are incorporated and the abrasiveness of the filler material [25]. In the present study, the glass fibers generated appreciably higher roughening of the steel surfaces than the carbon fibers [Fig. 15(b)], so that the higher wear rate of the glass-fiber materials is not surprising.

In principle, the same mechanisms are also valid for the glass or carbon fiber-reinforced thermosets. They are, however, modified by the fact that the fibers are continuous in length and that, due to the higher volume fractions of reinforcement, a much higher fiber density in the wear surface exists (Fig. 16). Hence, wear in these composites is primarily determined by the mechanisms associated with the fibers, i.e. wear thinning of fibers, fiber cracking and pulverization. A slightly different wear surface appearance is, however, found for the aramid woven fabric reinforced epoxy. Instead of broken and pulverized fibers typical for glass and carbon fiber composites, parts of the aramid fibers are torn out of the dense fiber bundles, which form the woven structure in the contact area. During this process, the fibers rupture in a ductile manner before their ends are splintered into many fine fibrils, which still adhere to the worn composite surface [Fig. 16(b)]. In this way, broken fibers do not immediately enter the wear debris (as broken glass or carbon fibers do) so that the amount of wear is lower in this material. This seems to be of special importance when local surface cracking can become a very dominant micro-mechanism of material removal.

3.3. Mathematical descriptions

3.3.1. General remarks

There have been made several attempts in the past to describe mathematically the wear behavior of composite materials against different counterparts. Some relationships are based on modified rules of mixture, describing either the wear rate of the composite or its wear resistance as a function of the corresponding properties of the individual composite elements and their volume fractions [6,22,26]. Others have correlated the wear properties with mathematical terms combining several mechanical properties and surface parameters of the materials in contact [24,27–35]. Both of the different approaches possess certain advantages; for example, the rule of mixture relationship leads to a better, more fundamental understanding of the role of the individual components and related wear mechanisms in the composite's wear behavior. The other method of description reconciles an idea of which material properties are worth improving in order to get optimum wear resistance. Therefore, an attempt will be made to describe some of the results of this investigation in both ways.

3.3.2. Rule of mixture approach

Figure 17 represents the experimental wear rates of two short glass fiber-reinforced thermoplastic matrix systems (GF–PES, GF–PA 6.6) as a function of fiber volume fraction and pv level, respectively. In general, PA 6.6 materials wear much slower than PES composites for a given pv value. Higher values of pv result in higher wear rates. The most spectacular effect is observed for PES, in which even a relatively small volume fraction of glass fibers (12 vol.%) improves the wear resistance by nearly two orders of magnitude. An improvement is also observed for PA 6.6 in this range of fiber content, but only by a factor of about two. This is mainly due to the fact that the neat polyamide material already exhibits a wear rate about 10^3-fold smaller than the corresponding PES value. A further enhancement of fiber loading in both thermoplastics results only in minor reductions of the wear rate. Both observations lead to the conclusion that, in the range of lower fiber content, wear behavior of the composite is dominated by the friction and wear properties of the matrix, whereas at higher fiber volume fractions, wear mechanisms associated with the fibers determine much more effectively the composite wear rates. Finally, this can even lead to an increase in wear rate with further enhancement of fiber reinforcement, which is clearly seen in the results measured for GF–PA 6.6 (indicated by V_f^*). The probability for such a transition in tendency seems to be highest, i.e. at a relatively low level of fiber loading, the higher the applied pv value. In addition, it must be assumed that the exact position of V_f^* is dependent on the properties of the individual components of the composite, the fiber/matrix bond quality, and orientation effects.

In order to be able to describe this kind of wear behavior by a modified rule of

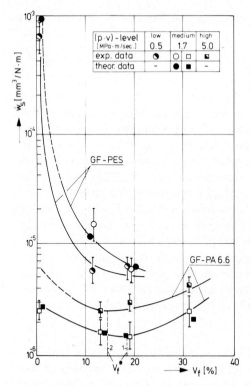

Fig. 17. Specific wear rates of PA 6.6 and PES composites tested at different pv levels as a function of fiber volume fraction. Some of the experimental results are compared with data theoretically calculated by a modified rule of mixture [eqn. (8)].

mixture, it has to be considered which particular wear mechanisms occur in the contact area and how they reduce the wear resistance of the composite, $\dot{w}_{s_C}^{-1}$. The evaluation of the worn surfaces had shown that 4 individual mechanisms dominate (more or less as a function of fiber material and volume fraction) the process of material removal [cf. Fig. 14(a)]. These are (a) matrix wear ($\dot{w}_{s_M}$), (b) fiber sliding wear ($\dot{w}_{s_{F_S}}$), (c) fiber cracking, and (d) wear by fiber/matrix separation at the interface. It can be assumed that mechanisms (c) and (d) occur sequentially and they can therefore be considered as a combined process of fiber cracking and interfacial separation. The partial value of this process, $\dot{w}_{s_{Fci}}$, is difficult to determine, but it is obvious that this mechanism becomes more important (on the cost of fiber sliding) the higher the fiber volume fraction and the pv product. Under these circumstances, more broken fiber particles, which are dug out of the surface, can act as third-body abrasives, thus inducing more sites of fiber breakage from which subsequent interfacial separation takes place.

As a rough estimate for the latter conditions, a relationship, successfully used for the correlation of wear test results with particulate filled thermoplastics [26], can be applied in the modified form

$$\dot{w}_{s_C}^{-1} = \left(\frac{1 - A_F}{A}\right)\dot{w}_{s_M}^{-1} + a\left(\frac{A_F}{A}\right)\dot{w}_{s_{Fs}}^{-1} + b\left(\frac{A_F}{A}\right)\dot{w}_{s_{Fci}}^{-1} \tag{6}$$

where A_F/A is the ratio of exposed area of filler to the apparent contact area. Its value is a function of fiber orientation with respect to the sliding surface, being a little bit lower for normal compared with antiparallel or parallel orientation. In a first-order approximation, it can be stated, however, that A_F/A is nearly equal to the fiber volume fraction, V_f. The sum of coefficients a and b must be equal to 1 and it is assumed that they consider the relative contribution of fiber sliding wear and fiber cracking/interfacial wear in the form

$$a = 0.5(1 + cV_f) \tag{7a}$$

$$b = 0.5(1 - cV_f) \tag{7b}$$

where c is a factor which considers how fast the transition from one mechanism to the other occurs ($c \geq 1$). It should be a function of the pv level applied and of the fiber orientation and geometry (aspect ratio, orientation factor, etc.).

Using these relationships, the equation for the wear resistance of the composite ($\dot{w}_{s_C}^{-1}$) as a function of the contribution of the individual mechanisms can be rewritten as

$$\dot{w}_{s_C}^{-1} = (1 - V_f)\dot{w}_{s_M}^{-1} + 0.5(1 + cV_f)V_f\dot{w}_{s_{Fs}}^{-1} + 0.5(1 - cV_f)V_f\dot{w}_{s_{Fci}}^{-1} \tag{8}$$

As a comparison of the mathematical description of the composite's wear resistance with the experimental reality, two sets of theoretically calculated wear rates were included in Fig. 17. For their calculation according to eqn. (8), data of the pure polymer matrices at a medium pv level ($\dot{w}_{s_{PA6.6}} = 2.6 \times 10^{-6}$ mm^3 N^{-1} m^{-1}; $\dot{w}_{s_{PES}} = 10^{-3}$ N^{-1} m^{-1}) as well as literature results of glass fiber material against steel counterparts ($\dot{w}_{s_{Fs}} = 10^{-6}$ mm^3 N^{-1} m^{-1}) obtained from measurements with unidirectionally continuous fiber-reinforced composites of very high glass fiber content [22] were used. Values of $\dot{w}_{s_{Fci}}$ and c were derived in combination with the first derivative of eqn. (8) to V_f, by using the transitional volume fraction V_f^* of the curve for GF–PA 6.6 and the corresponding values of the composite's wear rate under the assumption that c is constant for both materials at a given pv level of 1.7 MPa m s^{-1} ($\dot{w}_{s_{Fci}} = 0.175 \times 10^{-6}$ mm^3 N^{-1} m^{-1} for PA 6.6 composites; $\dot{w}_{s_{Fci}} = 9.75 \times 10^{-6}$ mm^3 N^{-1} m^{-1} for PES composites; $c = 3.5$). Once these values are generated, they should also be of the same order of magnitude for similar composite systems, provided always that a good fiber/matrix bonding is given, otherwise, changes in $\dot{w}_{s_{Fci}}$ and c must be expected. In the present case, however, the agreement in tendency between the measured wear data and the theory for both composite systems is encouraging.

3.3.3. Correlation of wear rate with other material properties

As already mentioned in connection with Fig. 4, wear of a material as a result of contact pressure and relative motion of a solid counterpart, is based mainly on different adhesive and abrasive wear mechanisms. As is known from other investigations on the abrasive wear of metallic materials, abrasion can be divided into three fundamental processes of material removal, i.e. microplowing, microcutting and microcracking [36,37]. Especially if the latter process becomes prominent, quite unpredictable changes in wear rate must be expected from what one would predict from the material hardness.

From microscopic observations of polymer matrix surfaces worn by steel counterparts, it can be concluded that the main contributions to wear are due to a sequence of plastic deformation and rupture in a thin surface layer, plowing in a larger material volume, and subsequent cutting of plastically deformed regions. The mathematical description of such behavior should follow the correlations given in eqns. (3) and (4), i.e. the wear factor $k^* = k_{pl}^* (\doteq \dot{w}_s)$ is inversely proportional to the hardness of the worn material (resistance of the material against indentation of counterpart asperities). In addition, k^* depends, through the wear coefficient, k, and the geometry of the surface morphology, on two types of physical properties of the material: the surface energy, which is related to adhesion and frictional forces (reflected by the frictional coefficient μ), and those properties of the material that are responsible for the mechanisms by which matter is separated from the surface. These are basically mechanical properties such as ultimate tensile strength, σ_f, and elongation to fracture, ϵ_f ($\sigma_f \epsilon_f =$ deformation energy). In fact, using the original Ratner–Lancaster approach [31,32] of the form

$$\dot{w}_s = \frac{1}{H} \mu \frac{1}{\sigma_f \epsilon_f} \tag{9}$$

has lead to a reasonably good correlation between the mechanical properties of the polymer matrices used here and their wear test results [38]. In this equation, the term $(\sigma_f \epsilon_f)^{-1}$ can be considered as a factor which determines the relative contribution to wear by abrasive mechanisms. It should be proportional to the abrasion factor, f_{ab}, defined by Zum Gahr and Mewes [39] as the ratio of the volume of wear debris to the volume of wear groove produced ($0 \leq f_{ab} \leq 1$)

$$f_{ab} = C_1 (\sigma_f \epsilon_f)_m^{-1} \tag{10}$$

where C_1 is a proportionality constant (dimension MPa), and the subscript m relates to the matrix polymer.

If fibers are incorporated, the hardness of the material increases and μ is modified. As the fibers support a great deal of the applied load, this should also result in a large reduction in the abrasion factor, f_{ab}, of the composite relative to the neat matrix. The effectiveness in the variation of f_{ab} is assumed to be a function of the topography of the contact area (influenced by the fiber arrangement in the composite), the fiber volume fraction, and the fiber material (low-friction material superior to high-friction fiber material; ductile fibers superior to brittle ones). An empirical approach for the description of the results obtained in this study was successful in the form

$$\dot{w}_s = f_{ab}\mu\frac{1}{H} \tag{11a}$$

with

$$f_{ab} = C_1(\sigma_f\epsilon_f)_m^{-1}\exp(-\xi V_f) \tag{11b}$$

where $C_1 = 5 \times 10^{-3}$ MPa. The exponent ξ was chosen to be 10, 11 and 12 for the glass, carbon, and aramid woven fabrics, and 20 and 22 for the short glass fiber- and carbon fiber-reinforcement, respectively (Fig. 18). The values of ξ chosen result from recalculations of the wear data measured for the short fiber composites and the fiber woven fabrics under different pv conditions. Hence, they can only be considered as a first-order approximation towards the conclusion that, due to their rougher topographies, only at higher fiber loadings are the woven fabric composites as effective in wear resistance improvement as the short-fiber reinforcements already are at lower fiber contant. The interwoven structure seems to favor more sites of

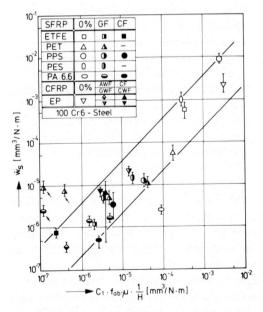

Fig. 18. Correlation of specific wear rates of the different composites tested against 100 Cr 6-steel with the product of reciprocal hardness, frictional coefficient (varies for all the materials between 0.13 and 0.21), and abrasive factor, f_{ab}, as defined by eqn. (11b).

fiber breakage than the smoothly wear-thinned short fibers so that a higher contribution of wear debris by pulverized fiber material is generated in the reinforced thermosets.

The correlation of eqns. (11a) and (11b) only hold for conditions where no important contribution of the third micromechanism of abrasive wear, i.e. the formation of brittle wear particles immediately after crack initiation by plowing counterface asperities, exists. The probability for these wear events increases with increasing counterface roughness. The latter can be initiated by roughening of a smooth counterface as a result of the abrasiveness of the reinforcing elements. This particular case is found for the PET– and PA 6.6–glass composites of higher volume fraction, for which such a transition from mild to severe abrasive wear mechanisms with V_f explains the deviation of the wear data from the general tendency given in Fig. 18 (arrows).

4. Abrasion by hard abrasive particles

4.1. Influence of external testing conditions

The external parameters chosen for the severe abrasive wear experiments and mentioned in Sect. 2.2 are the result of some pretests with a short glass fiber–PET matrix system in the antiparallel direction against Al_2O_3 paper of $D = 70$ μm grain size. It was of special interest to know how load, velocity and specimen size influence the wear behavior under the attack of abrasive particles. Figure 19(a) illustrates, for samples with a quadratic cross-section of apparent contact area $A = 10$ mm^2, that an increase in load (or contact pressure, p) results in a linear increase of the dimensionless wear rate, $\dot{w}$, according to eqn. (3). As the experiments were performed at a constant sliding velocity of $v = 300$ mm min^{-1}, this leads to the conclusion that the materials wear factor, k^*, (or specific wear rate $\dot{w}_s$) for these particular testing conditions is constant and independent of the pv product ($k^* \stackrel{\wedge}{=} k_0^*$). Its value is, of course, much higher that that measured against smooth steel counterparts. The same conclusion can also be drawn from the trend of wear rate vs. sliding velocity [Fig. 19(b)]. For a given contact pressure, the wear rate, $\dot{w}$, is independent of v in the range between 100 and 500 mm min^{-1}.

A very important piece of information, which must always be considered when comparing the results of laboratory tests with the wear behavior of a material in field application, is the effect of the specimen size exposed to the abrasive counterpart [Fig. 19(c)]. Specimens with a large contact area, A, exhibit lower wear rates than those of smaller size. The effect is more pronounced the higher the apparent contact pressure, p, is. The explanation is based on the fact that the abrasive particles are also worn by the glass fibers in the composite material tested, so that the abrasiveness of the particles is permanently reduced as long as they are in wear contact with the sample. As the sliding distance for each abrasive particle is longer for larger samples, the latter are, on average, in contact with grains of lower abrasiveness, thus exhibiting lower wear rates. Similar observations to those reported here for short

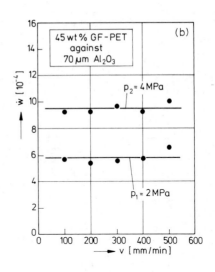

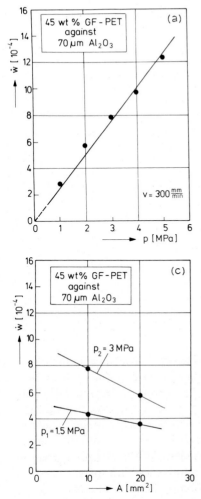

Fig. 19. (a) Linear increase of the dimensionless wear rate, $\dot{w}$, with applied load, shown for 45 wt.% GF–PET against 70 μm Al_2O_3 paper (AP direction; $A = 10$ mm^2). (b) Independence of the severe abrasive wear rate on the sliding velocity v (in the range $0 < v < 500$ mm min^{-1}; $A = 10$ mm^2). (c) Reduction in the dimensionless wear rate, $\dot{w}$, with increasing specimen size as shown for two different contact pressures. In both cases, the sliding velocity was $v = 300$ mm min^{-1}.

glass fiber–PET have also been made recently with continuous carbon fiber–epoxy resin composites [40].

4.2. Effect of composite microstructure

4.2.1. Specific wear rates

The favorable condition of wear rate reduction as a result of fibers in a polymer matrix does not always occur. The opposite is observed, for example, if steel

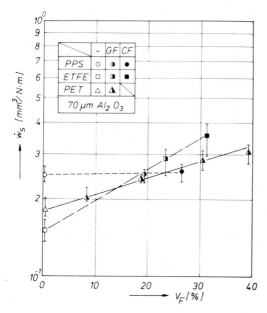

Fig. 20. Specific wear rate $\dot{w}_s$ vs. volume fraction of fibers V_f for various reinforced thermoplastics against 70 μm Al$_2$O$_3$ paper ($p = 2.22$ MPa; $v = 300$ mm min^{-1}).

counterparts possess a very high roughness, as reported by Lancaster for single-pass sliding conditions [17], or if wear is caused by erosion with hard quartz particles [41]. In both examples, changes in wear rate with fiber reinforcement became especially deleterious in composites with a more ductile (e.g. polyamide, PA) instead of a relatively brittle matrix, such as EP. In fact, these observations were also made in this study when the composites were worn under severe abrasive conditions. Sliding against abrasive paper containing hard Al$_2$O$_3$ particles (70 μm in size) caused an increase in wear rate of the thermoplastic matrix composites with increasing short fiber content (Fig. 20). In spite of the fact that this occurred only within one order of magnitude for the given range of fiber loadings, the increase was clearly detectable. The effect was more pronounced in the more ductile matrices (ETFE and PET) compared with the brittle PPS matrix.

Figure 21 shows that effects of fiber orientation in the short fiber-reinforced thermoplastics (polyamides) can be neglected with respect to severe abrasive wear conditions. It can be noticed, on the other hand, that the geometry of the reinforcing component may affect the wear resistance of the material much more. For a given volume fraction of reinforcing elements, bigger dimensions relative to the grain size of the abrasive seem to be more favorable in improving the composite's wear resistance.

Data measured under the same testing conditions for continuous fiber thermosets (Fig. 22) indicate that the latter reinforcement structures seem to be more effective in improving abrasive wear resistance than short fibers. Although in both cases the

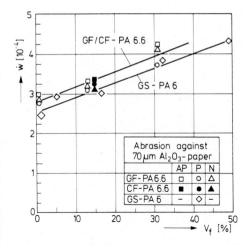

Fig. 21. Dimensionless wear rate, $\dot{w}$, of glass- and carbon fiber-reinforced PA 6.6 and glass sphere-filled PA 6, showing, against 70 μm Al_2O_3 paper, an increase with increasing filler content.

specific wear rates of the composites are higher by five orders of magnitude relative to the sliding wear experiments against steel, in the material group of Fig. 22, a reduction in wear rate with V_f is still observed. The relative efficiency in wear hindrance by the different fiber materials and arrangements seems to be the same as

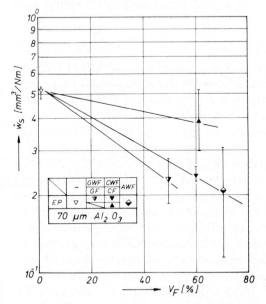

Fig. 22. Specific wear rate $\dot{w}_s$ vs. volume fraction of fibers for reinforced thermosets worn by 70 μm Al_2O_3 paper.

262

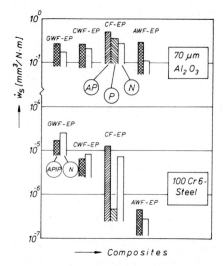

Fig. 23. Effect of fiber orientation on specific wear rate under two extremely different wear conditions for various reinforced thermosets: GWF–, CWF–, and AWF–EP = glass-, carbon-, and aramid fiber woven fabric–epoxy; CF–EP = unidirectionally oriented carbon fibers in EP resin.

that already described for the sliding wear conditions. It should be mentioned in this respect that the most effective wear resistance in both cases was detected for aramid fibers oriented normally to the wear surface (Fig. 23). This result is in agreement with others found for the wear behavior of aramid–epoxy composites against steel counterparts [23,28].

4.2.2. Microscopy of severe abrasive mechanisms

As mentioned in Sect. 3.3.3 and shown in Fig. 4, material removal from a surface by the attack of hard abrasive particles can be caused by three distinct micro-wear mechanisms, i.e. microplowing (combined with highly plastic deformation), microcutting and, eventually, microcracking [36,37]. By the first two mechanisms, longitudinal furrows in the direction of sliding are formed on the abraded surface. The amount of wear particles generated by these processes should be small when the resistance of the material against interprenetration of the asperities is high and when a high yield stress (resistance to plastic deformation) and a high work hardening capacity exists. The additional process of microcracking, which may only occur when the material is relatively brittle, is preceded by the development of individual cracks on the abraded surface. These cracks can propagate until wear particles break completely out of the surface without notable plastic deformation. An example of the efficiency of the different micromechanisms of material removal is clearly seen on Fig. 24(a) and (b), showing a comparison between the abraded surfaces of the thermoplastic PA 6.6 matrix (with high toughness and strength) and the brittle EP resin. The ductile PA 6.6 material has a highly plastically deformed surface interspersed with many plowed furrows parallel to the wear direction; material sep-

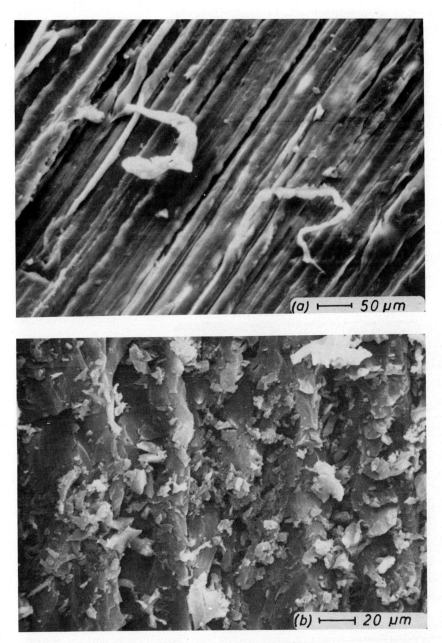

Fig. 24. Comparison of (a) a ductile PA 6.6 and (b) a brittle EP matrix worn by coarse abrasive paper (70 μm Al$_2$O$_3$).

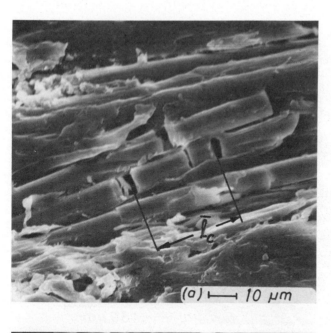

Fig. 25. Typical features on the surfaces of worn short fiber-reinforced thermoplastics. (a) Fibers oriented parallel to the sliding direction are frequently broken before being removed from the surface ($\bar{l}_c$ = average length of microcracked fiber pieces). (b) Size of an abrasive 70 μm Al_2O_3 grain (G) in comparison with fiber ends (F) broken out of the highly damaged surface (GF–PET, N direction).

aration and, thus wear debris formation, occurs in the form of long chips as a result of microcutting of the vaulted edges of the wear furrows by the sharp abrasive grains. The opposite, brittle polymer fragments due to multiple microcracking events, is observed on the worn surface of the epoxy.

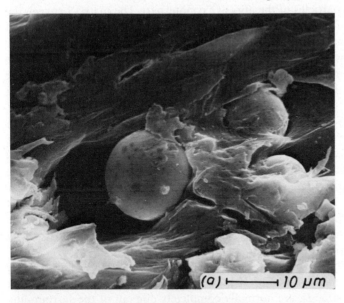

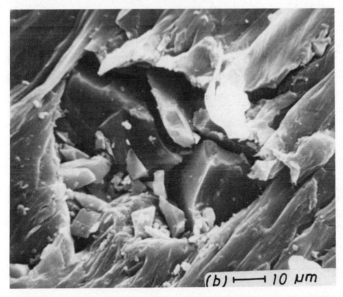

Fig. 26. Worn surfaces of glass sphere-filled thermoplastics characterizing the effect of sphere size on the mechanism of material removal by hard abrasive particles ($D = 70$ μm). (a) Small glass spheres ($d = 15$ μm) in PET. (b) Large glass spheres ($d = 70$ μm) in PA 6.

A more pronounced effect of the latter micro-wear mechanism is found, however, when the polymer matrices contain hard but brittle, short filler inclusions. Figure 25(a) illustrates for a parallel fiber orientation in thermoplastic PET that fibers in the wear surface are immediately broken into many short pieces when they come in contact with the harder particles of the abrasive counterpart. The matrix between the fibers is highly plastically deformed by an additional microplowing mechanism of the sharp abrasive grains before being cut and removed together with broken fiber ends. Figure 25(b) shows that a minor difference in wear mechanism exists if the fibers are normally oriented. Fiber pieces of a length-to-diameter ratio of about 1 are slicewise broken out of the surface and carried away together with plowed and cut matrix patches.

Concerning the short-fiber composites, this process is most destructive, i.e. wear rate is highest, the deeper the abrasive can penetrate into the surface (low matrix hardness and big abrasive particle size) and the more of the composite components are broken and immediately removed from the surface (favored by a high volume fraction of rigid short fibers). These observations explain the highest increase in wear rate with fiber volume fraction measured in the ETFE compounds. For the neat matrices, on the other hand, cracking mechanisms are only effective in PPS, thus having a higher wear rate than PET, PA 6 or PA 6.6, which are mainly worn by microplowing and microcutting events.

The different mechanisms of wear as a function of glass sphere diameter are illustrated on the SEM micrographs of Fig. 26. Glass spheres of diameters smaller than the abrasive grain size, D, are normally completely dug out of the surrounding polymer matrix [Fig. 26(a)]. If, however, the glass sphere diameter is bigger than D, only small parts of the glass spheres are destroyed by the harder abrasive grits, whereas the rest of them remain in the worn surface. Thus they can still contribute to the wear resistance of the composite [Fig. 26(b)]. Schematically, the dominant mechanisms of severe abrasive wear of the short filler-reinforced thermoplastics are summarized in Fig. 27.

Considering now the wear of the other group of composites, i.e. the continuous fiber-reinforced epoxies, the neat matrix behaves similarly to the brittle, thermoplas-

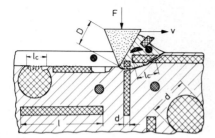

Fig. 27. Schematic illustration of severe abrasion mechanisms in reinforced thermoplastics as a function of filler orientation and geometry ($\tilde{l}_c$ = critical length of broken filler segments when entering the wear debris).

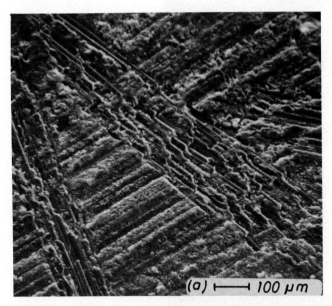

Fig. 28. (a) Low magnification micrograph of a glass fiber bundle-reinforced thermosetting resin worn by 70 μm Al$_2$O$_3$ paper. Although damage of the bundles has occurred, they still remain in the surface and contribute further to the wear resistance of the composite.

tic PPS. But, in contrast to the increase in abrasive wear rate with fiber loading in case of the short fiber-reinforced thermoplastics, the thermosetting systems exhibit a reduction in wear rate with V_F. This is assumed to be mainly due to the continuous fiber length and the arrangement of fibers in the form of big bundles. Individual abrasive particles, usually being much smaller in size than the bundles, can only destroy locally some of the fibers in the bundles when scratching the surface. As the fibers are very long, the major part of them still remains in the bundle (especially if a woven fabric structure predominates), thus contributing for a longer time to the wear resistance of the composite than individual, very short fibers can. The onservations are valid for the parallel, antiparallel and normal orientations and are very similar in both glass as well as carbon-fiber woven fabrics (Fig. 28).

Finally, it should be mentioned that the brittle cracking mechanism is completely eliminated if a continuous aramid woven fabric reinforcement is used. Though the surface of this material shows highly damaged regions of fibrillized and ruptured parts of the interwoven fiber bundles (Fig. 29), final material removal from the microstructure, even by a severe abrasive attack, seems to be most difficult. This is especially valid if the fiber bundles are oriented mostly normal to the wear surface. In fact, the lowest abrasive wear rates were measured for this type of material and orientation.

Fig. 28. (b) and (c) Antiparallel-oriented bundles of glass and carbon fibers, respectively, which were locally destroyed by the harder wear particles.

4.3. Counterpart effects

In addition to the influence of the microstructure of the composites investigated, the effective contributions of the three basic abrasive wear mechanisms to the

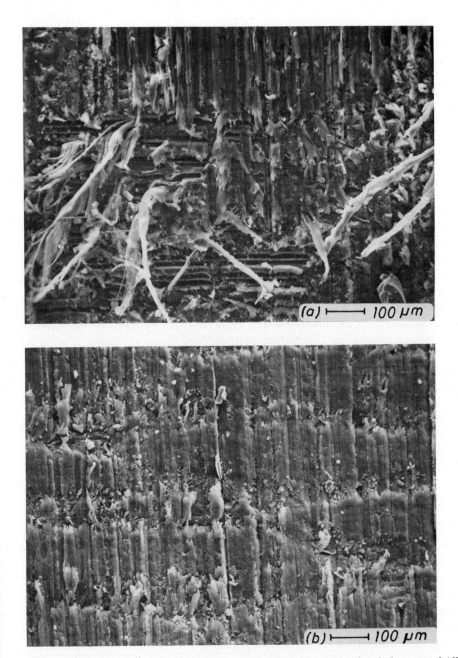

Fig. 29. SEM illustration of the fiber orientation effect on surface destruction during wear of AWF–EP by 70 μm Al$_2$O$_3$ paper. (a) AP/P-orientation and (b) normal orientation.

TABLE 5

List of the abrasive counterparts used (grinding papers) and the corresponding values of hardness and grain size

No.	Grinding material	Hardness, H_{abr} (MPa)	Grain diameter, D (μm)
1	SiC	2500	70
2a			7
2b	Al_2O_3	1800	70
2c			100
3	Flint	900	70
4	Glass	680	100

amount of material removed per sliding distance are a function of several properties of the counterpart asperities, in particular the shape, size, sharpness, hardness and density per unit area. In order to simulate the influence of these parameters on the wear of a particular composite material, different kinds of grinding paper can be used for the wear studies. In this part of the report, results of this kind of wear experiment carried out with thermoplastic matrix composites containing different kinds of reinforcement are discussed. The types and specifications of the different grinding papers chosen as counterface materials are listed in Table 5. Abrasive grains of four different hardnesses, H_{abr}, were used. As is known from similar studies on metals, the amount of wear should increase with increasing hardness of the abrasive counterpart [42]. An additional variation of the different abrasive papers is given by different grain sizes varying from $D = 7\,\mu$m up to 100 μm.

In Fig. 30, the wear rate of short fiber-reinforced polyethylene terephthalate is plotted against the grain size of the aluminium oxide particles of the abrasive papers used. For each of the three different sliding directions with respect to the fiber orientation in the specimen (AP = antiparallel, P = parallel and N = normal orientation) an increase of wear rate with the size of the abrasive grains was found, as expected from other studies [43]. There is only a very small effect concerning the sliding direction, showing higher values of wear rate for in-plane fibers (AP, P) compared with the case where fibers are oriented normal to the sliding plane. The orientation effect was not observed for the non-reinforced matrix and those filled with glass beads. That means that, in the latter cases, an isotropic wear behavior predominates, is opposed to the anisotropic condition of the fiber-reinforced material.

Figure 31 illustrates the effect of the shape of the reinforcement components (expressed by the aspect ratio l/d where l is the fiber length and d is the fiber diameter) on the wear rate of the materials under the influence of the three different abrasives already mentioned in connection with Fig. 30. For a given sliding direction and filler weight fraction, the wear rate increases slightly with the aspect ratio of the reinforcing element. In addition, the diagram indicates that the unreinforced poly-

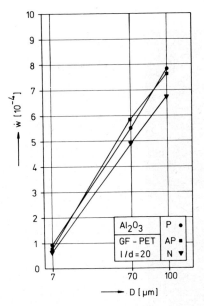

Fig. 30. Wear rate of GF–PET ($l/d = 20$) as a function of Al$_2$O$_3$ abrasive grain size, D, and the effect of fiber orientation.

mer exhibits the lowest wear rates under the influence of the three different abrasive counterparts.

Scanning electron micrographs taken from worn surfaces of the different com-

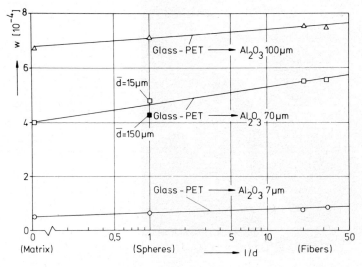

Fig. 31. Effect of E-glass filler inclusions and filler aspect ratio on the wear rate of thermoplastic polyethylene terephthalate against three Al$_2$O$_3$ papers of different grain size.

272

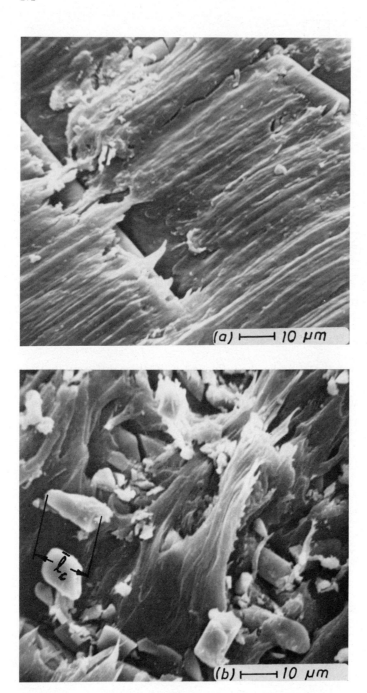

Fig. 32. Effect of abrasive Al_2O_3 particle size, D, on the appearance of worn GF–PET samples. (a) $D = 7$ μm, (b) $D = 70$ μm.

posite materials give an idea of the particular mechanisms involved in the wear process [Fig. 32(a), (b)]. Under the attack of very rough abrasive paper, the individual grains penetrate deeply into the surface of the material investigated, subsequently removing material from the surface by an extensive micro-plowing process. During this procedure, the polymer matrix is highly plastically deformed before being separated due to additional micro-cutting so that wear debris is formed. However, if there are additional rigid-filler particles (usually with about a twofold higher density compared with the polymer matrix) the amount of plastic deformation of the matrix is reduced at the cost of another wear micro-mechanism. Undeformed wear particles are formed due to micro-cracking of the brittle filler elements by the harder and sharper grains of the abrasive counterpart.

It can be assumed that the average distance, $\bar{l}_c$, of cracking events of in-plane fibers before being removed from the surface as a complete fiber segment is smaller than the diameter of the abrasive grains. This assumption is in agreement with observations of broken fiber pieces on the worn surfaces. Their length varied, in the cases of 70 and 100 μm papers, between 10 and 50 μm, with $\bar{l}_c = 30$ μm [cf. Figs. 25(a) and 32(b)]. If, however, the size of the part of the filler being momentarily in contact with the abrasive grain is smaller than the size of the abrasive particles, then the micro-cracking distance is a function of the penetration depth. This, in turn, is dependent on the relative hardness of filler and abrasive, the apparent pressure and the filler density per unit area. For the material filled with small glass beads, it can be estimated from the worn surfaces that the penetration depth is equal to the bead diameter [cf. Fig. 26(a)]. The same is valid for the N orientation of the fibers (estimated from the length of fiber segments having been cut slicewise in the case of normal fiber orientation (cf. Fig. 25(b); $\bar{l}_c \approx d$). The last mentioned conditions

TABLE 6

Comparison of wear data obtained in the P direction of selected PET composites against abrasive papers of different hardness and/or grain diameter. The symbols of each combination correspond to those used in Fig. 38(a)

Composite	Hardness, H (MPa)	Fracture energy, G_{Ic} (MPa m)	Reinforcement			$\dot{w} \times 10^{-4}$ and (μ)				
			l (μm)	d (μm)	$\bar{d}$ (μm)	SiC (1) [a]	Al$_2$O$_3$ (2b)	Al$_2$O$_3$ (2c)	Flint (3)	Glass (4)
GF–PET	140	5.69×10^{-3}	200	10		7.8 (0.68) ◑	5.5 (0.64) ◔	7.9 (0.72) ◐	6.1 (0.69) ○	6.5 (0.73) ◑
GS–PET	129	1.93×10^{-3}			15	7.1 (0.66) ▬	4.7 (0.62) ▣	7.1 (0.67) ▢	4.8 (0.67) □	6.2 (0.68) ▯
						4.2 (0.58) [b]				
GS–PET	139	1.79×10^{-3}				■				

[a] The numbers of the abrasives correspond to those in Table 5.

explain that there is (1) a minor, smaller amount of wear in the normal compared with the antiparallel or parallel fiber-orientation, (2) a lower wear rate for the glass-sphere filled material, and (3) a general increase in the wear rate with increasing hardness of the abrasive particles.

An additional explanation for the differences in the wear rate of the neat resin and the glass bead- and fiber-filled composites and for the different fiber directions arises from two other facts.

(1) A generally lower amount of volumetric wear of the neat matrix material as a result of the non-existent micro-cracking mechanism (opposite to the glass-filled cases).

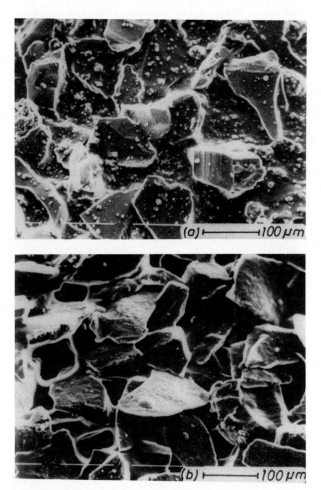

Fig. 33. Microscopic comparison of the grain density per unit area (f^*) of (a) Al_2O_3 paper (grain size 70 μm; $f^* = 0.48$) and (b) flint paper (grain size 70 μm; $f^* = 0.75$). The other abrasive papers used has f^* values of 0.70 (glass), 0.58 (Al_2O_3, $D = 100$ μm), and 0.65 (SiC), respectively.

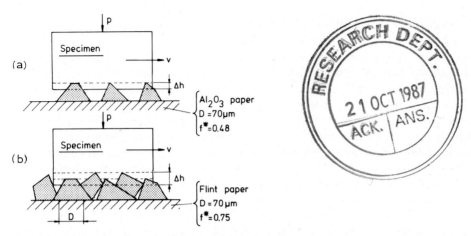

Fig. 34. Schematic of the attacking effectiveness of abrasive papers as a function of grain density per unit area, f^*. A relatively low value of f^* (a) leads to a lower amount of wear per unit time than a higher packing density of the abrasive grains (b).

(2) The lower exposed area of filler in the apparent contact area ($V_{\text{fexp}} = A_F/A$) in the cases of normal fiber orientation and in the glass-sphere filled case compared with the antiparallel in-plane fiber arrangement, although the filler volume fraction was supposed to be equal in all of these cases ($V_f = 18.5\%$; exposed area of filler: GF–PET, $l/d = 20$, x–y plane $V_{\text{fexp}} = 27\%$, x–z plane $V_{\text{fexp}} = 22\%$; GF–PET, $l/d = 32$, x–y plane $V_{\text{fexp}} = 26\%$, x–z plane $V_{\text{fexp}} = 19\%$; GS–PET, $\bar{d} = 15$ μm, $V_{\text{fexp}} = 17\%$; GS–PET, $\bar{d} = 150$ μm, $V_{\text{fexp}} = 25\%$).

Considering the effect of the hardness of the abrasive particles, a comparison of the data of the wear rate given in Table 6 on the basis of the same particle size shows that there is no systematic increase of the values with increasing abrasive hardness. It indicates that there seems to be another parameter of the abrasive papers which superimposes the effects of hardness and particle size. Analyzing the SEM micrographs of fresh abrasive papers carefully shows that there are differences in the density of the particles per unit area, $f^* = A_p/A$ (Fig. 33). Differences in the particle density per unit area can influence the wear process in two different ways. On the one hand, a higher density of the abrasive particles brings more of them in a position in which the attacking effectiveness of their sharp edges and tips is higher than in cases where particles are completely separated from each other (Fig. 34). The other point is that, in the case of low packing density, the time necessary to break up the fibers into segments of a length appropriate to remove the segments out of the surface is longer than in the densely packed condition [44].

4.4. Correlation of severe abrasion rate with the properties of the materials

Examining the correlation between the sliding wear rate and the properties of the wear couple [eqns. (2) and (11a)], i.e.

$$\dot{w} = f_{\text{ab}} \mu H^{-1} p \tag{12}$$

the predicted reduction of wear rate $\dot{w}$ with increasing material hardness (i.e. with V_f in case of fiber-reinforced plastics) at a given contact pressure, p, is not always true, even under sliding conditions against steel counterparts. The experimental results have shown that deviations from this rule occurred mainly above certain filler volume fractions V_f (V_f^*; cf. Fig. 17) and/or if the applied pressure, p, reaches some critical value (cf. Fig. 9). It was quoted that deviations in the wear data from the trend normally expected are the result of changes in the effective wear mechanisms, in particular a transition from adhesive and mild abrasive wear to severe abrasion, characterized by a remarkable amount of brittlely fractured surface fragments. The contribution of the latter mechanism, w_f, to the total wear of the composite should be the higher, the more the critical pressure (p_{crit}) for cracking brittle particles out of the surface is exceeded. It can be estimated from fracture mechanical considerations [45,46] that p_{crit} is a function of the material's hardness and fracture toughness (or fracture energy), i.e. both material properties should have an influence on this transitional wear behavior

$$p_{crit} \propto \frac{K_{IC}^2}{H} \tag{13}$$

where K_{IC} is the fracture toughness of the material tested and H its hardness. For a given apparent contact area, A, the value of p_{crit} is locally reached earlier the higher the load, F, thus the apparent contact pressure, p, and the lower the effective contact area, A_{eff}, according to

$$p_{eff} = p \frac{A}{A_{eff}} \tag{14}$$

This area should, on the other hand, be lower the higher the hardness H, and the higher the roughness (or the sharper the tips) of the hard counterpart asperities. This means that the probability that p_{eff} reaches p_{crit} (and micro-cracking events occur) is especially given by a material having high hardness and simultaneously low fracture toughness and being worn by a very coarse abrasive partner (Fig. 35). Schematically, the possible courses of $\dot{w}$ vs. H as a result of changes in the wear mechanisms are represented in Fig. 36. Transition II in this diagram indicates how a simple increase in p (for a material with given hardness H) may cause an over-proportional change in wear rate. The case of an increase in H at constant pressure p yielding a reduction in p_{crit} (from the condition of $p < p_{crit}$ to $p > p_{crit}$) associated with additional wear by micro-cracking events, is represented by transition I. If both an increase in H, for example by additional fibers in a polymer matrix, and an enhanced load, F, at a given counterpart roughness become effective, the probability for wear dominated by micro-cracking mechanisms should be highest. As a very demonstrative example for this transitional behavior, Fig. 37 illustrates a comparison of two wear furrows produced on polished surfaces of 18.5 and 39 vol.% GF–PET, respectively, by scratching with a diamond needle (tip angle 120°) under different

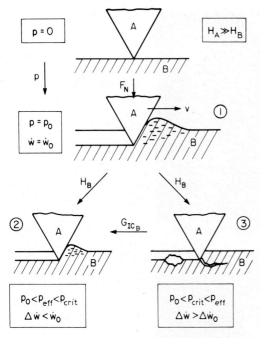

Fig. 35. Schematic representation of surface deformation and different wear mechanisms as a function of pressure, hardness of the materials and fracture energy.

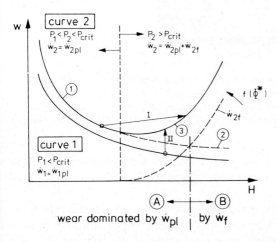

Fig. 36. Possible courses of abrasive wear rate $\dot{w}$ vs. hardness H of the material to be worn. Curve 1 represents the well known reduction of wear rate with hardness, whereas curve 2 shows a change in this trend as soon as wear by microcracking ($\dot{w}_f$) becomes effective. Curves 1–3 correspond to the numbered parts of Fig. 35.

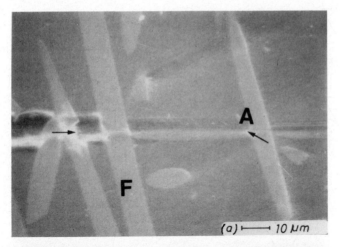

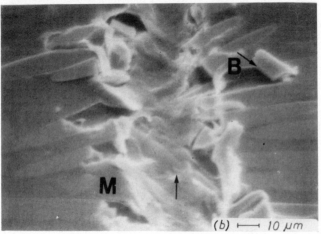

Fig. 37. Comparison of wear furrows on polished GF–PET surfaces as produced by scrartching with a diamond needle. The arrow indicates the sliding direction (AP). (a) 18.5 vol.% GF–PET, load 25 p, showing at (A) hinderance of matrix plowing by the fibers (F), i.e. $p < p_{crit}$. (b) 39 vol.% GF–PET, load 50 p, with a high contribution of lateral fiber breakage (B) out of the surrounding matrix (M), i.e. $p > p_{crit}$.

load. Doubling the load from $25p$ to $50p$ results in an increase in volumetric wear by a factor of about 40 as a result of extensive fiber breakage when a certain load level is exceeded.

The latter type of surface damage was also observed in multiple form on the worn surfaces of all the composites tested in this study against coarse abrasive papers. Under these circumstances, the most effective mechanism of material removal is a spontaneous cracking of brittle fiber inclusions and their subsequent transportation out of the contact zone by the hard abrasive grains. Based on Hornbogen's [47] and Zum Gahr's [45] approaches to this problem in metals and a more general discussion on wear and fracture mechanics by Rosenfield [48], a simplified correlation of material properties and wear rate as a result of two separate contributions can be established in the form [11,12]

$$\dot{w} = \dot{w}_{pl} + \dot{w}_f \tag{15a}$$

$$\dot{w} = \left(k^*_{pl} + k^*_f \right) p \tag{15b}$$

$$\dot{w}_s = \frac{\psi^*}{H^a} + \frac{\Phi^* H^b}{G^c_{IC}} \quad \text{with } \Phi^* > 0 \text{ if } p > p_{crit} \text{ and } \Phi^* = 0 \text{ if } p < p_{crit} \tag{15c}$$

where ψ^* is equal to the wear coefficient k [$\psi^* = k = \mu f_{ab}$ in accordance with eqns. (4) and (11)] and Φ^* is a characteristic term of the abrader/wear surface combination. As a first-order approximation, the exponents a, b, and c can be assumed to be 1, 1/2 and 1, respectively. p_{crit} is the threshold value above which wear particles are formed by microcracking. If it is drastically exceeded by p, the second term of eqn. (15c) can become of much more importance with respect to the final specific wear rate of the material than the first (cf. Fig. 36).

The expression of the abrasive wear coefficient Φ^* is basically a combination of geometrical and frictional properties of the wear partners, but should also include some information about the probability that wear particles are formed by microcracking. The latter is especially considered in Zum Gahr's idea for the abrasive wear of metals [45] and, after appropriate modifications, his approach can also be made applicable to polymer composite materials

$$\Phi^* = \left(H_{abr} \right)^\alpha D^\beta f^* \mu^\gamma \Omega^* \tag{16}$$

where the first three terms are related to the abrasive counterpart (H_{abr} = hardness, D = grain size, f^* = particle density per unit area). The frictional coefficient, μ, is a combined term of both partners and the factor Ω^* (= probability factor for microcracking) mainly depends on the properties of the composite material investigated and on the loading conditions

$$\Omega^* = V_{f_{exp}} \left(1 + \frac{2\bar{l}_c}{d} \right) \left[1 - \exp\left(-\sqrt{p/p_{crit}} \right) \right] \tag{17}$$

Assuming that, for the given testing conditions, $p \gg p_{crit}$, the second bracket of eqn. (17) approaches 1 and some simplified version of Φ^*, or Ω^*, can be identified as

$$\Phi^*_1 = \Omega_1 \propto \left(H_{abr} \right)^{1/2} Df^* \mu^2 V_{f_{exp}} \left(1 + \frac{2\bar{l}_c}{d} \right) \tag{18a}$$

(with $\alpha = 1/2$, $\beta = 1$, and $\gamma = 2$) or

$$\Phi^*_2 = \Omega_2 \propto D\mu^2 V_f \left(1 + \frac{2\bar{l}_c}{d} \right) \tag{18b}$$

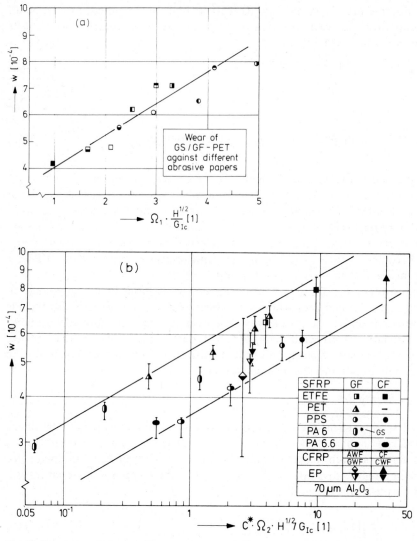

Fig. 38. Correlation of dimensionless wear rate, $\dot{w}$, of different composites with their fracture energy, G_{IC}, their hardness, H, and the modified abrasive wear coefficients, Ω_1 and Ω_2 [according to eqns. (18) and (19)]. (a) GF/GS–PET against various abrasive papers of different hardness and/or grain size. Symbols of each combination correspond to those used in Table 6. (b) All other composites tested against 70 μm Al$_2$O$_3$ paper. Factor C^* represents those counterpart parameters which were held constant (f^*, H_{abr}), i.e. $C^*\Omega_2 = \Omega$. (Values of Ω varied between 0.587 and 0.850.)

with the prerequisite that H_{abr} and f^* are constant and $V_{f_{exp}} \approx V_f$. In this form, both expressions, Ω_1 as well as Ω_2, can be considered as helpful quantities for estimating the effect of composite and counterpart parameters on the wear rate under severe abrasive conditions. As the latter is mainly characterized by the second term of eqn. (15c), a correlation of the form

$$\dot{w}_s \propto \Omega \frac{H^{1/2}}{G_{IC}} \tag{19}$$

should lead to a good agreement between the experimental wear data and the theory (corresponds to region Ⓑ on Fig. 36). In fact, this is actually found for the results discussed in Sect. 4.3 (with $\dot{w}_s \propto \Omega_1 H^{1/2}/G_{IC}$), as well as the results reported in Sect. 4.2 (with $\dot{w}_s \propto \Omega_2 H^{1/2}/G_{IC}$). Both relationships are represented in Fig. 38(a) and (b). But it should be noted that the results obtained with the neat matrices and those obtained with the composites against very fine abrasive paper (7 μm Al_2O_3) were not included. Since, in both cases, it is more likely that wear follows the mixed expression of eqn. (15c) rather than being dominated by its second term, the assumptions made for the above approach are no longer quite acceptable here.

5. Concluding remarks

A recently published review by Lancaster [25] on current achievements and future prospects of composites for increased wear resistance identifies in a systematic way how reduction in wear rate of composites against steel surface is achieved by a variety of microstructural parameters. Considering his conclusions, as well as the results obtained in this study on material properties in their correlation with wear, the following rules can be derived for a systematic development of a polymer composite with generally good wear resistance.

(1) The polymer matrix should possess a high resistance against viscoelastic/plastic deformation, which is usually the case if the material has a high hardness, H, and yield stress, σ_y.

(2) At the same time, a tendency of the polymer to deform plastically with a high work to rupture, $\sigma_f \epsilon_f$, is required.

(3) In addition, a combination of low frictional coefficient, μ, and

(4) a high thermal stability are desirable.

(5) For steady-state sliding wear conditions, e.g. against steel counterfaces, the addition of short or continuous fibers results in improvements to mechanical properties and significant reductions in wear. The latter is mainly due to preferential load support of the reinforcement components, by which the contribution of abrasive mechanisms to the wear of the material is highly suppressed.

(6) This is especially effective if the structure of the fiber arrangement yields a smooth wear surface topography.

(7) Ductile fiber materials are superior to brittle ones, especially if they possess a low frictional coefficient and low abrasiveness to the steel counterpart.

282

(8) On the other hand, a certain abrasiveness can be beneficial as long as a smoother steel surface is generated and thus wear is reduced.

(9) The formation of third-body films either as a result of material transfer during wear or by the introduction of solid or fluid lubricants usually favors the wear endurance of the composite.

(10) The addition of lubricants, on the other hand, degrades those mechanical properties of the material which are of special importance if wear is of a severe abrasive type, e.g. as a result of sliding against grinding paper containing very hard abrasive grains. Under these circumstances, it is necessary that,. in addition to the requirements mentioned above, the property ratio $H^{1/2}E/K_{IC}^2$ of the composite material with a given fiber volume fraction V_f is minimized.

(11) With respect to the reinforcement structure, it seems to be promising that the ratio of the asperity size, D ($\hat{=}$ size of the abrasive grains), to the dimensions of the filler components, d, is small, that the fibers are continuous in length and are interwoven rather than unidirectionally oriented. An orientation of tough fiber bundles normal to the sliding interface is very effective in hindering of the most destructive microcracking mechanism, which acts during severe abrasive wear loading conditions.

Usually, all these requirements cannot be fulfilled simultaneously, either as a result of the particular choice of the composite composition or other necessities, for instance with respect to a high load-carrying capacity of the material. The latter are at least of the same importance for material selection, but are not always in agreement with those for optimum wear resistance. The solution to the problem usually ends in a compromise, which combines the most important factors of both requirement profiles. As a final result of all these general conclusions, in Fig. 39 a hypothetical microstructure of a polymer composite material is suggested, which is assumed to have both a generally high resistance against different wear loadings and,

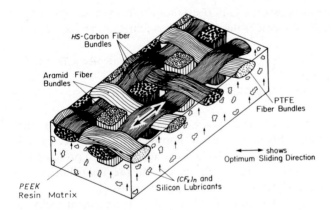

Fig. 39. Hypothetical model of a composite having 3D weave structure of different fiber materials in a modified PEEK matrix, which is assumed to have very high load-carrying capacity and a generally high wear resistance.

simultaneously, a good load-carrying capacity. It consists of

(I) a three-dimensional weave of high-strength carbon fibers, providing (a) high load-carrying capacity [49], (b) less sensitivity to counterface roughness when compared with high-modulus carbon fibers [25], and (c) higher resistance against severe abrasion relative to a unidirectional fiber orientation;

(II) additional aramid fiber bundles, in-plane perpendicular to the sliding direction and normal to the wear interface, acting mainly as crack-stopping elements under very abrasive wear loading;

(III) PTFE in filamentary form as part of the structure of the reinforcing fabric yields an easy formation of a third-body film when sliding against steel counterparts. The film cushions the intermediate spaces between the other fiber bundles and the counterpart asperities, smooths the wear surface topography, and thus reduces frictional coefficient and wear rate;

(IV) a thermoplastic polyetheretherketone (PEEK) matrix, which combines high strength and modulus with excellent toughness and thermal stability [50], even under wear loading [51]. The interspersion of the matrix with graphite fluoride, $(CF_x)_n$, [25] and silicon lubricants provides an optimum in frictional conditions, thus reducing wear still further.

Similar, but less complex, structures have been tested under various conditions in recent years [49,52–56] and have proved quite satisfactory in comparison with other composite systems. It can be assumed that the hypothetical structure suggested here, which combines the advantages of simpler microstructures, exhibits even better properties and thus can be considered as a further step in the direction of a maximum limit of general wear resistance attainable with an organic-based composite. Unfortunately, a material which preserved all these properties does not yet exist, so that final evidence for the hypothetical assumptions cannot be furnished at present.

List of symbols

A	apparent contact area (mm^2)
A_{eff}	effective area of contact during motion (mm^2)
A_F	exposed area of filler (mm^2)
A_p	exposed area of particles (mm^2)
a	exponent, factor
B	plaque thickness (mm)
b	exponent, factor
C_1, C_2, C^*	proportionality constants (variable)
c	exponent, factor
D	average size of counterpart asperities (μm)
d	fiber or sphere diameter (μm)
E	elastic tensile modulus (MPa)
F	load (N)

f_{ab}	abrasion factor
f^*	particle density per unit area (%)
G_{IC}	fracture energy of material tested (MPa m)
H	hardness of material tested (MPa)
H_{abr}	hardness of abrasive counterpart (MPa)
Δh	reduction in depth during wear (μm)
K_{IC}	fracture toughness of material tested (MPa m$^{1/2}$)
k	wear coefficient
k^*	wear factor (mm^3 N^{-1} m^{-1})
k_0^*	basic wear factor (mm^3 N^{-1} m^{-1})
L	sliding distance (m)
l	fiber length (μm)
$\bar{l}_c$	critical length of broken fiber segments during abrasive wear (μm)
Δm	mass loss during wear (g)
p	apparent contact pressure (MPa)
p_{crit}	threshold pressure for microcracking (MPa)
p_{eff}	effective normal pressure (MPa)
R_{tm}	average roughness of steel counterpart (between maximum peak and valley) (μm)
T	temperature (°C)
t	testing time (min)
V_f	filler volume fraction (%)
v	sliding velocity (m s^{-1})
$\dot{w}$	dimensionless wear rate
$\dot{w}_s$	specific wear rate (mm^3 N^{-1} m^{-1})
$\dot{w}_t$	time related (depth-) wear rate (μm h^{-1})
$\dot{w}^{-1}$	wear resistance
α	exponent
β	exponent
γ	exponent
ϵ_f	strain at fracture
ξ	exponent
μ	frictional coefficient
Ω^*	probability factor for microcracking
$\Omega_{1,2}$	modified abrasive wear coefficients (variable)
ρ	density of material tested (g cm^{-3})
Φ^*	abrasive wear coefficient (variable)
ψ^*	wear coefficient
σ_f	fracture stress (MPa)
σ_y	yield stress (MPa)

Abbreviations

Al$_2$O$_3$	aluminum oxide
AP	antiparallel

AWF	aramid fiber woven fabric
CF	carbon fiber
CWF	carbon fiber woven fabric
EP	epoxy
ETFE	ethylene tetrafluoroethylene
GF	glass fiber
GS	glass sphere
GWF	glass fiber woven fabric
N	normal
P	parallel
PA	polyamide
PEEK	polyetheretherketone
PES	polyethersulfone
PET	polyethylene terephthalate
PPS	polyphenylene sulfide
PTFE	polytetrafluoroethylene
SEM	scanning electron microscope
SiC	silicon carbide
v/o	volume fraction (%)
w/o	weight fraction (%)
f( )	function of
3D	three-dimensional

Acknowledgements

I am grateful to Prof. R.B. Pipes, Center for Composite Materials, University of Delaware, U.S.A., for suggesting this field of study. The valuable discussions on this topic in recent years with Prof. E. Hornbogen, Institut für Werkstoffe, Ruhr-Universität Bochum, F.R.G., are also highly esteemed. Part of the work reported here was funded by the Ministerium für Wissenschaft und Forschung des Landes NRW, F.R.G. The author acknowledges materials supplied by E.I. Du Pont de Nemours Inc., Wilmington, U.S.A.; LNP-Corporation, Malvern, U.S.A.; Fa. A. Krempel, Stuttgart, F.R.G.; and Fa. MBB, Ottobrunn, F.R.G. Finally, I appreciate the help of my colleagues and students at the Ruhr-Universität Bochum, Technische Universität Hamburg-Harburg, F.R.G., and the University of Delaware, U.S.A.

References

1 P. Beardmore, J.J. Harwood, K.R. Kinsman and R.E. Robertsoin, Science, 208 (1980) 833.
2 E.A. Margus, Pumps, Pompes, Pumpen, 164 (1980) 202.
3 J.H. Mallinson, Proc. 1st Ann. Techn. Conf. Reinforced Plastics/Composite Inst., The Society of Plastics Industry, 1976, Session 15-A, pp. 1–6.
4 F. Gooch, Pumps, Pompes, Pumpen, 171 (1980) 567.

5 G.C. Pratt, in M.O.W. Richardson (Ed.), Polymer Engineering Composites, Applied Science Publishers, London, 1977, p. 237.
6 E. Hornbogen, Fortschr.-Ber. VDI-Z. Reihe 5, 24 (1976).
7 J.H. Mallinson, Chem. Eng., May (1982) 143.
8 K.C. Ludema, Chem. Eng., July (1981) 89.
9 E. Hornbogen, Metall (Berlin), 12 (1980) 1079.
10 K. Friedrich, Fortschr.-Ber. VDI-Z. Reihe 18, 18 (1984).
11 K. Friedrich, in D.W. Wilson, R.C. Wetherhold, H.M. Cadot, R.L. McCullough and R.B. Pipes (Eds.), Composite Design Guide, Vol. 4, University of Delaware Press, Newark, DE, 1984, Chap. 4.9.
12 K. Friedrich and J.C. Malzahn, Proc. Int. Conf. Wear Mater., Reston, VA, 1983, p. 604.
13 J.F. Archard, J. Appl. Phys., 24 (1953) 981.
14 J.C. Anderson, Polymer Materials for Bearing Surfaces, Selection and Performance Guide, National Centre of Tribology, Warrington, U.K., 1983.
15 J.K. Lancaster, Tribology, 12 (1973) 219.
16 LNP Internally Lubricated Reinforced Thermoplastics, Liquid Nitrogen Processing Corporation Bull. 254–782, Malvern, PA, 1982.
17 J.K. Lancaster, J. Phys. D, 1 (1968) 549.
18 K. Tanaka, J. Lubr. Technol., 10 (1977) 408.
19 B. Briscoe, Tribology, 8 (1981) 231.
20 G. Erhard and E. Strickle, Kunststoffe, 62 (1972) 2, 232, 282.
21 H. Voss and K. Friedrich, Proc. Int. Conf. Wear Mater., Vancouver, 1985.
22 H.M. Hawthorne, Proc. Int. Conf. Wear Mater., Reston, VA, 1983, p. 576.
23 N.-H. Sung and N.P. Suh, Wear, 53 (1979) 129.
24 H. Czichos, Wear, 88 (1983) 27.
25 J.K. Lancaster, Proc. Int. Conf. Tribology in the 80's, NASA, Conf. Publ. 2300, Vol. 1, Cleveland, OH, 1983, Session 1–4, p. 333.
26 S.V. Prasad and P.D. Calvert, J. Mater. Sci., 15 (1980) 1746.
27 T. Tsukizoe and N. Ohmae, J. Lubr. Technol., 10 (1977) 401.
28 T. Tsukizoe and T.N. Ohmae, Proc. 4th Natl. 1st Int. Conf. Composite Mater., Centro Materiali Compositi, Naples, 1980, Paper 11.
29 B.J. Briscoe and D. Tabor, Br. Polym. J., 3 (1978) 74.
30 S.B. Ratner, in D.I. James (Ed.), Abrasion of Rubber, McLaren, London, 1967.
31 S.B. Ratner, I.I. Faberova, O.V. Radyukevich and E.G. Lur'e, Sov. Plast., 12 (1964) 37.
32 J.K. Lancaster, Proc. Inst. Mech. Eng., 183 (1968–69) 98.
33 J.M. Thorp, Tribology 4 (1982) 59.
34 J.P. Giltrow, Wear, 15 (1970) 71.
35 M.A. Moore, in D.A. Rigney (Ed.), Fundamentals of Friction and Wear of Materials, American Society for Metals, Metals Park, OH, 1981, p. 73.
36 K.H. Zum Gahr, Met. Prog., 9 (1979) 46.
37 M.A. Moore and F.S. King, in K.C. Ludema, W.A. Glaeser and S.K. Rhee (Eds.), Wear of Materials, ASME, New York, 1979, p. 275.
38 K. Friedrich, Proc. Int. Conf. Wear Mater., Vancouver, Canada, 1985.
39 K.H. Zum Gahr and D. Mewes, in K.C. Ludema, W.A. Glaeser, S. Bahadur, S.K. Rhee and A.W. Ruff (Eds.), WEar of Materials, ASME, New York, 1983, p. 130.
40 M. Cirino, private communication, 1984.
41 G.P. Tilly and W. Sage, Wear, 16 (1970) 447.
42 J. Larsen-Basse and B. Premaratne, Proc. Int. Conf. Wear Mater., Reston, VA, 1983.
43 J.C. Roberts and H.W. Chang, Wear, 79 (1982) 363.
44 K. Friedrich and M. Cyffka, Wear, 103 (1985) 333.
45 K.H. Zum Gahr, Fortschr. Ber. VDI-Z. Reihe 5, 57 (1981).
46 J. Lankford, J. Mater. Sci. Lett., 1 (1982) 493.
47 E. Hornbogen, Wear, 33 (1975) 251.

48 A.R. Rosenfield, in D.A. Rigney (Ed.), Fundamentals of Friction and Wear of Materials. American Society for Metals, Metals Park, OH, 1981, p. 221.
49 M.N. Gardos and B.D. McConnell, Development of a High Load, High Temperature Self-Lubricating Composite, Parts I–IV, ASLE Prepr. 81-3A-3–81-3A-6 (1981).
50 F.N. Cogswell, SAMPE Q., 7 (1983) 33.
51 M.P. Wolverton, J.E. Theberge and K.L. McCadden, Mach. Des., 2 (1983) 111.
52 J.K. Lancaster, ASTM-STP 769, (1982) 92.
53 T.O. Bautista, Proc. 35th Annu. Tech. Conf. Reinforced Plastics/Composites Inst., The Society of Plastics Industry, Session 5B, 1980, p. 1.
54 K.E. Jackson, NASA Tech. Pap. 2262, Tech. Rep. 83-B-7, (1984).
55 J.C. Anderson, Tribology, 10 (1982) 255.
56 M.N. Gardos, Tribology, 10 (1982) 273.

Chapter 9

Mild Wear of Rubber-Based Compounds

JAN-ÅKE SCHWEITZ and LEIF ÅHMAN

Department of Materials Science, Uppsala University, S-751 21 Uppsala (Sweden)

Contents

Abstract

Some fundamental mechanical properties of rubber are discussed, with particular emphasis on surface mechanics and the strong dependence of rubber friction on mechanical bulk properties. Theories of fracture and tearing are briefly reviewed. Parameters influencing the wear process are discussed for various set-up conditions (rubber specimen, counterface, running conditions, environment). The resulting wear process is divided into three phases of wear: (1) predetechment phase (deterioration of material strength), (2) detachment phase, and (3) postdetachment phase (transportation of wear debris). Deterioration effects occurring during the predetachment phase are discussed, with particular reference to modified surface layers, ageing, fatigue, crack formation and growth. The micro-mechanical detachment events are described in terms of a brittle-fluid mechanism spectrum and the formation of various types of wear patterns is discussed. It is noted that the mode of transportation of wear debris in the contact region may influence the friction and wear properties of the system considerably and the transportation process is subdivided into various modes of loose debris transportation, transfer to counterface, and back transfer. Some special wear patterns are discussed in order to elucidate the relationships between operating wear mechanisms and resulting wear patterns. Finally, a few suggestions are made concerning important fields of future rubber wear research.

1. Introduction

What is included in the term "wear of rubber" and, more specifically, what is ment by "mild" wear of rubber? What features, other than a lower wear rate, distinguish mild wear of rubber from severe wear? And why pay special attention to mild wear cases; would not more be gained by solving problems concerning severe wear?

The OECD Committee on Terms and Definitions in Tribology has agreed on the definition of wear [1] as "the progressive loss of substance from the operating surface of a body as a result of relative motion at the surface". This definition is in agreement with intuitive notions as well as common experience. It should be noted that the term "relative motion" includes not only sliding motion parallel to the surface, but also motion perpendicular to it. The latter type of motion is particularly important in many cases of wear of rubber, where periodic or intermittent loading of mainly compressive character generates physical or chemical deterioration processes, eventually resulting in mass loss from the surface. It should also be pointed out that, although central in any definition of wear, the actual mass loss events are merely a minor part of the whole wear process. The detachment mechanisms are indeed important elements of the process, but usually they represent neither its initial nor final steps. A wear process, especially one of mild wear, may have been initiated by strain-induced deterioration of the material long before the actual mass loss events take place. And once detached, the loose wear debris in the contact region may be of crucial importance to the evolution of the wear process and to the resulting wear rate. Hence, the "predetachment" and "postdetechment" phases are clearly as important to the wear process as the detachment mechanisms themselves.

Now, which are the distinguishing features of *mild* wear of rubber? Obviously, a low enough wear rate is the most determining feature. In practice, wear rates are often measured in terms of volume loss (or thickness loss) per unit time, e.g. mild wear may be defined as the wear regime $\leq 10^{-2}$ mm h^{-1}, which is a normal rate interval for tyre wear. From a scientific viewpoint, however, it is often better to express wear rates in terms of mass loss per unit contact surface and unit movement, e.g. mild rubber wear occurs in the interval $\leq 10^2$ μg cm^{-2} m^{-1}. Naturally, wear rate definitions can be further elaborated in terms of linear intensity, specific intensity, etc. [2].

The mechanisms of mild wear differ from those of severe wear mainly as regards the importance of the predetachment phase. A slow rate of material removal allows time-consuming processes like ageing and fatigue to play important roles. All mechanisms of severe wear (micro-cutting, micro-tearing, roll formation etc.) may also be present, to various degrees, in mild wear cases, but they occur under conditions and circumstances which have been modified during the predetachment phase.

As to the question of why mild forms of wear are particularly interesting to study, the answer is obvious. In practice, most cases of rubber wear are mild. In severe cases, alternative wear materials, or alternative designs, are chosen in order to reduce the wear to acceptably low rate levels. Furthermore, from an analytic point of view, it is often easier to identify and study fundamental wear mechanisms under conditions of mild wear when the picture is not obscured by, for instance, the serious thermal effects usually present in more severe cases.

A few comments concerning the use of the terms "compound" and "composite" are in order. Commonly, a rubber compound is considered to be an intimate union of microscopic constituents, i.e. the polymeric matrix with filler particles, while a rubber-based composite usually denotes a union of macroscopic constituents as, for instance, in a cord-reinforced tyre. In other fields of materials science, for example metallic materials, the use of terminology is sometimes the opposite. Be that as it may, a filler-reinforced rubber compound is obviously an intimate union of clearly distinguishable components and as such it can be considered a member of the composite family.

2. Mechanical behaviour of rubber

2.1. Visco-elasticity and rubbery behaviour

The most spectacular mechanical properties of elastomers are their extremely high fracture strains of several hundred percent and their exceptional resilience. These rubbery properties arise from a particular combination of molecular interactions, which allows the material to behave both as a solid body and in a fluid-like manner at the same time. In unvulcanized rubber, the interactions between the molecules are weak and the material behaves in a viscous manner. In the vulcanization process, strong intermolecular cross-links are formed, and the material attains a more solid nature; the more the density of cross-links is increased, the stiffer the rubber becomes. The molecular chains are extremely long and entangled. The links between the molecular segments, however, can be turned in such a manner that the angles of the covalent bonds are unchanged, but the linear extension of the molecule is greatly increased in the direction of an applied tensional stress. This is basically a statistical process and the elastic forces seeking to restore more probable molecular configurations are of an entropic nature, rather like the pressure exerted by a gas. In thermodynamical terms, the line tension, J, can be expressed as (see, for example, ref. 3)

$$J = \left(\frac{\partial U}{\partial L}\right)_T - T\left(\frac{\partial S}{\partial L}\right)_T \qquad (1)$$

The fact that the entropy term dominates over the internal energy term for strains in the interval 100–400% leads, in some cases, to thermomechanical properties opposed to those of other materials. For instance, if rubber is heated in a strained state, it gets stiffer, whereas steel gets softer. Thus, an extended rubber band will tend to shorten when heated, and elongate when cooled. This is called the Gough–Joule effect [4,5]. Furthermore, the elasticity of rubber is very strain-rate sensitive. If the strain rate applied requires the molecular segments to move faster than their natural retarded mobility allows, the viscous resistance will cause the rubber to stiffen. During the deformation, heat will be generated by internal friction in the structure, and the temperature will increase. In unloading, some heat will be reabsorbed. A load cycle always exhibits a more or less pronounced hysteresis effect. Stress-softening due to scission or rearrangement of molecular chains, fracture of weak entanglements, and structural changes of carbon aggregates have been reported [6,7]. The effect decreases as the number of stress cycles increases. Thus, the hysteresis effect is pronounced during the first stress cycles but gradually vanishes (see Fig. 1). For non-crystallizing rubbers, the generation of deformation heat is most pronounced in a certain strain-rate interval, characteristic of the particular rubber compound. Hence, mechanical vibrations will be more effectively damped in a certain frequency interval than in others. This rate sensitivity of rubber, caused by its fluid–solid double nature, is termed *viscoelasticity*. It should be emphasized that rubber does

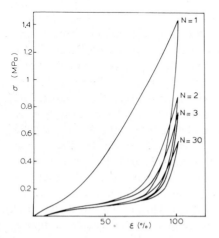

Fig. 1. Influence of the number of stress cycles on the hysteresis of NR filled with 40 phr carbon black [7].

not obey Hooke's law and that therefore no simple elastic modulus can exist for rubber.

For large strains, some types of rubber may crystallize to a certain degree (see, for example, ref. 3). This means that the rubber transforms into an ordered, solid phase of markedly lower energy and tougher mechanical properties. Crack propagation in natural rubber (NR), for instance, is hampered by crystallization effects at the crack tip, where the local strains are extremely large. In crystallizing gums, the hysteresis effect is mainly due to crystallization and not to internal viscosity.

2.2. Bulk and surface mechanics

One striking difference between elastic–plastic materials and viscoelastic materials is their different responses to a *tangential surface force* (i.e. a shearing force attacking the surface of the body). Suppose that a surface region of a body is subjected to a tangential surface force. Furthermore, suppose that the initial force is zero and that it then gradually increases in strength.

In a typical *elastic–plastic* material, for instance steel, the following will happen. The growing surface force will first generate purely elastic fields of stress and strain, which, in principle, are extended throughout the whole volume of the body. Hence, the initial elastic deformation is a typical bulk phenomenon. Eventually, as the surface force grows stronger, the yield limit of the elastic–plastic material is reached, whereupon it plasticizes. As the surface layer starts to plasticize, large amounts of energy are absorbed in a small volume (thin layer) near the surface and the elastic field component will not grow further; all energy will be expended (stored or dissipated as heat) by the growing plastic field component. Upon continued increase of the surface force, the elastic field energy will soon become negligible in comparison with the energy stored or dissipated by the plastic field in the surface layer. This is the reason why friction is considered to be a *surface phenomenon* in steel; the energy absorption mainly occurs in, or very close to, the contact surface. Figure 2(a) gives a qualitative idea of stress and strain versus depth below the surface.

In a *visco-elastic* material such as vulcanized rubber, the picture is completely different. This material does not exhibit a yield limit, i.e. the strain field will grow throughout the volume of the body until the fracture limit is eventually reached at some location. Consequently, in this material, the attack of a surface force results in a pronounced *bulk phenomenon*, Fig. 2(b). The bulk character of rubber friction is further enhanced by the damping effect previously mentioned, which causes the energy of a dynamic stress field to "leak out" gradually as heat. Of course, this behaviour does not prevent the occurrence of certain surface phenomena, such as viscous effects in a thin surface layer caused by rapid, sliding friction. In fact, surface phenomena of thermo-chemical origin turn out to be of great importance in many cases of mild wear, as we shall see later.

In conclusion, and somewhat idealized, we can make the following statement concerning the response of the two materials to the action of tangential surface forces: the plastic nature of steel will result in energy dissipation near the surface

294

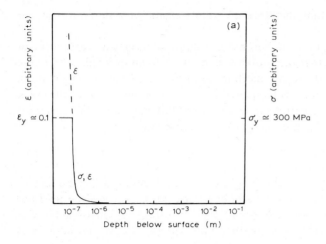

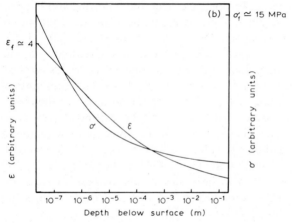

Fig. 2. Qualitative comparison of the stress and strain gradients below a surface subjected to a frictional force for (a) steel and (b) rubber.

(surface effect), while the visco-elastic nature of rubber results in energy dissipation in extended parts of the volume (bulk effect). This behaviour of rubber has several important consequences for the friction and wear processes.

Firstly, the coefficient of friction loses some of its significance as a characteristic parameter of interfacial sliding, since it is implicitly dependent on the volume and shape of the rubber specimen. Secondly, the surface free energy will also lose much of its usefulness as a mechanical parameter. It may always be related to the energy of the unsaturated bonds in the surface, but it is not so directly related to the crack formation energy as is the case in the classical Griffith crack theory for high modulus, brittle materials [8]. Instead, the tearing energy, defined as the total strain energy dissipated per unit area of crack growth, emerges as a fundamental crack growth parameter. Finally, the bulk dependence of rubber friction and wear makes it possible to affect the wear behaviour of a rubber by changing its mechanical bulk properties through fibre reinforcement, e.g. cord reinforcement of a pneumatic tyre.

As mentioned earlier, this bulk dependence does not prevent the occurrence of certain surface phenomena causing alterations of the material in a relatively thin surface layer. These alterations are usually of a thermo-chemical origin, but are of great importance to the mechanical properties, e.g. the mild wear properties, of the surface layer. Hence, the wear properties of rubber can be said to be governed by its fundamental visco-elastic bulk properties, superimposed by surface effects of mainly thermo-chemical origin.

2.3. Fracture and tearing

The basic mode of mechanical failure in rubber is tensile rupture. Shear fracture cannot occur in rubber, simply because there are no natural shear planes in the molecular structure. Naturally, applied shear stresses can result in failure, but a close investigation will always reveal that the crack has propagated at right angles to the maximum, local tensile stress. Figure 3 shows a SEM micrograph of a strained rubber surface. The horizontal surface texture (the light streaks), indicating the principal axis of straining, has been produced by mild etching with argon ions in

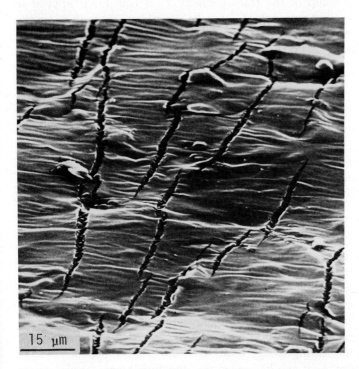

Fig. 3. Ion etched surface of a strained rubber specimen showing cracks at right angles to the direction of straining [74].

vacuum. It is seen that the microcracks are about 1 μm wide and oriented traversely to the applied tensile stress. Plastic deformation is prevented by the strong cross-links between the molecules in vulcanized rubber. Thus, in this sense, mechanical failure in rubber may actually be caused by brittle fracture, as indicated by the appearance of a relaxed fracture surface.

The classical theory of crack propagation in hard, brittle materials is due to Griffith [8]. In this theory, the critical stress of crack growth can be expressed as

$$\sigma_G = \left(\frac{2\gamma_s Y}{\pi c} \right)^{1/2} = \left(\frac{G_c Y}{\pi c} \right)^{1/2} \tag{2}$$

where σ_G is the tensile strength of the material, γ_s is the surface free energy, Y is the elastic modulus, $2c$ is the crack length, G_c is the critical (fracture) value of the strain energy release rate defined by Irwin [9]

$$G_c = -2\left(\frac{\partial E}{\partial A} \right)_{\text{fract}} = 2T \; (= 2\gamma_s \text{ in the Griffith theory}) \tag{3}$$

E is the total strain energy of the system and A is the fractured surface. Equation (3) is also a defining relationship of the tearing energy T. If eqn. (3) is applied to imperfectly elastic, non-linear materials, G_c can be complemented with a term ϕ accounting for the work irreversibly dissipated by plastic deformation or internal friction in a region close to the crack tip [10]

$$G_c = 2\gamma_s + \phi = 2T \tag{4}$$

This surface energy is defined [11] as fracture surface energy. For glassy polymers, the values of the irreversible work, ϕ, are generally a factor 10^3 larger than the surface free energies, γ_s, which hence are of small, if any, importance to the tearing energy.

In highly elastic materials like rubber, the tearing energy can be calculated from a generalization of Griffith's formulation [12]

$$T = 2kcE_d \tag{5}$$

where k is an empirical, slowly varying, strain-dependent function, and E_d is the bulk strain-energy density. The critical tearing energy, T_0, related to naturally occurring flaws in the rubber has been shown to be a factor 10^2 greater than γ_s of rubber [13], and is not directly related to the macroscopic tensile strength [14].

In principle, the tearing energy must also be implicitly dependent on the volume and shape of the rubber specimen. As it turns out, however, several macroscopic methods to determine the tearing energy under different test geometries yield the same value [15]. Furthermore, it can be argued that the microscopic tearing events taking place in an abrasion process must, in practice, be fairly independent of the macroscopic volume and shape of the rubber specimen, even though the mechanical bulk properties *must* be taken into account.

3. Conditions and parameters influencing the wear process

Traditionally, the mechanisms of rubber wear are said to be four: abrasive wear, fatigue wear, wear by roll formation, and pyrolytic wear. This traditional classification, however, is of a phenomenological character; it is based on the appearances of the worn surfaces rather than on the fundamental physical and chemical mechanisms operating at a microscopic level. On closer study, it is found that each traditional category can be decomposed into a number of more basic mechanisms.

In Fig. 4, an attempt has been made to summarize the major factors and effects of importance to the wear process in rubber and their interdependencies. The figure is intended to illustrate the enormous complexity of a rubber wear system and to highlight the fact that a change in any parameter anywhere in this system will influence practically all other parameters of the system, directly or indirectly. In the figure, set-up conditions and wear process phases are accounted for separately. The set-up conditions are divided into external influences (environmental and running conditions) and materials and design factors (composition, properties and designs of the rubber specimen and its counterface). The phases of wear resulting from these set-up conditions are described in terms of three main stages involved in the loss of substance from the operating rubber surface: deterioration of material strength properties (involving no actual mass loss), detachment of wear particles, and transportation of wear debris in the contact region. In the present section, a brief survey is given of some of the most important set-up conditions and their influences on the wear process. In the subsequent section, the three phases of the wear process are surveyed in greater detail.

3.1. Rubber specimen

The basic mechanical, chemical and thermal behaviour of a rubber specimen is determined by its chemical composition, although its behaviour is also affected by many other factors (cf. Fig. 4). The type of polymer, or blend of polymers, of the rubber matrix determines some fundamental properties of the rubber compound, for instance its ability to strain-crystallize, but the general wear properties of a compound are usually even more influenced by the choice of curing system, various additives, type of filler reinforcement, etc.

The influence of filler reinforcement, particularly carbon black, has been investigated by many workers over the years and a comprehensive review of this complex subject has recently been published by Rigbi [16]. It is well known that carbon black fillers increase the hardness, tear strength, and abrasion resistance of the compound and that the reinforcing effect is determined by the activation energy, grade, density and dispersion of the black in the rubber matrix.

298

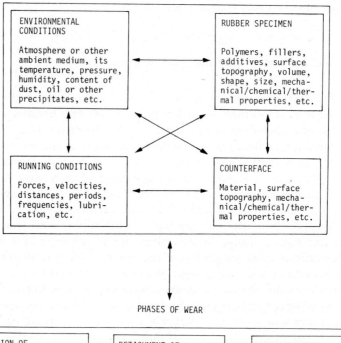

PHASES OF WEAR

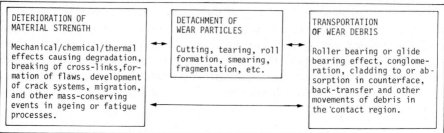

Fig. 4. Interdependencies in a rubber wear system.

Other additives of importance to the wear properties are anti-ozonants, anti-oxi-dants, anti-fatigue agents, light stabilizers etc. Naturally, these additives are of particular importance to wear mechanisms of a chemical or mechano-chemical nature, primarily ageing and fatigue. A recent article by Kuzminsky [17] reviews the fatigue mechano-chemistry and the influences of various additives.

Design factors, such as surface finish, fibre reinforcement, other load-carrying structures, volume, shape factor, etc., are of obvious importance to the overall wear properties of a rubber specimen. It would lead far beyond the scope of this article to discuss the influences of all these factors. The theory and applications of cord–rubber tyre composites have been rather comprehensively reviewed by Walter [18] and the influence of short-fibre reinforcement of elastomers has recently been reviewed by

Goettler and Shen [19]. It is quite clear that much work still remains to be done concerning the relationships between various types of fibre reinforcement and resulting wear properties. It should also be noted that no relationships of general validity exist between design factors and wear properties. A smooth surface topography, for instance, may be better or worse than a rough one, depending on the type of counterface, running conditions, ambient medium and temperature, and, of course, on the direction in which the wear process develops.

3.2. Counterface

To some extent, the wear process is influenced by the surface chemistry of the counterface, especially if it contains some constituent which is chemically aggressive to rubber. The chemical composition of the counterface may also be of some importance to the adhesive properties. Various types of adhesion components involved in the surface energetics of polymers have been reviewed and discussed by Lee [20]. Normally, however, the wear process is mainly affected by the physical and geometrical characteristics of the wear partner. Its hardness and roughness are of obvious importance to the removal of surface material, its thermal conductivity is of crucial importance to the accumulation of frictional heat in the contact region, and its ability or inability to capture or transmit wear fragments will determine the mode of transportation of wear debris.

3.3. Running conditions

Among the running conditions, we include parameters or conditions which are directly controlled by the user of the system, for instance load, velocity, running time or distance, frequency, cooling, lubrication, etc., but we do not include environmental conditions which would also prevail in the absence of the wear system.

The two most fundamental operating parameters are load and velocity. The normal load, or surface pressure, will determine the real contact area and the size and distribution of the deformed zones. The sliding velocity, in combination with the load, will determine the frictional force and coefficient of friction, which, in turn, determine the operating temperature. The latter will retroact on the coefficient of the continued friction and affect the material strength and wear properties as well. Some strength properties are, to a certain degree, also determined directly by the sliding velocity due to their strain rate dependence. The influence of various running parameters on the friction and wear of rubber have been treated by numerous authors over the years [e.g. refs. 2,21–26].

The operating temperature is, of course, not only determined by the generation of frictional heat, but is also affected by the ambient temperature, bulk hysteresis losses in the rubber specimen, thermal conductivity of the counterface, cooling arrangements, etc. The temperature in the contact region is one of the basic factors controlling wear. It has been found that the wear rate of pneumatic tyres, for example, increases by some 2% for each degree Centigrade temperature rise [25]. The well-known difficulty of finding reliable relationships between various running parameters and wear is, in many cases, due to the fact that the contact temperature has not been properly controlled or monitored during the experiments.

3.4. Environmental conditions

By environmental conditions we mean external influences that are not controlled by the user of the system, i.e. atmosphere or other ambient medium, its temperature, pressure, humidity, content of dust, oil or other precipitates, etc. These external parameters are very important to the wear process, particularly in cases of mild wear.

As further discussed in Sec. 4.1.2, oxygen and ozone have a direct effect on the strength properties during wear, oxygen by thermal or mechanical oxidative break-down of the hydrocarbon molecules and ozone by so-called ozone cracking due to rupture of strained double bonds. The latter effect is particularly marked at low tearing energies [27]. Ozone plays a major role in determining fatigue lives longer than one megacycle, whereas oxidation is the major cause of shorter fatigue lives [28]. As mentioned earlier (Sect. 3.1), both effects can be considerably reduced by the use of appropriate protective agents.

Relative humidity has a large effect on wear [29,30] and the varying tendency of different rubbers to absorb moisture can be a serious problem in comparative wear rate measurements, unless special arrangements are made [30].

4. Three phases of wear

As mentioned in Sect. 3, a rubber wear process can be subdivided into three stages: (1) deterioration of the material strength, (2) detachment of surface material, and (3) transportation of wear debris [31]. Of course, in a steady-state wear situation, all three phases occur simultaneously all over the contact surface. From the viewpoint of an individual wear particle, on the other hand, the three stages follow each other in a logical time sequence. They are not independent of each other, however, and they are all influenced by the set-up conditions, as illustrated in Fig. 4. In the following subsections, we will discuss them one by one.

4.1. Predetachment phase. Deterioration of material strength

4.1.1. Modified surface layers

Though facilitating wear, deterioration in itself does not yield mass loss. When exposed to wear, the surface material may become altered, resulting in a *modified surface layer* with properties quite different from those of the bulk. Such a layer is commonly acknowledged by people working with wear of steels [e.g. ref. 32] but has so far seldom been explicitly described for rubbers. For filled NR, Schallamach and others [29,33–35] have observed thin viscous layers which seem to cause decreased wear by smearing. The layers are suggested to be produced by oxidative and thermal degradation and to have properties similar to those of unfilled vulcanizates. The rate of chemical attack is accelerated when combined with generation of frictional heat, which accumulates near the surface due to the low heat conductivity of rubber. A four-zone layer model has recently been proposed by Cong et al. [36], but further work is clearly needed. It has also been noted that the electrical resistance decreases upon deformation, indicating a breakdown of the carbon black structure in a top layer of 0.1 mm thickness [21,37]. The strains and stresses in the surface region also affect migration of waxes, anti-ozonants, etc. from the bulk to the surface [17], as well as the surface temperature and the rate of oxygen access. With increased temperature, chemical reactions accelerate and oxidation, additional cross-linking, or decomposition may occur, depending on the environment, on the chemical composition, and on the nature of the polymer [35,38,39]. The modulus and hardness of the rubber also changes.

Hence, although bulk strength properties are necessary in describing the field of stress and strain in the bulk and at the surface, it may be misleading to use bulk data in the treatment of the micromechanisms of wear close to the surface.

4.1.2. Ageing and fatigue

The importance of the chemical degradation of rubber was early recognized and much work has been done, focused particularly on the role of oxygen and ozone [15,17,27–29,38–41]. In early work, fatigue was considered to be of an entirely chemical nature, but today it is well known to be a mechano-chemical process. There seems to be some confusion though, regarding the difference between ageing and fatigue [42], in fact some authors denominate ageing as "static fatigue". *Ageing* is a weakly time-dependent process causing gradual deterioration of the material properties due to chemical, thermal, and/or mechanical exposure under static conditions (note that creep does not necessarily involve ageing). If this deterioration is further enhanced by making the load cyclic or intermittent *with unchanged average value,* by definition we have a superimposed element of *fatigue.* The periodicity of a load may promote deterioration, for instance by hysteresis effects, by "pumping" effects influencing the access and partial pressure of oxygen and ozone, and by subjecting the material to a whole range of strains and strain rates.

Ageing is caused by reactions between oxygen or ozone and rubber leading to rupture of chains and cross-links [38,43]. Addition of anti-oxidants and anti-ozonants decreases the degradation rate, probably by neutralising the free radicals [17]. Ozone starts to attack stretched double bonds at a critical strain of about 3% [27,28,44], causing initiation of surface cracks. Low strains result in a few large cracks whereas high strains give many small cracks. The ageing process is accelerated by heat and light.

Fatigue is caused by a mechano-chemical process under dynamic load conditions. Ratner and Lure suggested the wear rate, $\dot{w}$, to depend on the activation energy, a_0, for oxidative degradation and on the applied load, L, according to [45]

$$\dot{w} \propto \exp\left[-(a_0 - kL)/RT\right] \tag{6}$$

i.e. the rate of chemical degradation increases in the presence of stresses. It has been suggested that fatigue includes changes in the cross-link density [46], but recent studies have shown that no major change in the network structure occurs [47,48]. Nor does ageing affect the fatigue strength [28]. The shape and amount of fillers have been reported to affect the resistance to mild wear [49] and fatigue [46]. This will be further discussed in the next subsection.

Today, most researchers [15,22,35,40,50,51] agree with Lake and Lindley [28] and Lake and Thomas [27] that fatigue is caused by mechano-chemical crack growth originating from small natural flaws present in all rubber vulcanisates. According to their model, the number of cycles, N, to failure is given by

$$N = \frac{1}{Bc_0(2kE_d)^2} \tag{7}$$

where c_0 is the size of the small natural flaws, E_d is the strain energy density, and B and k are constants.

4.1.3. Crack formation and growth

By introducing the concept of tearing energy, T, into fracture mechanics, it has been shown that, under dynamic stresses, cracks develop from naturally occurring flaws [15,27,28,40,44,52,53], the size of which has been estimated (perhaps overestimated) from eqn. (7) to be 40 ± 20 μm, almost irrespective of the type of polymer or rubber blend. The flaws are suggested to originate from voids, irregular cross-link density, stress concentrations around filler particles, clusters of sulphur or oxygen, and dust [17,48,53]. A transmission electron microscopy (TEM) study has indeed shown voids to form around the filler particles as a rubber specimen is elongated [54]. The number of voids was observed to increase with increasing strain. An example of a tearing event induced by a void is seen in Fig. 5. It shows a fracture surface from a fatigue test. A small, spherical void (~ 8 μm diam.), situated below the surface, has induced a micro-tearing event.

Investigations have shown that the correlation between tearing energy and crack growth is equal in tension and compression, hence confirming the idea that this is a fundamental material parameter [55,56].

The tearing energy was defined in Sect. 2. Experiments with tear test pieces of the trouser type and tensile strips have given the results summarized in Fig. 6 [27,28]. The figure shows the rate of crack growth in NR under repeated loading versus peak value of the tearing energy, T. For low values of T, the cracks grow at a constant rate, determined entirely by the ozone concentration, q (i.e. independently of the number of stress cycles), as

$$\frac{dc}{dt} = 0 \qquad T < T_{oz} \tag{8}$$

$$\frac{dc}{dt} = k_i q \qquad T \geqslant T_{oz} \tag{9}$$

where T_{oz} is the tearing energy corresponding to the critical strain for ozone cracking

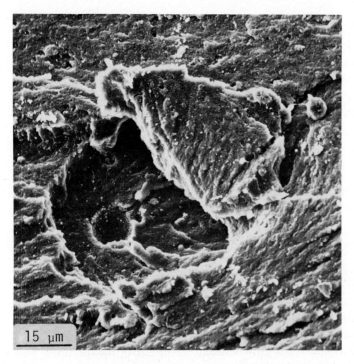

Fig. 5. Micro-tearing event induced by small, spherical void.

($\sim 3\%$, cf. Sect. 4.1.2). For NR, T_{oz} is about 10^{-3} J m^{-2}. At a critical value T_o, oxidative crack growth, promoted by the mechanical stresses [according to eqn. (6)], causes an abrupt increase in the crack growth rate. The value of T_o has been determined to be 40–80 J m^{-2}, depending on the cross-link density, in good

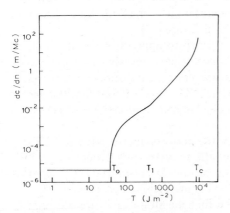

Fig. 6. Cut growth rate versus maximum tearing energy for NR under cyclic load [27].

agreement with theoretical predictions based on the assumption that T_0 is dependent on the strength properties of individual chains [15,27]. In the region just above T_0, the crack growth is linear in T. For $T = T_1$, a kink occurs in the curve of Fig. 6. The origin of the kink is not known, but a change from mainly oxidative to mainly mechanical rupture may explain this behaviour. Above T_1, the experimental data have been adapted to the expression

$$\frac{dc}{dn} = BT^\alpha \qquad (10)$$

where α is a parameter depending on the type of rubber, varying from 2 for NR to 4 for unfilled SBR and other synthetic rubbers. The constant B, the same as in eqn. (7), has been named the crack growth constant and is considered to be a fundamental material parameter, although its physical significance is as yet unclear. For instance, the units of B are not well-defined since, as α varies, so also does the dimension of B. It must therefore be kept in mind that eqn. (10) is a mathematical expression that has been adapted to experimental data and no physically well-founded theory exists today.

In practice, eqn. (10) is used in the whole range of T values, though truly significant in the region $T_1 < T < T_c$ only [27,52]. Eventually, for a high enough tearing energy, T_c, catastrophic crack growth results in rupture of the specimen. This final process is of a purely mechanical nature.

Figure 6 applies qualitatively to both non-crystallizing and crystallizing rubbers [15,22,28], but the crack growth rate is generally considerably lower in a crystallizing rubber. [For NR and SBR, the values of B in eqn. (10) are 10^{-14} m $(J\ m^{-2})^{-2}$ and 10^{-18} m $(J\ m^{-2})^{-4}$, respectively.] This is explained by the ability of some rubbers to form crystallites in areas of high strain. In a crystalline region, the crack growth rate is virtually zero. If the relaxation between the stress cycles is complete, the crystallites will melt and, on reloading, the crack grows until the rubber at the crack tip crystallizes. If the relaxation is not complete, the crystallites do not melt and no crack growth occurs upon reloading. The temperature in a crack tip region has been shown to be 40–50°C higher than the bulk temperature [17].

The results discussed above have been confirmed to apply in wear situations as well. Tests with a razor blade abrader [15,35,40,44] have revealed that the theory agrees well with the experimental data. For non-crystallizing rubbers, however, a constant misfit by a factor 0.63 has been reported [44]. It is believed that this factor is due to filler particles causing the cracks to deviate from a direction orthogonal to the surface.

In the razor blade test and other wear tests, the performance of NR and SBR are often surprisingly similar [35] in spite of the strain-crystallizing properties of NR, which should result in improved wear resistance due to crystallization in contact regions with extreme local deformation [50]. This anomalous behaviour is not yet fully understood, but one explanation may be that the strain rates are too high to allow NR to crystallize, thus eliminating the main difference between NR and SBR.

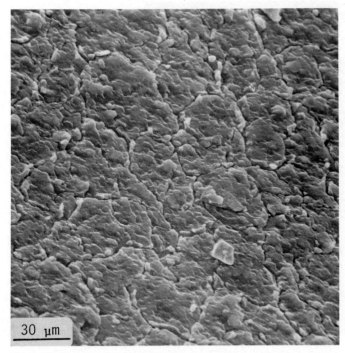

Fig. 7. Crack system in a NR–BR slurry pump lining.

In recent years, crack systems have been observed and studied in SEM [47,48,57–59]. Figure 7 shows an example of a network of fine cracks which has developed in the surface of a rubber lining in a slurry pump. The network is of an isotropic character, indicating a stress-free state in the surface.

Figure 8 shows a sectioned surface of a weathered and repeatedly strained, but unworn, part of a used rubber tyre. The coarse crack system has been initiated by ozone cracking and further aggravated by weathering. The cracks are about 0.2 mm deep, extending through a layer of modified surface material, indicated by lighter contrast in the picture. In general, the cracks do not extend down into the unmodified material, with the exception of a few, occasional cracks. Such an exception, about 0.5 mm deep, is shown in higher magnification in Fig. 9. It is seen that the appearance of the crack is altered at the transition from modified to unmodified material. In the modified surface layer (M), the crack surfaces exhibit a rough, weathered character, whereas the crack segment in the unmodified zone (U) has a "cleaner", fractured appearance. The latter crack segment is probably caused by fatigue. Another interesting feature is seen in the region to the left of the fatigue crack. Rather extensive deformations in the material are indicated by "flow lines", which seem to focus at a point P in the crack surface. This suggests that the fatigue crack was not initiated by the ozone crack in the modified layer, but at a stress concentration *below* this layer. From this point of initiation, the fatigue crack has

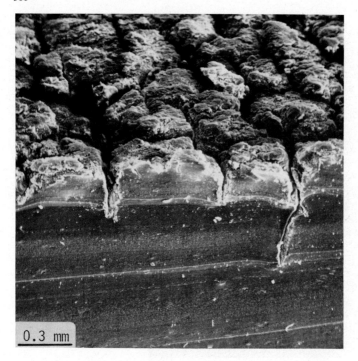

Fig. 8. Crack system and cross-section of an unworn part of a used tyre.

propagated upwards, eventually coalescing with the ozone crack, and some 0.1 mm downwards, where it has been halted by some kind of obstacle or stress concentration of unfavourable orientation.

Figures 10–14 show examples of cracks and crack systems initiated by flexing fatigue. A specimen with a fatigue crack is mounted for SEM studies according to Fig. 10. In the flexed position, several different crack systems are visible. In region A, a system of long, parallel, and fairly wide (~ 50 μm) cracks are seen (Fig. 11). Region B displays a dense structure of smaller cracks (≤ 10 μm; Fig. 12). The presence of these crack systems is not discernible in the relaxed specimen, which serves to demonstrate that surface examinations of prestressed rubber specimens in relaxed positions can be very misleading. Figure 13 shows the crack bottom (or tear front, region C) of the fatigue crack. It is clearly seen that the tearing takes place by tensile rupture of thin fibrils of rubber. Figure 14 (region D) illustrates that the catastrophic fatigue crack does not propagate through solid, undamaged material, but rather through a system of smaller cracks, which progressively develops at some distance in front of the catastrophic crack.

4.2. Detachment of surface material

The micro-mechanisms of material detachment from a rubber surface during a wear process are most readily described in terms of two extreme cases: detachment

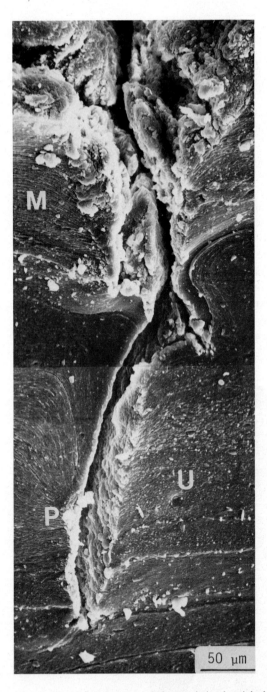

Fig. 9. Partial enlargement of the crack in the right-hand part of Fig. 8. M is modified subsurface material, U is unmodified rubber, and P is the site of crack nucleation.

308

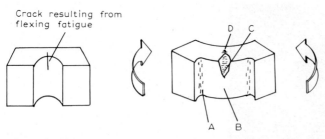

Crack resulting from
flexing fatigue

Fig. 10. Sketch illustrating the location of different crack systems in a rubber specimen subjected to flexing fatigue.

by brittle fracture of well-defined particles and detachment by smearing of a non-particulate substance. On a micro-scale, brittle fracture occurs by tensile rupture of thin and severely strained strings of rubber (Fig. 13). This is true in cases of micro-cutting, micro-chipping, micro-tearing, micro-flexing, etc. The opposite extreme case, smearing, occurs by chemical (e.g. oxidative), thermal, and perhaps also mechanical [35] degradation of the polymeric structure, rendering the surface layer a viscous nature. In between these two extremes (brittle fracture and smearing), exists a continuous spectrum of more or less viscous fracture mechanisms, in which the tensile rupture is combined with molecular degradation and a lowered degree of cross-linking.

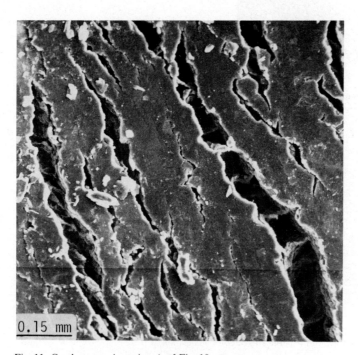

0.15 mm

Fig. 11. Crack system in region A of Fig. 10.

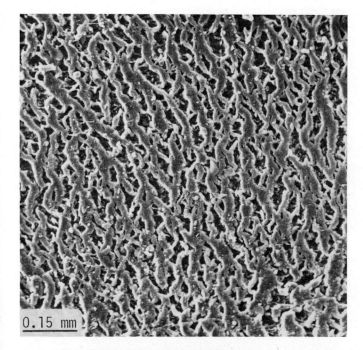

Fig. 12. Small surface cracks in region B of Fig. 10.

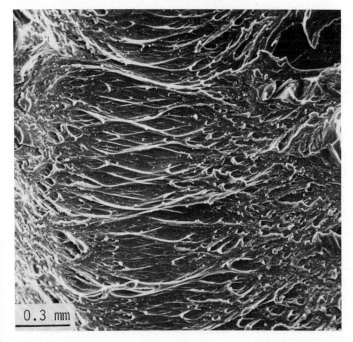

Fig. 13. Crack bottom (at C in Fig. 10) with severely strained rubber fibrils.

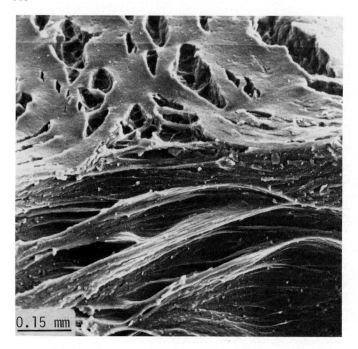

Fig. 14. Close-up of a fatigue crack tip (at D in Fig. 10) showing crack system in front of propagating main crack.

The above discussion of a brittle-fluid spectrum of mechanisms concerns the actual detachment events. These events are always preceded by processes of local straining and deformation of the surface material. Although the deformation processes are initiated in the predetachment phase, they are of a mechanical character and determine the local geometric conditions of the detachments. Schallamach [60] and others [61,62] have studied the contact pattern for rubber sliding against a smooth glass surface under high friction but negligible wear rate. The regions of contact are found to be parallel lines, traversely oriented to the sliding direction and travelling like wave-fronts in the sliding direction at a velocity exceeding that of the sliding. The non-contact regions are compressed and buckled away from the counterface, while the contact regions are strained.

Abrasion patterns, similar to the Schallamach waves, are often found in rubber surfaces exposed to more severe wear. These Schallamach patterns [21] are described in the literature as ribs, bands, ridges, etc. [31,33–35,44,53,58,59,63,64]. Although formed under different conditions of friction and wear, both Schallamach waves and Schallamach patterns are most probably the results of periodic fluctuations between compression and tension, which may occur along the contact surface under certain sliding conditions. As material is removed from the contact regions by wear, the conditions of compression–tension are shifted, and the parallel lines of the Schallamach abrasion pattern will wander along the surface at a velocity lower than the sliding velocity.

The appearances of these patterns may vary, depending on material properties, surface topographies, and sliding conditions. In some cases, the pattern does not constitute long, parallel ridges, but rather a dense structure of small, irregular, and incoherent tongues or scales. If the pressure is increased, the scales grow larger and gradually coalesce into long, coherent edges, the spacing between the parallel edges increasing with increasing pressure [21,31].

The detachment of wear particles in razor blade abrasion experiments has been discussed by Gent and Pulford [15,35,65,66]. The primary wear events occur at the root (or tear tip) of a tongue, as indicated in Fig. 15. This is where the actual "digging" into the rubber surface takes place by a combination of tearing and detachment of microscopic wear particles. Most of the mass loss, however, does not occur in the primary zones, but by detachment of small or large fragments from the edges of the strained tongues.

Depending on the type of rubber blend, the edges of the scale structure may or may not develop "edge rolls". Figure 16 shows edge rolls formed in a case of wear of NBR against a polyester gauze. Such rolls are often detached in fairly large pieces. In the SBR case of Fig. 17, on the other hand, edge rolls have not been formed, and material is detached from the edges in small fragments. Figure 18 shows a NR–BR case with strong elements of smearing. As further discussed in the next subsection, the size and consistency of the detached particles are of great importance to the transportation mechanisms and to the resulting wear rates.

In the general case, an abraded rubber surface does not exhibit a well-defined scale structure, or any other structure of regular appearance, but instead displays the irregular topography of Fig. 19. The detachment of fragments from such a surface occurs by a complex interaction of tearing, cutting, flexing, plucking, etc. and the sizes of the detached fragments can vary over a considerable range. In cases of so-called fatigue wear, for instance mild cases of tyre wear, the worn surface exhibits

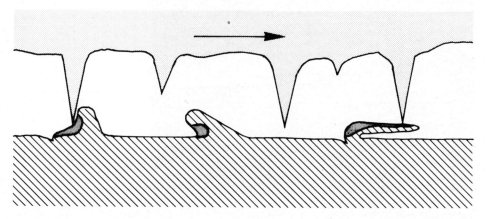

Fig. 15. Schematic cross-section of a rubber surface worn by sharp asperities. The primary wear regions are shadowed.

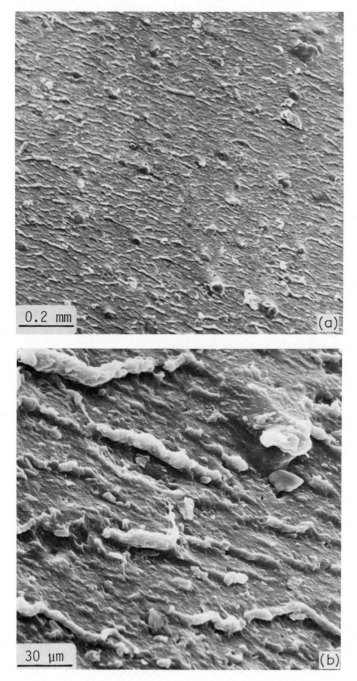

0.2 mm

(a)

30 μm

(b)

Fig. 16. Edge rolls formed during wear of NBR on a polyester gauze [31]. Wear direction: upwards. (a) Overall wear pattern. (b) Detail of (a) showing edge rolls.

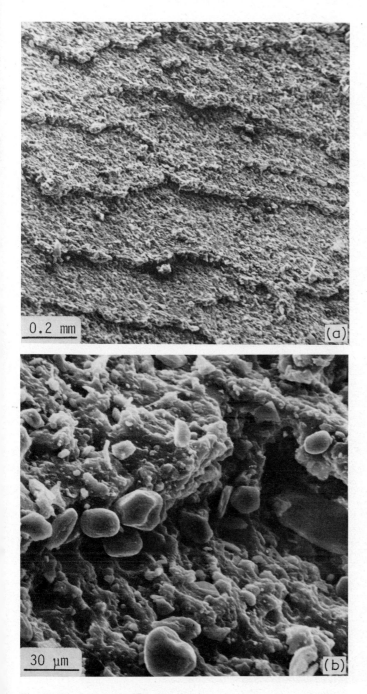

Fig. 17. Wear pattern in SBR after sliding on a polyester gauze [31]. Wear direction: upwards. (a) Overall view. (b) Detail of (a) showing fragmentation.

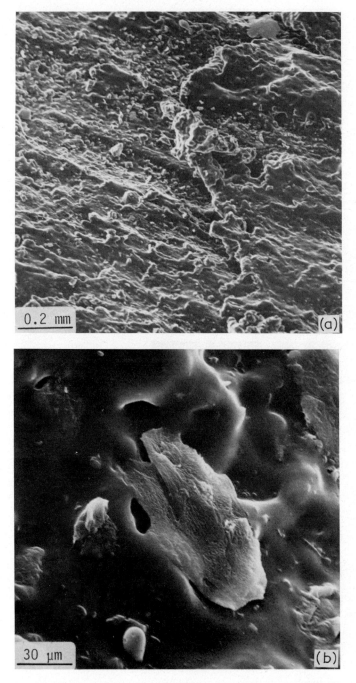

Fig. 18. Smearing of NR–BR after sliding on a polyester gauze [31]. Wear direction: left to right. (a) Scales and furrows are seen in low magnification. (b) Detail of (a) showing smooth surface with traces of tensile fracture.

Fig. 19. General appearance of complex abrasion.

a topography that looks smooth to the naked eye, rather like that of an unworn surface. In higher magnification, however, the surface assumes the irregular, gravelly appearance that is characteristic for cases of complex abrasion. The sizes of the detached fragments are, in this case, about $1-10$ μm.

4.3. Postdetachment phase. Transportation of wear debris

Once detached from its original site in the rubber surface, a wear particle can meet many different fates depending on the properties of the rubber surface and its counterface and on the prevailing sliding conditions. The mode of transportation of wear debris in the contact surface may influence the friction and wear properties of the system considerably; a change in transportation mechanism may, in fact, change the wear rate by as much as two orders of magnitude [31].

The transportation of detached material in the contact region can be subdivided into different modes according to an extension of the discussion in ref. 31 (δ is some characteristic depth measure of the surface topography).

(A) Loose debris transportation
    (1) Fragments smaller than δ
    (2) Fragments of size $\sim \delta$ (e.g. detached edge rolls)
    (3) Fragments larger than δ (e.g. roll conglomerates)

(B) Transfer to counterface
  (1) Particle penetration of counterface
  (2) Particle deposition in counterface
  (3) Smearing onto counterface
(C) Back transfer

In the above classification of transportation modes, we have not included some extreme cases; for instance, the extreme pyrolytic effects taking place in an aircraft tyre at the moment of touch-down.

The class A modes of transportation deal with fragments of a loose particle character which cannot immediately escape from the contact surface, but must travel along it for a while as sliding proceeds. This will happen in a large variety of practical wear cases where the detached fragments are hard enough, and the loading conditions are mild enough, to allow the fragments to maintain their particle character throughout the transportation process.

If the fragments are smaller than the characteristic measure, δ, of the topographic variation (mode A1), their transportation will not, in general, retroact much on the detachment or predetachment mechanisms and hence not influence the wear rate. However, if the density of fragments in the contact surface becomes high enough, small fragments may tend to agglomerate into larger particles and a change in transportation mode from A1 to A2 or A3 can occur. This has been seen to happen in cases of mild wear of SBR in a track previously contaminated by wear fragments, as illustrated by Fig. 20 [31].

In cases where the detached fragments are about the same size as the topographic variation δ (mode A2), an effect on the wear rate can sometimes be seen. This happens, for example, in a case of mild wear of NBR where detached edge rolls (Fig. 16) act as a *roller bearing* in the contact surface, lowering the friction and the wear rate [31]. In this case, the transportation mode clearly retroacts on the detachment, and probably also on the predetachment, mechanisms. If the wear track is contaminated with detached edge rolls from previous passages, larger roll conglomerates will form and a change in mode from A2 to A3 occurs.

When roll conglomerates considerably larger than δ are formed (mode A3), the effect on the wear rate is quite drastic provided the rolls possess a load-bearing capacity good enough to maintain a separation between the wear partners during sliding. With a change in transportation mode from A1 to A3, the roller bearing effect can lower the wear rate by as much as two orders of magnitude [31].

Roll formation phenomena in rubber wear are being paid quite a lot of attention in the literature [2,15,59,64,66–69] and roll formation has even been nominated one of the main mechanisms of rubber wear. It should be noted, though, that roll formation, when it occurs, is merely one link in the chain of events constituting a wear process. Furthermore, its position in this chain may vary: roll formation may occur either *before* detachment, by formation of edge rolls (as discussed in Sect. 4.2), or *after* detachment, by agglomeration of small fragments into larger rolls. This distinction is usually neglected in the literature.

The class B modes of transportation deal with cases when detached material is

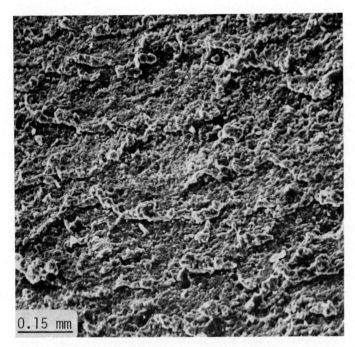

Fig. 20. Roll formation by conglomeration of wear fragments [31]. Wear direction: upwards.

captured or transmitted by the counterface. The latter case (mode B1) occurs, for example, in wear of SBR against a polyester gauze [31] when the detached fragments are small enough to penetrate the gauze structure. Thus, subsequent to detachment, the fragments are almost momentarily removed from the contact surface and a close contact between the rubber specimen and the gauze is maintained throughout the wear process, promoting a relatively high wear rate. If the wear takes place in a circular track, the gauze structure will sooner or later become saturated with wear fragments. A change in transportation mode from B1 to A2 or A3 may then occur, resulting in a considerably lowered wear rate.

If solid rubber fragments are captured by the counterface (mode B2), for instance in surface cavities or by some adhesive mechanism [20], the sliding will, to some extent, occur rubber-to-rubber. In these cases, one must expect either some type of seizure (e.g. stick–slip effects) or a change in transportation mode, for example to smearing or back transfer.

The third mode of material transfer, smearing (B3), will locally generate thin, viscous films in points of contact between the two wear partners. Since smearing is usually associated with high interfacial temperatures, disastrous wear rates could be expected. It turns out, however, that if the smearing element is not too strong, it may actually have a lubricating effect [29,34,66], which in analogy to the roller bearing effect can be denominated *glide bearing effect*. In this effect, thin, viscous rubber layers in the interface cause the rubber specimen to "float" along the counterface under low friction and wear. It is reasonable to assume that such an effect could only occur under low pressure conditions, since the load-bearing capacity of the viscous layers must be limited.

The class C modes of transportation concern the back transfer phenomenon, i.e. cases when loose debris, or material transferred to the counterface, is recaptured by the rubber surface. Back transfer effects and their influence on detachment mechanisms and wear rates are difficult to study experimentally. It has been suggested [70] that a radionuclide tracer technique would be one possible way to approach this problem.

In conclusion, in cases of mild, sliding wear, the postdetachment phase may sometimes be of decisive importance to the wear rate and must not be neglected in the analysis. In spite of this fact, transportation mechanisms are rarely discussed in the literature. Further studies of the postdetachment phase are clearly in order.

5. Special wear patterns

In previous sections, we have discussed the irregular surface topography in cases of complex abrasion (Fig. 19) and we have seen examples of smearing and scale patterns (with or without edge rolls) in cases of mild, sliding wear against a polyester gauze (Figs. 16–18). In the present section, we will discuss some special types of wear patterns, in order to elucidate the relationships, or lack of them, between operating wear mechanisms and resulting wear patterns.

Figure 21(a) shows an example of a classical Schallamach pattern produced by grinding on a fairly coarse abrasive paper. The pattern exhibits a wavy appearance, with wave-fronts traversely oriented in the sliding direction. In higher magnification [Fig. 21(b)], it is seen that the wear picture is rather complex, with signs of cutting, chipping, roll formation and detachment of small fragments (fragmentation). The main transportation mode is A1, with elements of A2, B2 and C. Hence, we may conclude that this Schallamach pattern is not a result of a particular, microscopic wear mechanism and probably not of a particular combination of mechanisms either. This conclusion supports the assumption of Sect. 4.2, that Schallamach patterns will form when certain frictional conditions are fulfilled in the surface, and that the detailed wear mechanisms involved in the friction process are of secondary importance in this case.

Figure 22 shows a wear pattern in a sand-blasted rubber specimen. The particle stream has hit the rubber surface at an angle from below in the picture. The resulting scale pattern is almost identical to the scale pattern of Fig. 17 (mild, sliding wear of SBR against a polyester gauze). This resemblance may seem surprising, considering the great difference in wear conditions. There are, however, certain features which may contribute to explain the striking similarity of the two wear patterns. In both cases, the detachment of wear particles occurs by small-particle fragmentation and, in both cases, the mode of transportation is B1, i.e. immediate removal of wear

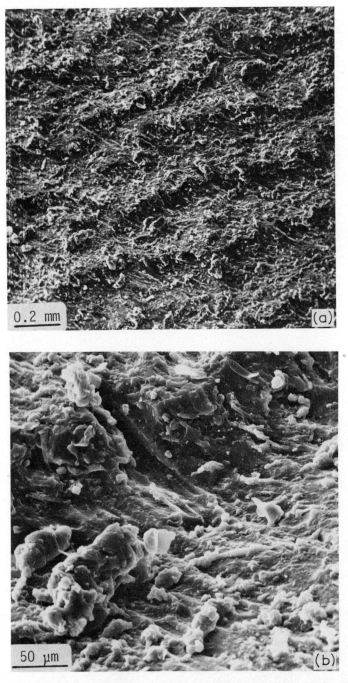

Fig. 21. Rubber surface topography after grinding with abrasive paper. Wear direction: upwards. (a) Schallamach pattern. (b) Detail of (a) showing cutting, fragmentation and roll formation.

320

7 μm   0.3 mm

Fig. 22. Scale pattern resulting from sand blasting. Wear direction: upwards.

fragments from the "interface". Scale patterns of similar appearance, due to resembling conditions of detachment and transportation, are also found in many cases of razor blade abrasion (cf. Sect. 4.2).

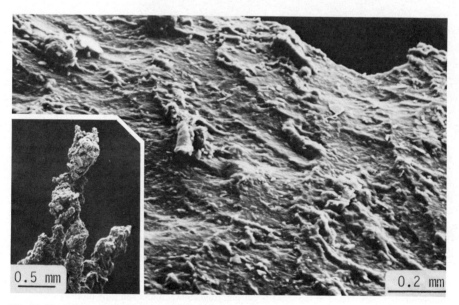

0.5 mm   0.2 mm

Fig. 23. Wear pattern resulting from rolling/sliding contact.

Figure 23 shows the surface of a small rubber wheel exposed to rolling contact with a larger, rotating steel cylinder. To introduce a certain degree of sliding in the contact region, the axes of the rubber wheel and the steel cylinder were tilted relative

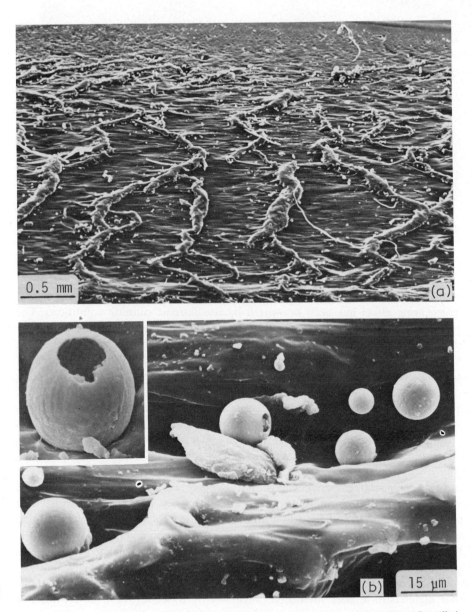

Fig. 24. Roll formation resulting from sliding impact [74]. Wear direction: left to right. (a) Overall view of the scratch. (b) Detail of (a) showing spherical shells.

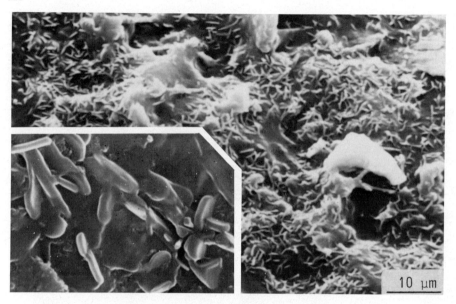

Fig. 25. Formation of small pins or platelets during pendulum scratching.

to each other. The resulting steady-state wear pattern looks rather like the surface of a stormy sea and exhibits strong elements of smearing. In this case, the occurrence of a glide bearing effect has been prevented by the extreme running conditions. The detachment of wear debris occurs by a markedly viscous fracture mechanism. Very large conglomerates are formed from extruded tacky wear debris.

Figure 24 shows an example of roll formation under impact sliding conditions. A protruding steel "tooth" at the end of a heavy pendulum arm has been brought to hit the specimen surface at a very low impingement angle. It is seen that roll formation can occur even under high energy rate conditions. Signs of thermal effects are seen in higher magnification [Fig. 24(b)]. Small, spherical shells (~ 10 μm diam.) are formed, possibly as a result of high, local flash temperatures. These shells resemble the rubber particles of impact polystyrene [71] and of ABS polymers [72,73].

In some cases, rubber surfaces have hard "skins", i.e. surface layers which have been hardened in the manufacturing process or by, for instance, treatment with sulphuric acid prior to adhesive bonding (cyclization). Such surfaces may display wear patterns which are not observed in "skinless" rubbers. Figure 25 shows the surface of a black-filled rubber after pendulum impact scratching. The surface seems to be covered with a rather dense structure of small pins, which in higher magnification are seen to be the edges of small platelets pointing out of the rubber surface. Similar platelets have been identified previously by Bascom [57] in microcracks of unworn cyclized surfaces and were found to contain high concentrations of zinc and sulphur. Hence, one cannot disregard the possibility that the platelet pattern is not

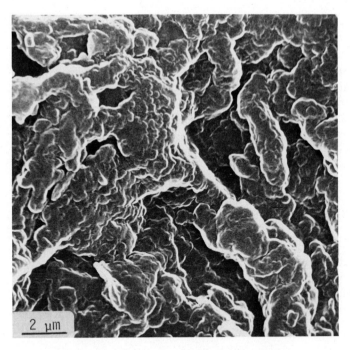

Fig. 26. High magnification micrograph of weathered surface.

really a "true" wear pattern, but rather some kind of internal structure of the skin which has been revealed by the wear process.

As discussed in Sect. 4.2, the characteristic wear pattern in cases of mild tyre wear (fatigue) is that of complex abrasion with particle sizes in the interval 1–10 μm (Fig. 19). Another interesting topographic feature of rubber tyres exposed to mild fatigue wear is found for very high magnifications ($\geqslant 10\,000\times$). The worn surface exhibits a very porous, coral-like structure (Fig. 26) with pore sizes in the submicron interval. Exactly the same type of structure is also found in samples taken from *unworn* parts of the same tyre, whereas a freshly cut surface, magnified to the same degree, appears to be very smooth and solid. This suggests that the coral structure of Fig. 26 is the result of ageing and fatigue, possibly combined with swelling and drying. Hence, in this particular case of mild tyre wear, the characteristic wear pattern is found in the 1–10 μm range, whereas a characteristic structure of ageing and fatigue is found in the submicron range. The extreme porosity of the submicron coral structure probably promotes the processes of thermo-chemical deterioration by facilitating access to the surface layer of atmospheric oxygen, moisture, etc.

6. Tendencies and suggestions

A simple, computer-based search for scientific publications on wear and fatigue of rubber revealed that over 1200 titles (patents excluded) have been registered during the last twenty years and that considerably more than half of that number were published in 1977 or later. In this vast flow of publications, clearly predominant tendencies in current rubber wear research are not easily discernible. It is, however, clear that the subject is approached in various manners in different parts of the world and that too many authors tend to neglect work performed in parts of the world other than their own.

A few current tendencies should be mentioned, though. It seems as if the attempts of the sixties and early seventies to develop extensive, and rather complicated mathematical relationships between phenomenological macro-parameters (like friction coefficient and wear rate; see, for example, ref. 2) are gradually being replaced by theories based on a more fundamental understanding of the physical and thermo-chemical processes operating on a micro-scale. This is a natural consequence of the increased use of SEM (and, to a limited extent, TEM) in the wear mechanism analysis.

Energy methods are attracting increasing interest, particularly in the development of theories for crack and tear growth, but also concerning other aspects of the wear process [20,62]. Further work is needed on the analysis of the energy expenditure in a wear process. How much is dissipated as non-destructive heat? How much is spent on predetachment deterioration? How much is spent on mechanical detachment? How does the postdetachment phase influence the total energy loss and its expenditure?

It seems inevitable that any theoretical treatment of a wear process, which is a statistical phenomenon, sooner or later must involve statistical methods. Today, such methods are utilized on a very rudimentary level, usually in the form of simple mean values. More detailed statistical distribution analyses, like the distribution analysis of rubber fatigue lives performed by DeRudder [46], will eventually be needed.

Further experimental and theoretical work should be directed towards the mechanical behaviour of modified surface layers. It is well known that the mechanical properties of modified layers may differ drastically from bulk properties, but quantitative theories are still lacking. This demanding analysis does not only involve changes in stress and strain characteristics due to deterioration, but also alterations imposed by extreme, local strain rates and temperatures in the surface layer. It seems probable that modern techniques of surface analysis can be utilized to a greater extent than today to determine the compositions of modified surface layers. Furthermore, it does not seem out of reach to study the internal structure of modified surface layers by means of the TEM technique, although certain problems concerning the preparation of specimens remain to be solved.

Finally, the possibilities of improving the wear properties by fibre reinforcement have not been fully explored (cf. Sect. 3.1). The relationship between wear properties and bulk reinforcement is by no means clearly understood and more complex systems, e.g. long-fibre reinforcement of the bulk material combined with short-fibre reinforcement of the surface layer, have not been investigated at all.

List of symbols

A	area
a_0	activation energy
B	crack growth constant
c	surface crack depth ($2c$ = internal crack length)
c_0	size of natural flaws
δ	characteristic depth of surface topography
E	total strain energy
E_d	strain energy density
ϵ	strain
ϕ	irreversible work dissipated during crack growth
G_c	strain energy release rate
γ_s	surface energy
J	line tension
k	empirical constant, or empirical strain-dependent function
L	length, or load
N, n	number of cycles
q	ozone concentration
R	gas constant
S	entropy
σ	stress
σ_G	critical stress of crack growth
t	time
T	tearing energy, or temperature
T_o	critical tearing energy (general, or for oxidative crack growth)
T_{oz}	critical tearing energy for ozone cracking
T_c	critical tearing energy for catastrophic crack growth
U	internal energy
$\dot{w}$	wear rate
Y	Young's modulus of elasticity

References

1 OECD, Glossary of Terms and Definitions in the Field of Friction, Lubrication and Wear, Paris, 1969.
2 G.M. Bartenev and V.V. Lavrentev, in L.H. Lee and K.C. Ludema (Eds.), Friction and Wear of Polymers, Elsevier, Amsterdam, 1981, Chap. 4, p. 111 and Chap. 6, p. 202.
3 L.R.G. Treloar, The Physics of Rubber Elasticity, Clarendon Press, Oxford, 3rd edn., 1975.
4 J. Gough, Mem. Lit. Philos. Soc. Manchester, 1 (1805) 288.
5 J.P. Joule, Philos. Trans. R. Soc. London, 149 (1859) 91.
6 L. Bateman (Ed.), The Chemistry and Physics of Rubber-like Substances, MacLaren, London, 1963.
7 H. Hirakawa, F. Urano and M. Kida, Rubber Chem. Technol., 51 (1978) 201.
8 A.A. Griffith, Philos. Trans. R. Soc. London, 221 (1921) 163.

326

9 G.R. Irwin, J. Appl. Mech., 24 (1957) 361.
10 E. Orowan, Rep. Prog. Phys., 12 (1949) 185.
11 J.P. Berry, J. Appl. Phys., 34 (1963) 62.
12 G.J. Lake and A.G. Thomas, Proc. R. Soc. (London) Ser. A, 300 (1967) 108.
13 L.H. Lee, J. Polym. Sci. Part A-2, 5 (1967) 1103.
14 G.J. Lake and P.B. Lindley, J. Appl. Polym Sci., 9 (1965) 1233.
15 A.N. Gent and C.T.R. Pulford, Dev. Polym. Fract., 1 (1979) 155.
16 Z. Rigbi, Rubber Chem. Technol., 55 (1982) 1180.
17 A.S. Kuzminsky, Dev. Polym. Stab., 4 (1981) 71.
18 J.D. Walter, Rubber Chem. Technol., 51 (1978) 524.
19 L.A. Goettler and K.S. Shen, Rubber Chem. Technol., 56 (1983) 619.
20 L.H. Lee, Polym. Sci. Technol., 5A (1974) 31.
21 A. Schallamach, Wear, 1 (1957/1958) 384.
22 J.K. Lancaster, Wear, 14 (1969) 223.
23 K.A. Grosch and A. Schallamach, Rubber Chem. Technol., 43 (1970) 701.
24 F.C. Weissert, Kautsch. Gummi Kunstst., 25 (1972) 571.
25 A.G. Veith, Rubber Chem. Technol., 46 (1973) 801, 821.
26 V.V. Lavrentev, in L.H. Lee (Ed.), Advances in Polymer Friction and Wear, Plenum Press, New York, London, 1974.
27 G.J. Lake and A.G. Thomas, Kautsch. Gummi Kunstst., 20 (1967) 211.
28 G.J. Lake and P.B. Lindley, Rubber J., 146 (1964) 30.
29 A. Schallamach, J. Appl. Polym. Sci., 12 (1968) 281.
30 L. Åhman, A. Bäcklin and J.-Å. Schweitz, in S.K. Rhee, A.W. Ruff and K.C. Ludema (Eds.), Proc. Int. Conf. Wear Mater., San Francisco, CA, 1981, ASME, 1981, pp. 161–166.
31 J.-Å. Schweitz and L. Åhman, in K.C. Ludema (Ed.), Proc. Int. Conf. Wear Mater., Reston, VA, 1983, ASME, 1983, pp. 610–616.
32 S. Hogmark, S. Söderberg and O. Vingsbo, Proc. 3rd Conf. Mech. Behaviour Mater., Cambridge, 1979, pp. 621–631.
33 A. Schallamach, Proc. NRPRA Jubilee Conf. 1964, pp. 150–166.
34 B.B. Boonstra and F.A. Heckman, Rubber Chem. Technol., 44 (1971) 1451.
35 A.N. Gent and C.T.R. Pulford, J. Appl. Polym. Sci., 28 (1983) 943.
36 K.C. Cong, W. Yang and Y.Y. Huang, in K.C. Ludema (Ed.), Proc. Int. Conf. Wear Mater., Reston, V, 1983, ASME, 1983, pp. 591–595.
37 P.L. Hurricks, Wear, 52 (1979) 365.
38 G.I. Brodskii, N.L. Sakhnovskii, M.M. Reznikovskii and V.F. Evstratov, Sov. Rubber Technol., 19 (1960) 22.
39 M. Tichava, Plaste Kautsch., 9 (1962) 539.
40 A.G. Thomas, J. Polym. Sci. Polym. Symp., 48 (1974) 145.
41 J.K. Lancaster, Trans. Inst. Met. Finish., 56 (1978) 145.
42 J.R. Beatty, J. Elastomers Plast., 11 (1979) 147.
43 L.R.G. Treloar, Introduction to Polymer Science, Wykeham, London, Winchester, 1970.
44 D.H. Champ, E. Southern and A.G. Thomas, in L.H. Lee (Ed.), Advances in Polymer Friction and Wear, Plenum Press, New York, London, 1974, p. 133.
45 S.B. Ratner and E.G. Lure, in D.I. James (Ed.), Abrasion of Rubber, MacLaren, London, 1967, p. 155.
46 J.L. DeRudder, J. Appl. Phys., 52 (1981) 5887.
47 N.M. Mathew, A.K. Bhowmick and S.K. De, Rubber Chem. Technol., 55 (1981) 51.
48 N.M. Mathew and S.K. De, Int. J. Fatigue, 5 (1983) 23.
49 H.E. Railsback, N.A. Stumpe, Jr. and C.R. Wilder, Rubber World, 160 (1969) 63.
50 E. Southern and A.G. Thomas, Rubber Chem. Technol., 52 (1979) 1008.
51 A.N. Gent and C.T.R. Pulford, Dev. Polym. Fract., 1 (1979) 155.
52 A.N. Gent, in F.R. Eirich (Ed.), Science and Technology of Rubber, Academic Press, New York, 1978, pp. 419–454.

53 C.T.R. Pulford, Ph.D. thesis, University of Akron, 1979.
54 J. Kruse, Rubber Chem. Technol., 46 (1973) 677.
55 P.B. Lindley and A. Stevenson, Rubber Chem. Technol., 55 (1982) 337.
56 A. Stevenson, Int. J. Fract., 23 (1983) 47.
57 W.D. Bascom, Rubber Chem. Technol., 50 (1977) 327.
58 A.K. Bhowmick, S. Basu and S.K. De, J. Mater. Sci., 16 (1981) 1654.
59 A.K. Bhowmick, Rubber Chem. Technol., 55 (1982) 1055.
60 A. Schallamach, Wear, 17 (1971) 301.
61 B. Best, P. Meijers and A.R. Savkoor, Wear, 65 (1981) 385.
62 M. Barquins, Wear, 91 (1983) 103.
63 A. Schallamach, Gummi Asbest Kunstst., 31 (1978) 502, 506, 510.
64 N.M. Mathew and S.K. De, J. Mater. Sci., 18 (1983) 515.
65 A.N. Gent and C.T.R. Pulford, Wear, 49 (1978) 135.
66 C.T.R. Pulford, J. Appl. Polym. Sci., 28 (1983) 709.
67 N.L. Sakhnovskii, L.I. Stepanova and V.F. Evstratov, Sov. Rubber Technol., 31 (1972) 26.
68 S.M. Aharoni, Wear, 25 (1973) 309.
69 D.F. Moore, in D. Dowson (Ed.), Proc. Leeds–Lyon Symp. Tribol., 3rd Meeting, 1976, Mechanical Engineering Publ., London, 1978, pp. 161–162.
70 N.S. Eiss, J.H. Warren and S.D. Doolittle, Wear, 38 (1976) 125.
71 H. Keskkula and P.A. Traylor, Polymer, 19 (1978) 465.
72 K. Kato, Kolloid Z. Z. Polym., 220 (1967) 24.
73 K. Kato, Polym. Eng. Sci., (January) (1965) 38.
74 J.-Å. Schweitz, S. Hogmark and L. Åhman, Res. Rep. UPTEC 81 104R, Institute of Technology, Uppsala University, Sweden, 1981.

Chapter 10

The Wear and Friction of Commercial Polymers and Composites

JOHN C. ANDERSON

National Centre of Tribology, UKAEA, Risley, Warrington WA3 6AT (Gt. Britain)

Contents

Abstract

For convenience, commercially available plastic bearing materials are divided into six groups. Methods of wear testing for polymer and composite materials are briefly discussed and basic wear data for representative materials in each group are given. The majority of the chapter is devoted to an analysis of the factors that affect, principally, the wear rates of materials. These factors include what may be described as "operating parameters" such as bearing pressure, temperature, sliding velocity, and counterface roughness. These parameters may possibly be changed at the design stage. In addition, some of the factors that influence the choice of materials are also discussed. These are mainly concerned with operating environments such as water and other fluids, vacuum and inert gases, and ionising radiation. Other, less obvious, factors such as trapped wear debris in conforming geometries, and the nature of the fillers that are included in composites, can also significantly affect friction and wear performance. At the end of the chapter, some guidance on the choice of polymer composites for sliding applications is provided with a suggested step-by-step selection procedure.

1. Introduction

In the U.K., there are over 70 polymer-based composites that are sold and used as bearing materials. World-wide, there may well be twice that number. Whilst unfilled polymers have been, and no doubt will continue to be, used in a wide variety of applications, polymer composites are also used, particularly for difficult applications, where a blend of properties is required. There are many sources of information on the friction and wear characteristics of commercially available polymer composites [1–4] and a good deal of research has gone into the development of specialised materials.

Polymer composites began to be used in bearing applications prior to the Second World War [5], but it was undoubtedly the advent of PTFE composites in the 1950s, together with other materials such as polyamides and polyacetals, that led to their use in diverse applications. Although there is a wide range of commercially available polymer composites, the great majority are based on one of six polymers, viz. polyamide, polyacetal, PTFE, polyethylene, polyimide, or phenolic resin. It is the diversity of fillers, reinforcements and composite construction that gives rise to the range of polymer composite bearing materials. Of course, there is some duplication with different firms producing similar materials with different trade names and this is particularly true for PTFE composites.

In order to discuss the subject of polymer bearing composites, it is essential to use some form of classification which puts them into categories describing their appearance, performance attributes, or some other characteristic. For the purposes of this chapter, six groups have been chosen. These are (1) unfilled thermoplastics, (2) filled thermoplastics, (3) high temperature polymers, (4) thin layer materials, (5) filled PTFE, and (6) reinforced thermosets.

Such a classification may be arbitrary, but is nevertheless useful. Unfilled thermoplastics are, of course, not composites at all but their inclusion is justified on the grounds that their performance can be compared with those of Group 3 composites, high temperature polymers, which includes those aromatic polymers, thermoplastic and thermosetting, capable of operating continuously at temperatures above 200°C. Thin layer materials (Group 4) are those in which a thin polymer or composite layer, usually less than 0.5 mm thick, is bonded by some means to a backing material. In most of these, the thin layer is based on PTFE, but their

TABLE 1
Examples of polymer materials within each group

Group	Material
1 Unfilled thermoplastics	Polyamide (nylon) Polyethylene (HD) Polyethylene (UHMW)
2 Filled thermoplastics	Carbon fibre-reinforced plastics Polyacetal + oil Polyamide + oil Polyurethane + fillers
3 High temperature polymers	Polyimide Polyether–etherketone Polyamide–imide
4 Thin layer materials	PTFE + filler/bronze sinter PTFE + filler/bronze mesh PTFE fibre/glass fibre PTFE flock/Nomex PTFE-based thin layer
5 Filled PTFE	PTFE/glass fibre PTFE/carbon–graphite PTFE/bronze PTFE/glass/MoS_2 PTFE/mica
6 Reinforced thermosets	Asbestos/phenolic Polyester Organic fibre/resin Cotton/phenolic

operating range and bearing applications differ significantly from the filled PTFE category. The latter is, however, numerically the largest single category of polymer composite bearing materials.

Some indication of the range of materials in each category is given in Table 1.

2. Applications

In most tribological situations, the objective is a design which will give an acceptable friction level and rate of wear. Usually, this objective is met by ensuring that a film of fluid separates the surfaces whilst the components are in motion. Frequently, this state of affairs cannot be achieved because ensuring adequate fluid lubrication is difficult, impossible, or even undesirable. Examples of this are where bearing loads are high and relative speeds low, when maintenance is infrequent or non-existent, or where a fluid could contaminate a product, as in the food industry.

TABLE 2

Tribological applications of polymers and composites

Group	Material	Seals	Gears	Abrasive conditions	High pressures	Water	Pivot bearings	Slideways	High temperature	Compressor rings	Nuclear
1	PTFE	✓						✓			
	Polyacetal	✓	✓			✓	✓				
	Nylon		✓	✓							
	UHMWP	✓	✓	✓		✓		✓			
2	Nylon + MoS_2		✓								
	Polyacetal + oil		✓				✓				
	Nylon + oil		✓				✓				
	Polyurethanes			✓		✓	✓				
3	Polyimide (filled)	✓	✓						✓		✓
	Polyamide–imide		✓						✓	✓	✓
4	PTFE + Pb/bronze				✓		✓	✓		✓	
	PTFE fibre/glass fibre				✓		✓	✓		✓	✓
	PTFE flock/Nomex				✓		✓	✓	✓	✓	✓
5	Glass fibre						✓			✓	
	Carbon/graphite	✓									
	Bronze										
	Glass MoS_2	✓						✓			
6	Polyester laminate		✓			✓		✓			✓
	Asbestos/phenolic		✓			✓	✓	✓			✓
	Cotton/phenolic		✓			✓	✓	✓			✓

Because polymer composites used for bearings will usually operate satisfactorily with no fluid lubrication, they are natural candidates for such applications. Many of the materials listed in Table 1 are used for a range of sliding applications including gears and seals. So diverse are the uses of polymer bearing composites that it is impossible to list them all. What Table 2 does is to use 10 arbitrarily selected, but relevant, applications. The absence of a tick does not necessarily preclude use in a particular situation. Ticks are shown where particular materials are known to have been used in the past. Gears and seals are included because the friction and wear data are relevant to these applications also.

3. Performance rating

As in many other areas of tribology, the performance of polymer bearing composites depends upon a complex mixture of parameters and wear processes and cannot adequately be described by one single index. However, the PV factor is frequently used since PV (the product of bearing pressure and sliding speed) is related to the rate of wear and also to the temperature rise at the sliding surfaces. Two kinds of PV factor are commonly given. A limiting PV factor is the upper bound to the useful operating region of a material, the limitation being temperature in many cases. Also, PV factors which equate to a certain wear rate [typically 25 μm $(100 \text{ h})^{-1}$] are also quoted [2]. But PV factors apply to one particular set of conditions. For example, they do not take into account the temperature of the environment, on which any increase in temperature due to frictional heating is superimposed. To overcome this, PV correction factors for various temperatures are sometimes plotted [6].

Because of the limitations of the PV approach, it has been suggested [7] that a better method would be to present wear rate data as a function of bearing pressure and temperature (PT). A typical PT–wear rate graph might look like that in Fig. 1. This method has been adopted in two publications [3,4] which give guidance on the selection and performance of dry bearing materials. A drawback of this approach is that an estimation of the temperature of sliding surfaces in a given application is required. This can become quite a complicated process [8], depending on the complexity of the design.

4. Basic friction and wear data

Before presenting basic friction and wear data for polymers and composites, it is appropriate to discuss briefly the methods of testing by which such data are generated.

Laboratory-scale testing is usually carried out either to rank the performance of candidate materials for an application or to investigate particular wear process and measure the wear rate of a selected material or materials. When the wear rate of a

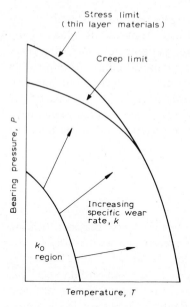

Fig. 1. Idealised pressure–temperature–wear rate curve for polymers and composites.

material is determined, one of two approaches is usually adopted. The test may be made as representative of the conditions of the intended application as possible and every effort made to reproduce the load, speed, contact geometry etc. of the application. Alternatively, an accelerated test method is used. Lancaster [9] has shown that accelerated tests can reduce the test time from weeks to hours whilst still producing the same ranking for the same group of materials tested under both accelerated and representative conditions. However, the danger with accelerated tests is that acceleration of the wear process can introduce wear mechanisms which would not normally be present in a practical situation and which can jeopardise the accuracy of the results. Accelerated tests can be used to rank materials, but the wear rates derived from such tests are likely to be different from those pertaining in real applications.

Laboratory methods of wear testing and specimen geometries are described in other chapters and can be divided into two general categories, non-conforming contacts and conforming contacts. The former tend to be used for quick accelerated tests (e.g. crossed cylinders, pin-on-ring) whilst the latter, though more representative of real bearings, are used where reasonably accurate wear rates are required for life prediction.

Figure 1 shows, in a generalised way, how the wear rates of polymers and composites vary with bearing pressure and surface temperature. There is a region at lower bearing pressures and ambient temperature where rate is essentially independent of these parameters. Under these conditions, specific wear rate (the volume of material worn per unit applied load per unit distance of sliding) has been designated

TABLE 3

Units of specific wear rate and conversion

$\mathrm{mm^3\ N^{-1}\ m^{-1} \times 10^{-9} = m^3\ N^{-1}\ m^{-1}}$	
$\mathrm{cm^3\ kg^{-1}\ cm^{-1} \times 9.81 \times 10^{-6} = m^3\ N^{-1}\ m^{-1}}$	
$\mathrm{in.^3\ min\ ft.^{-1}\ lbf^{-1}\ h^{-1} \times 2 \times 10^{-7} = m^3\ N^{-1}\ m^{-1}}$	

k_0. From both a materials selection and design point of view, comparative k_0 values for materials are essential, since the life of a bearing is inversely proportional to specific wear rate. (Units of specific wear rate are given in Table 3.)

The specific wear rates for various materials within each category are shown in Fig. 2. The k_0 values here were derived from thrust bearing tests with a bearing

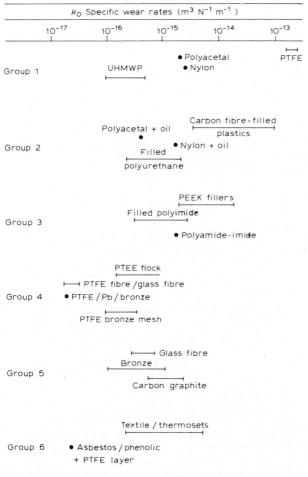

Fig. 2. Summary of k_0 values for material groups.

pressure of 1 MPa and a sliding speed of 0.03 m s^{-1} against a mild steel counterface. Under these conditions, all polymer materials are within their k_0 regions and the tests were not accelerated in any way. The figure demonstrates the wide variation of specific wear rates for these materials and also that some fillers are particularly effective in reducing wear rates, notably fillers in PTFE. A considerable amount of wear testing on similar materials has also been carried out in the laboratories at the

Group	Material	Specific wear rate ($m^3 N^{-1} m^{-1} \times 10^{-15}$)				
		NCT	British rail[a]			
		Thrust washer	Reciprocation	Continuous rotation	Fretting	Abrasion
Unfilled plastics	Polyacetal	2	0.4			
	Nylon 6,6	3	8.8	60	4.2[b]	90
	UHMWP	1.5	0.3	1	1.7[b]	1
Filled plastics	Polyacetal + oil	0.3	0.4	36	6.1	350
	Nylon + oil	0.2	7.3	1.9	24.4[b]	130
	PU/filler	2		11	10	
	PU/filler	1.5		5.5	9.3[b]	760
High temperature plastics	Polyimide + graphite	0.5	0.02	5.3	3.7[b]	3450
	PEEK	20	5.2		1.8[b]	360
	Polyamide-imide/graphite	1.5	1	1.4		1620
Thin layer materials	PTFE/lead/bronze	0.02	C.01		14.8[b]	1700
	PTFE fibre/glass fibre	0.03				
	PTFE flock/Nomex	0.5				
	PTFE/bronze mesh	0.1				
Filled PTFE	Graphite	0.6		23	0.26	470
	Glass fibre	0.5	1.1	14	0.38[b]	330
	Bronze/graphite	0.08	0.2	5.3	1[b]	440
	Bronze	0.5	0.3	3	0.66[b]	490
Textile thermosets	Cotton/phenolic + PTFE	8	3	8.4	83[b]	1740
	Asbestos/phenolic + filled PTFE surface layer	0.02	0.9			

Mild steel counterface exept where noted.

[a]Reproduced by kind permission of British Railways Board.

[b]Case hardened steel counterface.

Fig. 3. Wear rates of some polymer composites.

British Rail Technical Centre and some results are shown in Fig. 3. The abrasion test used consists of continuous sliding of the material in an oil/silica slurry and is considerably more severe than other test conditions. This comparison demonstrates the fact that the same group of materials show a different ranking order under different test conditions. For the purpose of selecting a material, basic wear rates derived from test conditions which most closely simulate the intended application should be used.

With the exception of abrasive conditions, the lowest specific wear rates are generally obtained with filled PTFE materials, either in bulk form (Group 5) or as thin layers bonded to a metallic backing (Group 4). Ultra-high molecular weight polyethylene (UHMWP) also gives a low rate of wear and is abrasion resistant. Carbon fibre fillers do not, in general, greatly improve wear resistance, but this topic

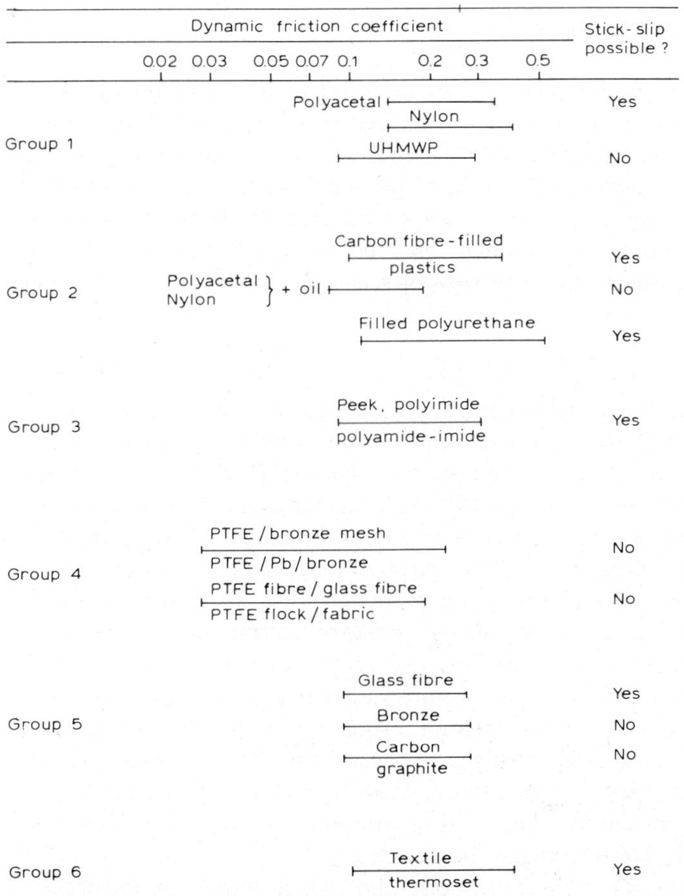

Fig. 4. Range of friction coefficients for material groups.

is discussed in detail elsewhere in this book. Textile reinforced thermosets, as a class, do not have such a good dry wear resistance as the filled PTFEs, despite the fact that some types contain graphite and molybdenum disulphide. Their principal advantage is that they perform well in wet conditions but, as with all materials, dimensional changes due to water absorption have to be taken into account. The one textile-reinforced thermoset with a good dry wear resistance has a thin (0.25 mm) layer of filled PTFE at its rubbing surface.

Figure 4 shows the ranges of dynamic friction coefficients for polymer bearing materials. In general terms, friction coefficients tend to decrease as the bearing pressure increases (particularly with PTFE composites) and increase as the surface temperature increases. Some indication of the tendency to stick–slip motion has also been included, although this phenomenon is also dependent on the dynamic response of the bearing assembly.

5. Factors affecting friction and wear

The various factors which influence the friction and wear rate of polymers and composites will be considered individually. They can be divided into "application" parameters such as bearing pressure, sliding speed, and temperature and "environmental" factors like the presence of fluids, inert gases, or vacuum, etc. Consideration of these factors individually is, of course, somewhat artificial because, in a given application, various combinations of them may be present. However, individual consideration is useful from the point of view of selection of materials by a process of elimination.

5.1. Bearing pressure

For unfilled polymers, wear rate is generally independent of bearing pressure up to some critical value which is typically one-third of the compressive strength of the polymer. The results from thrust bearing tests on polyacetal [Fig. 5(a)] confirm that this is the case. In addition, Lancaster [9] has collected wear rate data for this material obtained from a variety of test geometries which show that wear rate is essentially independent of pressure over two orders of magnitude of pressure.

The data for ultra-high molecular weight polyethylene [Fig. 5(b)] indicate that the same is also true for this material over a smaller range of bearing pressure. However, this material is more sensitive to creep than polyacetal so that a limiting bearing pressure is probably defined by creep limitations rather than wear performance.

In long-term wear tests on this material, using a pin-on-disc configuration, Dowson et al. [10] have reported changes in wear rate after prolonged sliding. Typically, they show that specific wear rates increase by a factor of between 2 and 4 after many kilometres of sliding. Similar discontinuities of wear rate have been observed in thrust bearing tests at higher bearing loads.

Under these conditions, a brown "third-body film" which formed on the polymer surface became disrupted after prolonged sliding and gave rise to an increase in wear

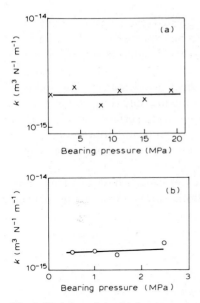

Fig. 5. Variation of k with bearing pressure for (a) polyacetal on mild steel and (b) UHMWP on mild steel.

rate. The data of this effect have been assembled in terms of the effect of applied load on the sliding distance to the onset of the wear rate transition (Fig. 6). This indicates that the sliding distance to the point of wear rate transition is strongly

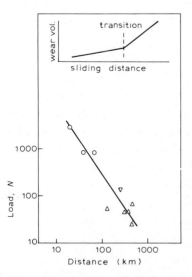

Fig. 6. Dependence upon load of wear rate transition of UHMWP against steel. ○, NCT; △, Dowson et al. [10]; ▽ Dowson et al. [23].

dependent on the applied load. If, as has been suggested, the transition marks the onset of a fatigue mechanism of wear, then it is to be expected that this would be strongly dependent upon load. However, the evidence of fatigue effects in this case is somewhat circumstantial.

A representative high temperature polymer composite (polyimide + 15% graphite) showed a continuous increase in wear rate with bearing pressure (up to 60 N mm^{-2}) under test conditions where the temperature at the bearing surface was kept below 50°C. At the highest pressure, the material deformed significantly during the test so the wear rate was determined by weight loss and was thus insensitive to the deformation.

From similar thrust bearing tests, the wear rate–pressure behaviour of two thin layer filled PTFE materials is presented in Fig. 7. Both exhibit a continuous increase

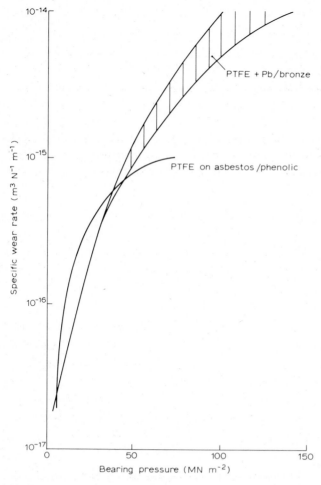

Fig. 7. Variation of wear rate with bearing pressure. Thrust bearing on En 3B mild steel.

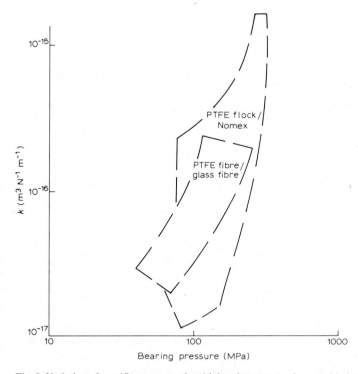

Fig. 8. Variation of specific wear rate, k, with bearing pressure for two thin layer materials (thrust bearing data).

in wear rate with bearing pressure and the rate of increase reduces at high bearing pressures. In both cases, the steel counterface was cooled so that the surface temperature did not exceed 50°C and the increase in wear rate was thus not due to temperature effects. With the PTFE/bronze sinter material, initial thrust bearing tests at high pressures showed that wear debris was contributing significantly to the high rates of wear, so a detailed investigation was undertaken. This is discussed in more detail later.

With thin layer materials based on PTFE fibres, either in continuous woven form or as chopped fibres, wear rate once again increased with bearing pressure (Fig. 8). In this graph, the bands of wear rate indicate the spread of results from a variety of test methods. This variation can be ascribed to differences in test geometry, counterface material and hardness, and type of motion (continuous or oscillatory). All these factors play a part in determining the tenacity and durability of the transfer film which is laid down by the material on the counterface. One investigation [11] has revealed something of the complexity of the films generated by woven PTFE fabric-type materials during dry sliding.

5.2. Temperature

The mechanical properties of all polymers and composites decrease as the

TABLE 4

Effect of method of temperature generation on the wear rate of some polymers

Material	Counterface	Sliding velocity (m s^{-1})	Temperature (°C)	Wear rate (m^3 N^{-1} m^{-1})
Polyacetal	En 3B	0.26	82	1.06×10^{-15}
Polyacetal	En 3B (heated)	0.013	86	1.54×10^{-15}
Polyacetal	18/8 stainless	0.16	86	1.27×10^{-15}
Polyacetal	18/8 stainless (heated)	0.013	88	3.05×10^{-15}
Polyimide + graphite + PTFE	En 3B	0.14	94	4.25×10^{-16}
Polyimide + graphite + PTFE	En 3B (heated)	0.013	90	3.25×10^{-16}

temperature is increased so that their wear rates would be expected to increase with temperature, other factors being equal. The actual variation of wear rate with temperature will also depend on the influence of fillers and the way in which they modify the counterface during sliding and on the formation of oxides or other films on the counterface.

The temperature at a bearing surface can rise due either to a high environmental temperature or to the generation of frictional heat at the sliding surface. In the latter case, sliding speed is particularly important. In addition to raising the mean temperature of the surface, high transient temperatures at asperity contacts (so-called flash temperature) can substantially affect wear rates at high sliding velocities.

In a study carried out at NCT [12], the wear rates of two materials, polyacetal and graphite-filled polyimide, were measured in experiments where the same surface temperature was generated in two ways: first by heating the counterface electrically, but keeping the rubbing velocity low to minimise frictional heating, and secondly by increasing the sliding velocity. The test results are summarised in Table 4. The inference from this work was that the way in which the surface temperature was generated did not influence specific wear rate. That this will be true only where the flash temperature rise is not significant, has been demonstrated by the work of Evans and Lancaster [20].

Figure 9 presents the results of a series of thrust bearing tests performed on a number of filled polyimide materials and filled or reinforced on a number of filled polyimide materials and filled or reinforced PTFE composites. All the tests were carried out at low bearing pressure (1 MPa) and low sliding speed (0.03 m s^{-1}). For each test, a polished mild steel counterface was electrically heated to raise its temperature to a preset level. The wear rates plotted are the steady-state values after running in.

The polyimide materials all exhibited a continuous increase in wear rate with temperature. This behaviour is consistent with results reported by Giltrow [13] for commercially available polymer composites, tested in a crossed cyliners geometry. However, Giltrow's wear rates were in some cases an order of magnitude lower and

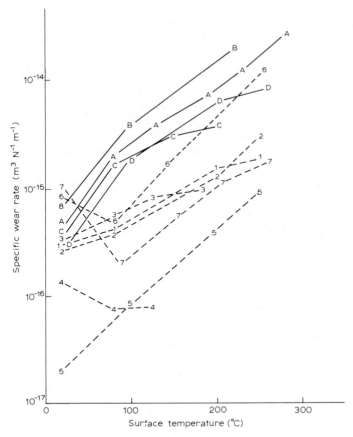

Fig. 9. Wear rate vs. temperature for some filled PTFEs and polyimides. Polyimides containing A, 40% graphite + 10% carbon fibres; B, 15% graphite; C, 40% graphite; D, 15% graphite + 10% PTFE. PTFEs containing 1, 2, mineral; 3, 60% bronze; 4, 15% carbon fibre; 5, PTFE/lead bronze sinter (metal backed); 6, 7, carbon/graphite.

this may be ascribed to the fact that he used a hardened steel counterface. Lewis [14] has also reported on the influence of temperature on the wear rate of one polyimide composite tested in a thrust bearing geometry. His results indicated no increase in wear rate with temperature up to 350°C. Above this, the wear rate increased rapidly. The reason for the discrepancy between the thrust bearing results may lie in the nature of the counterface and the influence this has on the formation of a transfer film.

Turning to the PTFE composites, the results shown in Fig. 9 are in reasonable agreement with those obtained by Evans [15] using a pin-on-ring test geometry. Evans conducted tests on PTFEs with inorganic fillers such as glass, asbestos and graphite and also on PTFEs with high temperature organic fillers (polyimide and aromatic polyamide fibre). The lower wear rates obtained with the latter composites at temperatures up to 150°C was ascribed to their low abrasiveness. However, the thrust bearing results indicates that this type of wear rate–temperature behaviour can also occur with PTFE composites containing inorganic fillers.

An interesting point emerges from these results for polyimide and PTFE composites. The latter are about half the cost of polyimides and their relatively low strength can be overcome by using, for example, a thin layer bonded to a metal backing. The implication is that, for temperatures up to 250°C, PTFE composites could be used in preference to polyimides unless particular properties such as high resistance to ionising radiation are required. Above this temperature, polyimides or polyether–etherketones are virtually the only contenders.

The wear rate–temperature results for the steel-backed PTFE/bronze sinter material agree will with those derived from results reported elsewhere [16] from journal bearing tests. This confirms the ability of this product to give reasonable wear life at temperatures up to 250°C.

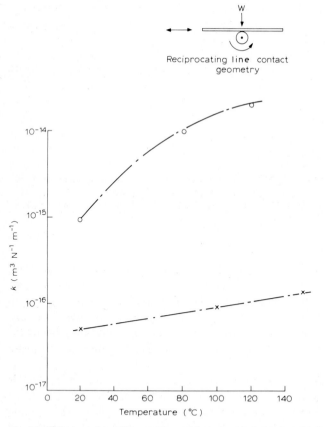

Fig. 10. Variation of specific wear rate with temperature for two thin layer PTFE materials (RLC test). O, Woven PTFE fibre/glass fibre; ×, PTFE flock/Nomex.

Some data are also available for the effect of temperature on the wear rate of thin layer PTFE fabric materials. Generally, however, their prime applications are at high bearing pressures and low sliding speeds, very often with infrequent operation, so that bearing temperature is rarely a problem. However, the wear rates of some of these materials at different temperatures have been measured both by the manufacturer and also in a non-conforming reciprocating line contact (RLC) test [17]. The variation of steady-state wear rate with temperature for woven PTFE fibre/glass fibre is shown in Fig. 10. The RLC results show higher wear rates and a greater increase with temperature than tests on spherical bearings. The reason for this is that the RLC test produces a much greater depth of local deformation in the bearing layer than a conforming geometry and this results in a much greater degree of disruption of the material which results in higher wear rates. The data for a PTFE flock/textile fabric material is also given in Fig. 10. The wear rate of this material is apparently independent of temperature up to 120°C and it appears to be insensitive to the increased local deformation in the RLC test.

5.3. Sliding velocity

The principal effect of sliding velocity on the wear rate of a polymer occurs through its influence on the temperature at the sliding surface. For a given bearing pressure and friction coefficient, the frictional heat generated at the bearing surface will increase in proportion to the sliding velocity. The steady state mean surface temperature attained, θ, will then depend upon the heat transfer properties of the bearing assembly. Thus

$$\theta = R\mu PV$$

where μ is the friction coefficient, P the bearing pressure, V the sliding velocity, and R the thermal resistance of the bearing assembly. Thus, on this argument, the influence of sliding velocity on wear rate can be converted to that of temperature.

The above applies to the *mean* surface temperature at the sliding surface. However, the frictional energy is dissipated as heat at the actual contacting asperities and, because of the poor thermal conduction of polymers, the instantaneous temperatures at the asperities (flash temperatures) can be significantly above the mean surface temperature. This was originally demonstrated by Archard [18]. At high Peclet numbers, flash temperature is more sensitive to sliding speed than load [19] and is also dependent upon thermal conductivity of counterface. Evans and Lancaster [20] carried out a series of pin-on-ring experiments with polyacetal pins sliding against different counterface materials. They demonstrated that, in each case, there was a critical sliding speed above which the polyacetal wear rate increased rapidly. This sliding speed corresponded to the point where the sum of the *calculated* flash temperature and ambient temperature reached the melting point of the polyacetal (175°C). In these tests, the duration of sliding was kept deliberately short to limit the mean surface temperature rise. Also, the critical sliding speed decreased if the counterface was heated externally.

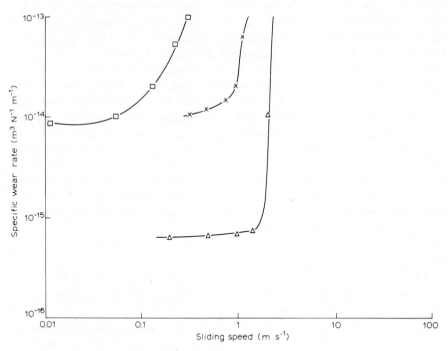

Fig. 11. Polyacetal on non-metallic counterface (pin-on-disc tests). □, Polyacetal/polyacetal [20]; ×, polyacetal/glass [20]; △, polyacetal/glass [21].

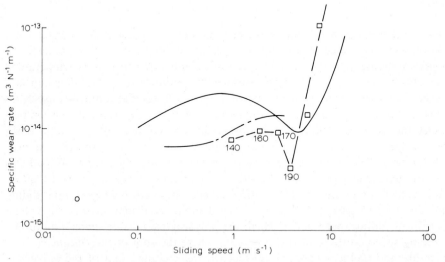

Fig. 12. Comparison of pin-on-disc tests for polyacetal sliding on steel. ○, k_0 wear rate; — — —, NCT pin-on-disc; ———, Evans and Lancaster [20] pin-on-ring; — · —, Tanaka,and Uchiyama [21] pin-on-disc.

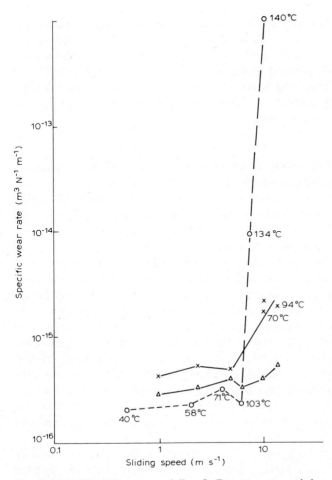

Fig. 13. UHMWP sliding on a steel disc. O, Dry; ×, water cooled; △, 95/5 water/oil emulsion.

The results of Evans and Lancaster [20] and of Tanaka and Uchiyama [21] with polyacetal on glass (Fig. 11) show good agreement on the critical speed. Figure 12 compares NCT and other pin-on-disc tests for polyacetal sliding against mild steel. Here, too, there is good agreement on the critical sliding speed. The temperatures shown against the NCT results refer to the sum of the measured mean pin temperature and the flash temperature obtained using Lancaster's formulae. Similarly, Fig. 13 compares the results of similar experiments with ultra-high molecular weight polyethylene sliding against steel.

Two points should be noted in connection with the effect of flash temperature. Firstly, there is no method of measuring the flash temperature rise in a polymer–metal sliding combination; they can only be estimated by calculation and the reliability of the estimation is critically dependent on the assumptions that are made. Secondly,

Lancaster's method assumes that the entire load is borne on a single asperity and that the resultant area of contact is plastically deformed. The flash temperatures calculated in this way are certainly optimistic though, as the results indicate, this method can be used to predict critical sliding speeds in pin-or-disc experiments with reasonable accuracy. In conforming contacts such as journal and thrust bearings, this method would greatly overestimate flash temperatures because the load is carried on more than one contact point.

5.4. Counterface roughness

Pressure, temperature, and sliding speed may all be described as "application parameters" which derive directly from the duty requirements of a bearing application. It may be possible for the designer to change these parameters at the design stage. Another parameter which greatly influences the wear rate of polymers and composites is the topography of the surface against which it slides.

Investigations of the influence of surface roughness on the wear rate of polymers is complicated by the fact that the initial roughness is modified either by transfer of the polymer or by the abrasive action of fillers in a composite. To overcome these effects, Lancaster [22] conducted experiments in which a polymer pin was loaded against a steel ring and continuously brought into contact with a fresh part of the ring. Under these conditions, the wear rates of a number of polymers are very sensitive to the centre-line-average surface roughness. The situation for high density and UHMWP in pin-on-disc tests with repeated traversals over the same track is shown in Fig. 14. Data from two publications [23,24] have been plotted on the same graph. Despite the fact that the polymer transfers during sliding, there is still a large increase in wear rate as the initial roughness increases from 0.3 to 8 μm. However, roughness values in the range 1–10 μm are non-typical of those normally encountered as bearing counterfaces.

A number of researchers have reported a minimum in the wear rate–surface roughness curve for high density polyethylene on steel. Dowson et al. [23] reported that the minimum occurred at around 0.03 μm Ra, whilst Gillis' work [25] indicated 0.15 μm, and that of Buckley [26], 0.3 μm. However, whilst the Ra value is frequently used, the work of Hollander and Lancaster [27] demonstrated that the radii of the asperity peaks plays a significant role in determining the wear rate of polymers. The latest digital profilometers give a range of surface profile descriptors and further work remains to be done on this topic.

The effect of counterface roughness on the wear rate of polymer composites depends markedly on the ability of the filler to modify the initial roughness. The more abrasive fillers such as glass are particularly effective in this respect and this is illustrated by the results of Briscoe and Steward [28] with PTFE containing either 10% graphite or 10% glass spheres. Graphite is considerably less abrasive than glass and consequently the composite wear rate is more sensitive to counterface roughness. As with unfilled polymers, a minimum in the wear rate–roughness curve occurs at around 0.35 μm Ra. By contrast, the wear rate of a PTFE composite containing small glass spheres is independent of roughness over the range 0.1–1.5 μm Ra.

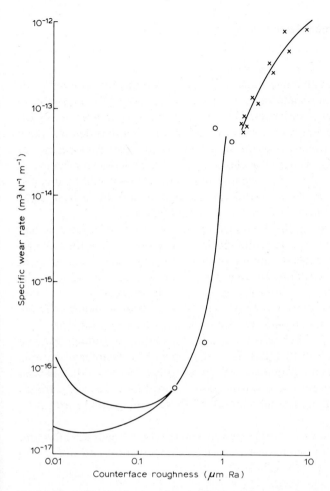

Fig. 14. The influence of counterface roughness on the wear rate of polyethylene. ○, Dowson et al. [23] UHMWP on En 5B pin-on-disc; ×, Bahadur und Stiglich HDP on steel.

High strength carbon fibres (type II) are considerably more abrasive than the high modulus variety (type I). Evans and Lancaster reported [29] that the wear rates of six polymer composites containing 30% type II fibres are much less sensitive to counterface roughness compared with composites of the same polymers containing type I fibres. From the data available, it is apparent that composites containing a filler of low abrasiveness are most sensitive to counterface roughness. The bearing geometry also appears to have an effect with composites containing glass fibre, tests with a "closed" conforming bearing show less sensitivity to roughness than "open" geometry such as pin-on-disc or line contact. If wear debris is retained within the bearing, it can presumably more effectively polish a counterface than if it escape.

6. Factors influencing material selection

6.1. The influence of fillers. Choice of counterface

It has already been noted that the large majority of non-metallic commercially available bearing materials are composites in which one or more filler is combined with a polymer and that filler abrasiveness plays a significant role in determining the influence of counterface roughness on wear rate. Other work [29] has demonstrated that a degree of filler abrasiveness is useful in enhancing the wear performance of PTFE composites operating in water. The choice of suitable materials for operation against relatively soft counterfaces (for example aluminium alloy) would be dictated by their abrasiveness. With all these considerations, it is important to have some method of assessing the abrasiveness of polymers and composites.

A convenient method of doing this is to measure the wear rate of a metallic ball loaded against a rotating polymer composite specimen. The wear rate of the ball at a fixed load and sliding speed depends, among other things, upon the abrasiveness of the composite and the ball hardness. In such concentrated contact conditions, the polymer does not form a transfer film (except when a large wear scar has been formed on the ball). Whilst the magnitude of the ball wear rates so determined are not, in themselves, significant, comparative values for different composites are.

Table 5 shows the results for such tests for a range of composites sliding against a brass ball [30]. The data confirm that glass fibre-reinforced PTFEs are considerably more abrasive than the carbon-, graphite-, and bronze-filled versions. Abrasiveness increases slightly as the glass fibre content increases. Petroleum cokes and carbons are used as fillers in PTFE and combinations of these with graphite are more abrasive than graphite fillers alone.

Of the two commonly used carbon fibre-reinforcements (type I high modulus and type II high strength), the former have a more graphitic structure and are conse-

TABLE 5
Relative abrasiveness of common polymer fillers

Filler	Relative abrasiveness [a] $(m^3 \, N^{-1} \, m^{-1} \times 10^{-17})$
Mineral/ceramic	1000–2000
Glass fibre	600–1600
Type II or AS carbon fibres	500– 600
Graphite/carbon	300
Textile fabric	200
Bronze powder	20
Type I carbon fibre	10
Graphite	1

[a] From the wear rate of a brass ball sliding against polymers with the above fillers under well-defined conditions.

TABLE 6
Influence of ball hardness on relative abrasiveness

Material	Ball (hardness, VPN)				
	Acrylic (30)	Aluminium alloy (120)	Brass (200)	316 stainless steel (300)	Ti alloy (400)
46% Carbon fibre-filled PES	10 000	860	190	2.4	5
25% Glass fibre-filled PES		20	8.3	1.0	0.08
PTFE + glass fibre	550	10	5	0.27	0.3
PTFE + carbon/graphite	100	20	3	0.04	

quently less abrasive than type II fibres. This effect seems to be largely independent of the type of polymer reinforced with carbon fibres. Type AS (intermediate) carbon fibres are similar in abrasiveness to type II fibres.

With some thin-layer materials, abrasive constituents (glass fibres) are present at the surface and so they have a high abrasiveness. Where no abrasive constituents are present at the surface, relative abrasiveness is low.

Table 6 shows how the abrasiveness of four composites decreases rapidly as the hardness of the ball is increased. When a hard carbon–chromium steel ball was used (700 VPN), none of these composites produced a wear scar.

The way in which the relative abrasiveness of composites dictates the choice of material to run against a soft counterface such as aluminium alloy is demonstrated as follows. The steady-state wear rates of aluminium alloy counterfaces have been measured in a thrust bearing test against composites of widely varying abrasiveness [30]. Figure 15 shows that the counterface wear rate increases rapidly above a critical abrasiveness. With highly abrasive composites aluminium pick-up took place with a high rate of counterface wear. These results indicate that low abrasiveness composites can be used in conjunction with aluminium alloys since, at light loads, the intrinsic oxide layer on the alloy is not disrupted. Alternatively, the hardness can be increased by hard anodising or other treatments if the use of abrasive composites is unavoidable.

6.2. Influence of the environment

Another major consideration in the selection of polymers and composites is the influence of the environment on their wear performance. This covers operation in water and other fluids, vacuum, inert gases, and exposure to ionising radiation and many polymer composites have been successfully used in these conditions. The effect of each of these environments will be considered in turn.

6.2.1. Water

There is a distinct difference in the effect of water on the wear rate of polymers and composites that rely on transfer films (e.g. filled PTFE, HDPE, UHMWPE) compared with those that do not (e.g. reinforced thermosets). The presence of water inhibits the formation of a transfer film and this results in higher wear rates. The

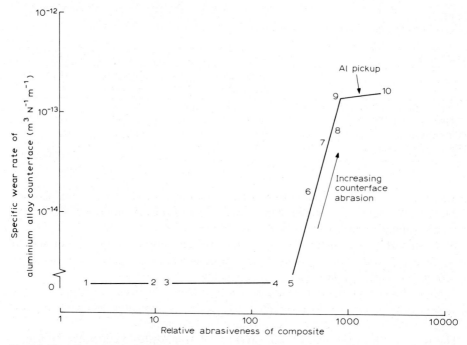

Fig. 15. The effect of various polymer fillers on the wear of aluminium alloy. 1, PTFE flock/fabric; 2, nylon 6,6 + 30% HM carbon fibre; 3, PTFE + 60% bronze; 4, PTFE + 22% carbon/graphite; 5, polyester/textile; 6, polyester + 30% AS carbon fibre; 7, PTFE + 15% glass fibre; 8, nylon + 40% AS carbon fibre; 9, PTFE + 25% glass; 10, PTFE mineral fibre.

TABLE 7

Effect of water on the wear rate of some polymers and composites

Material	Wear rate ($m^3 N^{-1} m^{-1} \times 10^{-15}$)	
	Dry	Wet
PTFE + graphite	1.3	14
PTFE/25% glass fibre	7.7	330
PTFE/25% asbestos	26	500
PTFE/mica	12.5	50
PTFE/graphite bronze	6.5	6
Polyacetal	20	20
Polyphenylene oxide	250	200
PTFE/carbon fibre	2	100
PTFE/polyimide	1	50
Polyurethane + fillers	9	80
UHMWP	0.048	4.5
Textile laminate	1.5	5.2
Cotton/phenolic + PTFE	5	4

results of thrust bearing tests indicate this effect; the wear rates of reinforced thermoset composites are only slightly affected by the presence of water (Table 7). The other materials, which rely on a transfer film all show a significant increase in wear rate when immersed in water.

6.2.2. Other fluids (excluding oils)

In a detailed study on the wear of polymers and composites boundary lubricated by a range of fluids, Evans [31] has reported as follows. For crystalline polymers such as polyacetal and PTFE, most fluids slightly decrease the wear rate below its unlubricated value. Many fluids also reduce the wear rates of amorphous polymers such as polyphenylene oxide and polymethyl methacrylate but very high wear rates occur in those fluids that are strongly absorbed, leading to stress cracking or crazing. This effect takes place when the solubility parameter (a measure of the intermolecular cohesion within the fluid) is close to that of the polymer. Finally, the wear rates of epoxy–carbon fibre and PTFE composites are increased by the presence of all fluids. This is due to the inhibition of transfer film formation.

Recent work [32] has shown that small quantities of perfluoropolyether fluids can significantly reduce the wear and friction of polyimides and PEEK at temperatures up to 200°C.

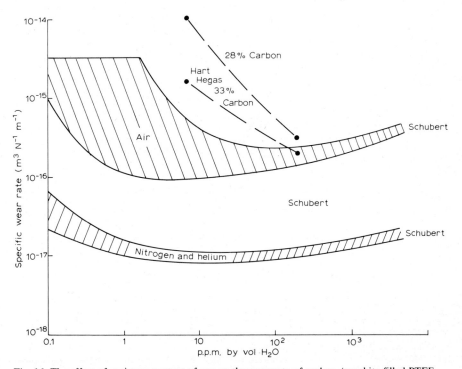

Fig. 16. The effect of moisture content of gas on the wear rate of carbon/graphite-filled PTFE.

6.2.3. Inert gases

The requirement of polymers and composites to operate in inert gas environments is particularly important in pump and compressor applications handling such gases. A number of investigations have been carried out on PTFE composites in inert gases to determine wear rates when the moisture content of the gas was varied. The moisture content of the gas appears to be particularly important with carbon/graphite filled PTFE (Fig. 16). Schubert [33] has shown that the wear rate of such composites increased as the moisture content was reduced below about 10 p.p.m. by volume. Hart [34] indicated a similar effect in helium, but Schubert's results showed a lower wear rate in dry helium and nitrogen than dry air. The lubricity of graphite is known to be critically dependent on the presence of moisture and it is to be expected that composites, including graphite, would show increased wear rates in dry conditions. The reason for the superior performance in dry nitrogen and helium is not clear. However, blends of carbon/graphite may well be less sensitive to moisture effects.

Schubert [33] and Fuchsluger and Taber [35] have tested bronze and glass fibre composites in air and inert gases with various moisture contents. Schubert's results (Fig. 17) showed the glass fibre composite to have a better wear resistance than

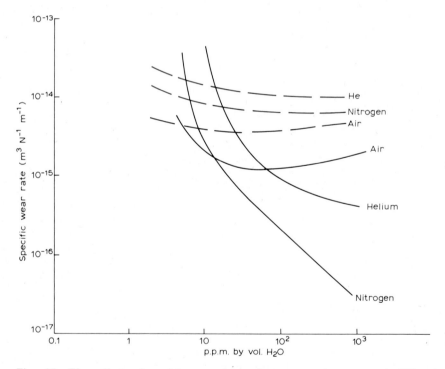

Fig. 17. The effect of moisture content of gas on the wear of 25% glass-filled (———) and bronze-filled (— — —) PTFE.

TABLE 8
Effect of vacuum on the wear rates of some composites

Material	Wear rate ($m^3 N^{-1} m^{-1} \times 10^{-15}$)	
	In air	In vacuum [a]
Unfilled polyimide	0.45	0.15
Polyimide + PTFE	0.75	0.075
Polyimide + graphite + PTFE	0.15	45
Polyimide + MoS$_2$	0.45	0.15
Polyimide + type II carbon fibres	1.5	15
Polyimide + type I carbon fibres	0.15	22
PTFE/glass fibres/MoS$_2$	0.19	0.015
PTFE + Pb/bronze sinter	0.15	0.3

[a] Less than 10^{-6} mbar.

bronze except in very dry conditions (below 20 p.p.m. moisture). Fuchsluger and Taber obtained the opposite result, although their wear rates were generally higher.

There seems, therefore, to be no discernible pattern in the wear of PTFE composites in inert gases. The moisture content plays a significant role with some composites, but has little effect with others.

6.2.4. Vacuum

The wear rates of a number of materials in air and in vacuum are compared in Table 8. Here, there is a distinct pattern in that the wear rate of graphite-filled composites, for example graphite-filled polyimide, increases dramatically in vacuum and this applies also to composites containing type I carbon fibres. Conversely, the wear rates of composites with molybdenum disulphide decreases in vacuum. This is consistent with the fact that the absence of condensable vapours in vacuum leads to "dusting" of graphite, whereas MoS$_2$ does not depend on adsorbed vapours for its ability to lubricate.

6.2.5. Irradiation

Polymers have a much lower resistance to ionising irradiation than, for example, metals [36,37]. The absorbed doses of gamma irradiation required to cause both moderate and severe reduction in the mechanical properties of typical dry bearing polymers are given in Table 9. This shows that aromatic polymers (for example polyimide and polyphenylene oxide), retain their mechanical properties at high absorbed doses compared with other polymers.

PTFE suffers a severe reduction in its mechanical properties at doses above 10^5 rad and so has usually been ruled out as a bearing material in conditions of high irradiation. Whilst this is the case for bulk PTFE containing dispersed fillers such as glass fibres, graphite and bronze, it is not necessarily true for metal-backed thin layer PTFE composites. Table 10 gives wear rates and running-in wear depths for a

TABLE 9

Effect of gamma irradiation on the mechanical properties of polymers

Material	Mild to moderate effect [a]	Moderate to severe effect [b]
Polyacetal	10^5–10^6 rad	Above 10^6 rad
Nylon	5×10^5–10^6 rad	Above 10^6 rad
Polyethylene	10^7–6×10^7 rad	Above 10^8 rad
PTFE	5×10^4–10^5 rad	Above 10^5 rad
Polyimide	10^8–10^{10} rad	Above 10^5 rad
Polyethylene oxide	10^8–10^9 rad	Above 10^9 rad
Polyester textile laminate	3×10^8–10^9 rad	Above 10^9 rad

[a] Less than 50% reduction in some mechanical properties.
[b] Greater than 50% reduction in mechanical properties.

PTFE fibre/glass fibre material from thrust bearing tests against polished stainless steel, for unirradiated specimens, and specimens irradiated to 10^6 and 10^8 rad. The main effect is that the running-in wear depth is increased, but the steady-state wear rate is essentially unaffected. The explanation is that, although the PTFE is degraded by irradiation, it can still form a transfer film and the load is carried by the resin and glass fibres.

Figure 18 presents the results of a series of tests on irradiated metal-backed PTFE lead/bronze sinter. The wear rate is unaffected at gamma doses of up to 10^8 rad. Above this, the PTFE/lead surface layer became extremely powdery and a higher wear rate resulted due to the inability to form an effective transfer film.

However, whilst the wear resistance of such PTFE composites is unimpaired at gamma doses up to 10^8 rad, it is possible that the breakdown products of PTFE could cause stress corrosion of metallic counterfaces and, for this reason, they have not found widespread use in radiation applications.

Of the thermoplastics, polyethylene has a relatively high resistance to irradiation. Severe reduction in mechanical properties occurs only above 10^8 rad. As Table 10 shows, the dry wear resistance of UHMWP is unaffected by gamma irradiation up to 10^8 rad. Artificial joint components made of this material are usually sterilised by

TABLE 10

Effect of gamma irradiation of the running-in wear and steady-state wear rates of a PTFE fibre/glass fibre composite

Gamma dose (rad)	Running-in wear depth (mm)	Steady-state specific wear rate ($m^3 N^{-1} m^{-1} \times 10^{-16}$)
0	0.062	4.5
10^6	0.08	5
10^8	0.1	6

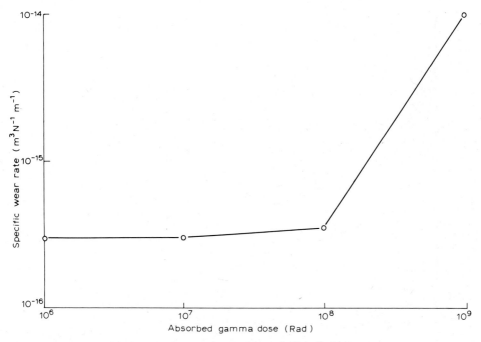

Fig. 18. PTFE + lead bronze vs. En 3B thrust washer at 1 MPa and 20°C.

gamma irradiation (normally to about 2×10^6 rad absorbed dose). It is therefore useful to know that its wear performance is unaffected by such dosage.

6.3. Bearing geometry. The effect of trapped debris

In typical engineering bearings of the journal and thrust type, the rubbing surfaces are highly conforming so that they are much less efficient in expelling wear debris compared with non-conforming geometries such as gears and cams. Wear rates obtained from thrust bearing tests might thus be affected by the trapping of wear debris, depending upon its nature. This section describes the experiments performed to assess the influence of trapped wear debris on the wear rate of certain polymers and composites [38].

The first indication of a wear debris effect became apparent during thrust bearing tests at NCT with the PTFE/bronze sinter material against smooth mild steel at high bearing pressures. These results indicated a very rapid increase in specific wear rate with bearing pressure, even though the surface temperature was kept below 50°C. Examination of the steel counterface during these tests revealed that, at best, the transfer film was patchy. As wear progressed, any transfer was removed by abrasive action of the wear debris particles.

In order to remove the wear debris, radial grooves were machined in the specimen. Initially, grooves were so machined but further tests were done with 16 and 32 grooves. The results obtained are presented in Fig. 19. The wear rate of this

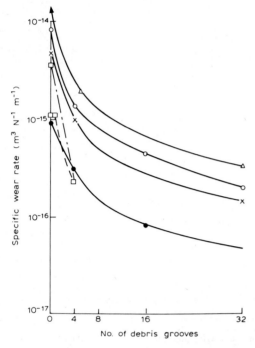

Fig. 19. The effect f debris grooves on the wear rates of PTFE + lead bronze (———) and PTFE + bronze mesh (— · —). △, 140 MPa; ○, 80 MPa; ×, 60 MPa; ●, 40 MPa.

material was reduced by an order of magnitude at high pressures in the test with grooved specimen. In these tests a thin, uniform PTFE transfer film was formed and further confirmation of a debris tapping effect is given by the results of a series of tests with specimens having 4 grooves (Fig. 20). With the same bearing pressure in each test, the specimen wear rate was measured as the amplitude of oscillation against the counterface was progressively increased. An angle of ±45° means a total sweep of 90° and that every part of the counterface was exposed to a debris groove at some stage during one oscillation. Thus, increasing the angle of oscillation improved the efficiency of debris removal which resulted in a reduced wear rate.

Two other thin layer PTFE composites reinforced with a bronze mesh have also been tested. Like other bronze-filled PTFE materials, the wear rates of both increase rapidly at pressures above 10 MPa when sliding against a mild steel counterface. Once again, the inclusion of 4 debris grooves results in a marked decrease in wear rate. For these materials, this reduction occurred at low, as well as at high bearing pressures (Fig. 19).

The PTFE fibre/glass fibre material produces glass wear debris which is extremely fine, so even though it is hard, it does not appear to act as an abrasive. The results of wear tests on this material, both with and without debris grooves, are given

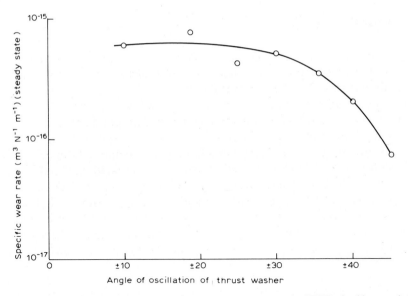

Fig. 20. The effect of the angle of oscillation on the wear rate of a PTFE + lead bronze thrust bearing with four debris grooves (90°). Bearing pressure = 82 MN m^{-2}.

in Table 11. These show that, even at very high stresses, the wear rate is unaffected by removal of debris.

6.4. Friction

In terms of operating parameters, the friction of polymers and composites tends, in general, to decrease with increasing bearing pressure and to increase as the temperature increases. The friction of the thin layer materials, however, tends to decrease with increasing temperature, whereas the friction of other classes is less sensitive to changes in temperature.

TABLE 11

Effect of debris grooves on the wear rate of PTFE fibre thin layer materials (thrust bearing tests)

Material	Bearing pressure (MN m^{-2})	Wear rate (m^3 N^{-1} m$^{-1} \times 10^{-17}$)	
		Plain bearing	4 debris grooves
PTFE fibre/glass fibre	82	7	5
PTFE fibre/glass fibre	140	17	10
PTFE flock/Nomex	140	20	18

7. Selection of materials

The preceding sections have provided information on the wear and friction of some commercially available polymer composites. The factors which influence performance, both operational and environmental, have been discussed. In any one application, it is unlikely that more than a few parameters will affect the friction and wear rate of the composite, but it is essential to identify the important parameters as part of the procedure for selecting a material.

A step-by-step procedure for selecting suitable polymer composites for sliding applications is shown in Fig. 21. It is a method by which candidate materials are chosen by a process of elimination and this type of method has been adopted in two publications [3,4]. This method is particularly suitable where application conditions are onerous by virtue of the environment or extremes of load, speed, temperature, etc. With "ambient" or "moderate" conditions, many materials may be suitable unless a particularly long life (low wear rate) is required. In such circumstances,

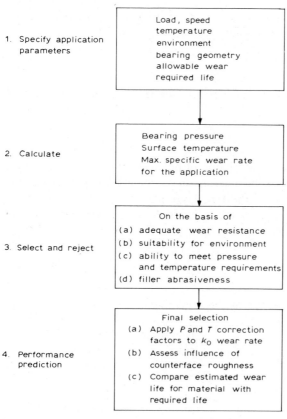

Fig. 21. Selection of materials.

other factors, such as cost, availability or suitability for volume production (e.g. injection moulding), will decide the final choice.

For a satisfactory outcome in terms of reliability and adequate life, all these factors must be taken into account at the design stage and a trial run in the application itself, or under conditions which closely simulate it [39], should be undertaken. Selection of polymer or composite materials merely from catalogue data, or even from past experience under supposedly similar conditions, may lead to problems later on.

References

1 J.K. Lancaster, Tribol. Int., December (1973) 219.
2 Eng. Sci. Data Item 68018, ESDU, London, 1968.
3 Eng. Sci. Data Item 76029, ESDU, London, 1976.
4 National Centre of Tribology, Polymer Materials for Bearing Surfaces, NCT, 1983.
5 General Discussion on Lubrication, Vol. 2, Inst. Mech. Eng., London, 1937.
6 Bearing Design, Polypenco Ltd., Welwyn Garden City, Herts., Gt. Britain.
7 A.B. Crease, Tribol. Int., 2 (1973) 15.
8 Eng. Sci. Data Item 78026–78029, ESDO, London, 1978.
9 J.K. Lancaster, RAE Techn. Rep. No 78046, 1978.
10 D. Dowson, J.R. Atkinson and K. Brown, in L.H. Lee (Ed.), Advances in Polymer Friction and Wear, Vol. 5B, Plenum Press, London, 1974, p. 533.
11 J.K. Lancaster, D. Play, M. Godet and A.P. Verrall, RAE Tech. Rep. No 78035, 1978.
12 J.C. Anderson and E.J. Robbins, in D. Dowson (Ed.), Wear of Non-metallic Materials, Proceedings of the 3rd Leeds–Lyon Symposium on Tribology, Mechanical Engineering Publications, London, Paper IV (iii), 1978.
13 J.P. Giltrow, RAE Tech. Rep. 72052, 1972.
14 R.B. Lewis, ASLE Prepr. 69 AM 56-2 (1969).
15 D.C. Evans, Tribology 1978, Mechanical Engineering Publications, London, 1978, pp. 1–5.
16 G.C. Pratt, in M. Richardson (Ed.), Polymer Composites, Applied Scientific Publishers, Barking, 1979, Chap. 5.
17 R.B. King, RAE Tech. Rep. 78663, 1978.
18 J.F. Archard, Wear, 1 (1958/1959) 200.
19 J.K. Lancaster, Tribology, 4 (1971) 82.
20 D.C. Evans and J.K. Lancaster, Discussion on paper quoted in ref. 12.
21 K. Tanaka and Y. Uchiyama, in L.H. Lee (Ed.), Advances in Polymer Friction and Wear, Vol. 5B, Plenum Press, London, 1974.
22 J.K. Lancaster, Proc. Inst. Mech. Eng. London, 183 (1968–1969) 98.
23 D. Dowson, J.M. Challen, K. Holmes and J.R. Atkinson, in D. Dowson (Ed.), Wear of Non-metallic Materials, Proceedings of the 3rd Leeds–Lyon Symposium on Tribology, Mechanical Engineering Publications, London, Paper IV (iv), 1978.
24 S. Bahadur and A.J. Stiglich, Wear, 68 (1981) 85.
25 B.J. Gillis, Ph.D. Thesis, University of Leeds, 1978.
26 D.H. Buckley, in L.H. Lee (Ed.), Advances in Polymer Friction and Wear, Vol. 5B, Plenum Press, London, 1974, p. 542.
27 A.E. Hollander and J.K. Lancaster, Wear 25 (1973) 155.
28 B.J. Briscoe and M. Steward, Tribology 1978, MEP, 1978.
29 D.C. Evans and J.K. Lancaster, in D. Scott (Ed.), Materials Science Series, Vol. 13, Wear of Polymers, Academic Press, New York, 1979.

30 J.C. Anderson and A. Davies, Lubr. Eng., 40 (1984) 41.
31 D.C. Evans, in D. Dowson (Ed.), Wear of Non-metallic Materials, Proceedings of the 3rd Leeds–Lyon Symposium on Tribology, Mechanical Engineering Publications, London, Paper III (i), 1978.
32 J.C. Anderson, L. Flabbi and G. Capporicio, in W. Bartz (Ed.), Synthetic Lubricants and Operational Fluids, TAE, Esslingen, 1984.
33 R. Schubert, Trans. ASME, 30 (1971) 206.
34 R.R. Hart, Proc. Inst. Mech. Eng., 181 (1967) 130.
35 J.H. Fuchsluger and R.D. Taber, J. Lubr. Technol., 95 (1971) 423.
36 M.H. Van de Voorde and C. Restat, Selection Guide to Organic Materials for Nuclear Engineering, CERN 72-7, Geneva, 1972.
37 J.W. Hitchon, AERE Rep. M3251, AERE, Harwell, 1963.
38 J.C. Anderson and E.J. Robbins, in K.G. Ludema (Ed.), Wear of Materials, American Society of Mechanical Engineers, New York, 1981, pp. 539–543.
39 J.C. Anderson and P.K. Williamson, in L.H. Lee (Ed.), Int. Symp. Polym. Wear Control, American Chemical Society, Washington, DC, 1985.

Chapter 11

Composites for Aerospace Dry Bearing Applications

JOHN K. LANCASTER

Materials and Structures Department, Royal Aircraft Establishment,
Farnborough, Hants. (Gt. Britain)

Contents

Abstract

Most bearings associated with flight-control applications in fixed wing aircraft and helicopters comprise polymer-based composites in the form of thin layers, ~ 0.3 mm thick, adhesively bonded to a metal substrate. The various types of composite used for this purpose are described and information given on how their friction and wear properties depend on the conditions of sliding; stress, temperature, counterface metal, contamination by fluids, type of loading, and various kinematic factors associated with the bearing design. A long-standing objective has been to relate friction and wear performance to composite structures and compositions and some results are given from recent research describing progress towards this end. Finally, brief mention is made of future trends and requirements for composites in dry-bearings, with particular reference to high temperature applications.

1. Introduction

During the past two or three decades, there has been an enormous increase in the usage of dry-bearings throughout industry in general and for aerospace applications in particular. A wide spectrum of materials is available ranging from ceramics to solid-film lubricants [1], but plastics-based composites offer an attractive solution for almost all applications except, of course, those involving very high temperatures. The introduction of fillers or fibrous reinforcements into plastics for bearing purposes originally arose from the need to improve mechanical properties governing load-carrying capacity, such as strength, stiffness, creep, etc. A large volume of

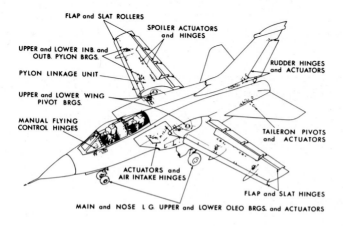

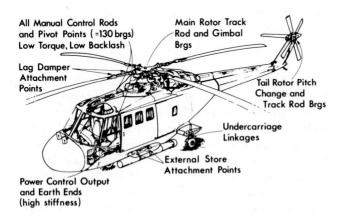

Fig. 1. Applications of thin composite bearing liners for flight control in fixed wing aircraft and helicopters.

literature now exists on these structural effects of reinforcements (see, for example, ref. 2). It was soon realized, however, that the presence of fillers and fibres in plastics could modify the friction and wear behaviour quite independently of their effects on mechanical properties. It is this aspect which is of primary concern in the present chapter. Discussion will be concentrated primarily on those composites which consist of thin bearing liners adhesively bonded to a harder substrate. Such bearing constructions are now very widely used for flight-control applications in both fixed wing aircraft and helicopters. Some of the specific areas involved are illustrated in Fig. 1.

2. Materials

Until about thirty years ago, all airframe control applications, such as those in Fig. 1, used bronze-steel bearings lubricated by grease and each requiring its own maintenance schedule. The costs of routine maintenance thus tended to become ever more expensive as aircraft increased in complexity. A major innovation occurred in the mid-1950s with the introduction of the first, thin-layer, composite material specifically designed to replace these lubricated metal bearings and so reduce maintenance [3]. The structure of this material is shown in Fig. 2(a) and there were three novel features. Firstly, the thickness of the composite bearing layer was deliberately kept small so as to increase the proportion of load support from its rigid substrate. Secondly, the lubricating constituent, PTFE, was introduced as a fibre rather than as particles because in fibre form it is appreciably stronger and stiffer than the bulk solid. Finally, the intricacies of the fabric weave were so arranged as to produce a contacting surface which was predominantly PTFE and a rear surface which was mainly of some other type of fibre in order to facilitate adhesive bonding to the metal substrate. Liners of this type are still in widespread use today. Glass is commonly used as the reinforcing, bondable fibre and it is also common for the PTFE-containing fabric to be backed by an all-glass fabric which, in turn, is bonded to the substrate. The adhesive used for bonding is frequently the same as the synthetic resin used to fill the interstices of the fabric.

Shortly after the marketing of the above type of bearing liner, and to avoid patent problems, an alternative type of thin liner was introduced in which particulate PTFE was dispersed throughout a fabric-reinforced resin [Fig. 2(b)]. The scope for modifications within these two basic formats is very great and a few of the various possibilities are listed in Table 1. Some of these began to be realized in the early 1970s with a "second generation" of liner types incorporating the latest developments in synthetic fibre and resin technology. Table 2 gives a partial list of the types of composite liners now available commercially, together with a general indication of their composition and structure; precise details are not usually disclosed by the manufacturers.

The thickness of almost all these composite liners is usually around 0.3–0.4 mm and their intrinsic moduli of elasticity are of the order of a few GPa. The overall

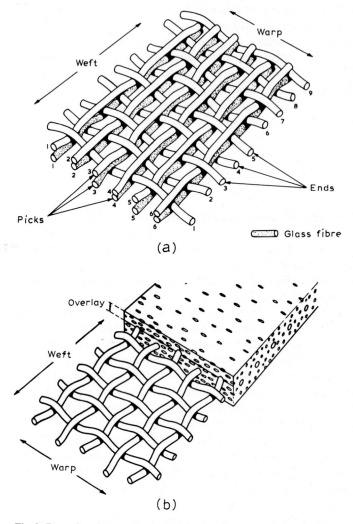

Fig. 2. Examples of two main types of dry-bearing composite liners. (a) PTFE fibre interwoven with glass fibre into fabric. (b) Plain-weave fabric with PTFE dispersed in resin.

stiffness of bearing assemblies incorporating these liners, however, tends to be appreciably greater than this because of the support provided by the metal backing. Load-carrying capacities under static, or low-speed conditions can be very high, $\simeq 400$ MPa, the exact value depending on the construction and the type of reinforcing fibre. The potential advantages of carbon fibres for such thin-layer, bearing composites were recognized at an early stage following fibre availability, and liner constructions then made incorporating helically wound, twisted carbon and PTFE fibres [4]. Initial performance data appeared very promising, but despite this

TABLE 1

Some possible variations in structure and composition for thin-layer, dry-bearing composites

Resin/adhesive	Fibre type	Fibre form	Solid lubricants
Phenolic	Polyimide-amide	Fabric	PTFE
Vinyl–phenolic	Aromatic polyamide	plain	particles/flock in resin
Epoxy	Aramid	satin	fibre interwoven in fabric
Polyimide	Polyester	twill	$(CF_x)_n$
	Glass	Gauze	Graphite
PPS	Carbon		MoS_2
PES	Metals	Filament winding	
PEEK	bronze	Single/multifilament fibres	
	nickel	Mixed multifilaments	
	boron	Fibre size and spacing	

there was no commercial development. Only more recently has interest again begun to emerge in using carbon fibres for high-strength bearing composites [5].

The synergistic effects of bronze and lead in improving the friction and wear performance of PTFE have been known for some time [6] and are attributed to chemical reactions between the Cu, Pb and degradation products of the PTFE [7]. Various composite bearing liners have therefore been made incorporating these elements. In particular, copper wires have been interwoven with PTFE fibres and filled PTFE has been reinforced with a bronze mesh (Table 2). The most successful of all these metal–PTFE composites, however, is undoubtedly that consisting of a

TABLE 2

Some proprietary types of thin-layer, dry bearing composites

Airflon	
Fiberslip	Interwoven PTFE/glass fibres + resin [as Fig. 2(a)]
Fabroid	
X1	As above but some PTFE warp fibres replaced by glass
Faftex	Interwoven PTFE/Aramid fibres + resin
Fibriloid	As above but with overlay of resin/PTFE flock
Faflon	Filament wound PTFE/Aramid fabric layer + resin
Duralon	80% PTFE/20% polyester fabric with filament wound glass/epoxy backing
Uniflon RR	PTFE flock in vinyl phenolic resin reinforced with Aramid fabric [as Fig. 2(b)]
Fraslip	As above but with polyester fabric
Kahrlon	As above but with granulated PTFE
Karon	Resin + fillers injected in bearing clearance
Fillmide	Moulded graphite fibre–polyimide resin + lubricating fillers
Unimesh	Filled PTFE reinforced by bronze mesh
Metalloplast	
Pydane	PTFE fibres interwoven with metal wires + resin
DU	Porous bronze on steel filled with PTFE/Pb

porous bronze layer sintered on to a steel backing and filled with a PTFE–Pb mixture. The maximum load-carrying capacity, ~ 120 MPa, is somewhat lower than that of many fabric-composite bearing liners because of the intrinsic weakness of the bronze particle-to-particle sinter bond in shear. This limitation has tended to preclude their widespread use for flight-control bearings. Variations on this type of construction include porous bronze filled with polyacetal, polyphenylene sulfide, polyvinylidene fluoride or polyether ether ketone. All of these, however, are intended for use in conditions of marginal lubrication rather than for dry sliding.

3. Performance testing

Because the reliability of flight-control bearings is so crucial to the operation of both helicopters and fixed wing aircraft, bearing procurement normally involves compliance with various standards and specifications. The most widely used are those provided by the U.S. Military and listed in Table 3. Selecting one example, MIL-B-81820 incorporates general requirements on materials, design, construction, dimensional tolerances, and ball-surface texture. In addition, there are specific requirements which involve performance testing: radial and axial static limit load; wear during oscillation under radial load at high temperature (325°F), room temperature, and low temperature (− 10°F); no-load, rotational, break-away torque;

TABLE 3
U.S. Military specifications and standards for self-lubricating bearings

Specifications	
MIL-B-8942 [a]	Bearings, plain, TFE-lined, self-aligning
MIL-B-8943 [b]	Bearings, sleeve, plain and flanged, TFE-lined
MIL-B-81819	Bearings, plain, self-lubricating, self-aligning, high-speed oscillation
MIL-B-81820	Bearings, plain, self-aligning, self-lubricating, low-speed oscillation
MIL-B-81934	Bearings, sleeve, plain and flanged, self-lubricating
MIL-B-81935	Bearings, plain, rod-end, self-aligning, self-lubricating
Standards	
MS 14101	Bearing, plain, self-lubricating, self-aligning, low-speed, narrow, grooved outer ring, − 65 to 325°F
MS 14102	Bearing, plain, self-lubricating, self-aligning, low speed, wide, chamfered outer ring, − 65 to 325°F
MS 14103	Bearing, plain, self-lubricating, self-aligning, low speed, wide, grooved outer ring, − 65 to 325°F
MS 14104	Bearing, plain, self-lubricating, self-aligning, low speed, narrow, chamfered outer ring, − 65 to 325°F

[a] Specifications replaced after 1978 for *new* designs by MIL-B-81820.
[b] Specifications replaced after 1978 for *new* designs by MIL-B-81934.

wear after exposure to various contaminating fluids; normal and circumferential bearing conformity; liner condition and bond integrity. Of most interest in the present context is the procedure for wear assessment. The sliding conditions to be used are an amplitude of oscillation of $\pm 25°$ at 10 c min^{-1}, together with a load which depends on bearing dimensions but which results in a nominal stress of approximately 140 MPa over the projected area of the liner surface. The maximum permissible wear also depends on the bearing dimensions, but for spherical bearings of bore sizes between 0.3125 and 1.0 in., it is 0.0035 in. after 1000 cycles and 0.0045 in. after 25 000 cycles.

Whilst these specifications play an essential role in maintaining an acceptable level of quality control, it is important to remember their limitations. The bearing performance demanded is, in general, fairly modest and can be satisfied by numerous types of composite liners. The duration of sliding, 25 kc, is far below that needed for most applications and qualification to specification cannot therefore provide any guarantee of acceptable long-term performance. In addition, because bearing performance depends so critically on the exact conditions of sliding imposed, data obtained from specification testing cannot, in general, be used for design purposes.

To obtain design information from which to predict the wear life of, for example, flight-control bearings, it is ideally necessary to use test equipment which can simulate the continuously varying spectrum of loads, speeds and temperatures characteristic of the flight profile of a particular bearing on a particular aircraft. Such tests are extremely expensive and, moreover, the data obtained is valid only for the conditions imposed; extrapolation to the performance of bearings in other areas or on other aircraft is difficult, if not impossible. A more versatile approach is to evaluate bearings at various fixed combinations of load, speed, temperature, etc. so as to build up a general pattern of behaviour from which extrapolation becomes more justifiable. For the extremes of high loads and low speeds, such tests involve the use of large-scale equipment operating for long periods of time. Figure 3 shows an example of a journal bearing test rig for evaluating 1 in. diam. bearings at nominal stresses up to 420 MPa and speeds of 10 c min^{-1}. In these conditions, the definition of a complete relationship between wear depth (or volume) and time can sometimes require several weeks of continuous operation. There is thus no prospect of being able to use full-scale bearing tests to explore the influence of all the many factors which can influence wear, such as counterface type and roughness, environmental contamination, etc. Full-scale bearing tests are also inappropriate in the early stages of a material's development programme where rapid feedback of information on performance is vital in order to influence the next stages of development. For both of these last two purposes, there is no real alternative but to use some form of accelerated wear-testing procedure.

For thin liner materials whose structure and composition may vary with depth, one appropriate accelerated wear test is to use a "pin-on-disc" geometry as shown schematically in Fig. 4(a). The surfaces of three pins, typically 5 mm in diameter, are lapped co-planar and discs punched from the liner are adhesively bonded to them. Self-alignment is provided by a spherical seating through which the load is applied

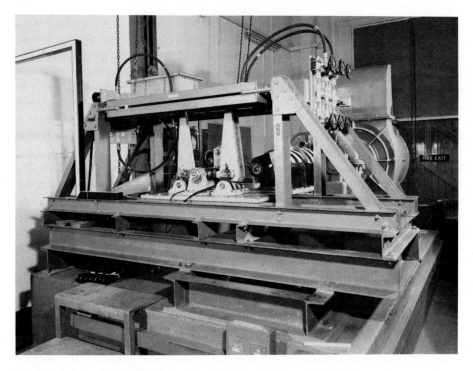

Fig. 3. Test rig for 25 mm diam. journal bearings at stresses up to 350 MPa.

and, with a suitable lever arm arrangement, it is possible to achieve mean stress levels commensurate with those in many applications, e.g. 150 MN m^{-2}. Because the apparent contact area is relatively small, the sensitivity of wear measurement is high. An alternative approach is to use a "pin-on-ring" configuration [8] in which a strip of the liner material is loaded in nominal line contact against the curved surface of rotating ring, Fig. 4(b). Line contact, and thus a high Hertzian stress, is maintained by slowly reciprocating the liner strip in a direction at right angles to the axis of rotation of the ring. The magnitude of the stress is calculated from elasticity theory, knowing the elastic modulus of the liner, and the latter can be derived from ball indentation experiments with the aid of a modified elasticity analysis appropriate to the case of a thin elastic layer on a rigid substrate [9]. Wear is determined from profiles of the worn groove in the liner using a low magnification profilometer.

The above "reciprocating line contact" arrangement bears no resemblance to a conventional cylindrical or spherical bearing geometry and it is therefore essential at the outset to demonstrate that it is able to establish trends in wear behaviour which are similar to those found from full-scale bearing tests. Two examples will suffice for illustration. Figure 5 shows that the general shapes of the wear depth–time relationships for two types of bearing liner are very similar in both bearing and reciprocat-

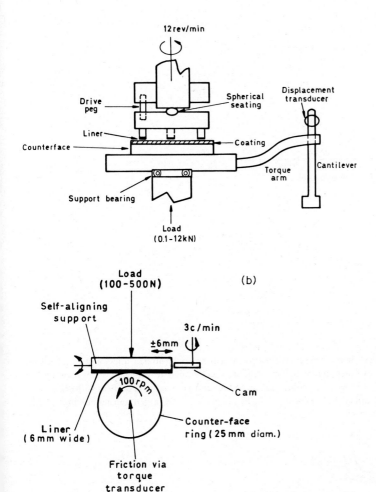

Fig. 4. Schematic arrangements of apparatus for accelerated wear tests. (a) Three pin-on-disc. (b) Reciprocating line contact.

ing line contact tests. The time scale involved in the latter is, of course, much shorter. The second example, in Fig. 6, shows a comparison of the depth–time relationships for nine experimental liner materials obtained in the two types of test. The bearings required some 18 months for evaluation, but the reciprocating line contact tests were completed in 3 weeks. Whilst there are some differences of detail, eight out of the nine materials rank in the same order in both tests. These, and other comparisons [8] have clearly established the value of accelerated wear testing for evaluating composite bearing liners on a comparative basis and for establishing performance trends. Some of the latter will now be reviewed.

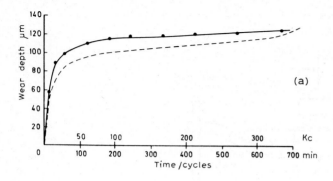

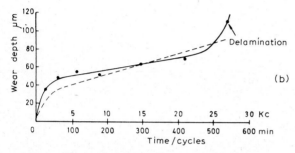

Fig. 5. Comparisons of wear depth–time relationships in bearing tests (broken lines) and reciprocating line contact (solid lines). (a) Liner similar to Fig. 2(a). (b) Liner similar to Fig. 2(b). Reciprocating line contact conditions: 450 N, 0.13 m s^{-1}, unidirectional motion, and Hertzian stress of (a) 65 MPa and (b) 85 MPa. Bearing tests: (a) Spherical bearings, ±25°, 10 c min^{-1}, 256 MPa; (b) journal bearings, ±25°, 40 c min^{-1}, 234 MPa. (From ref. 1.)

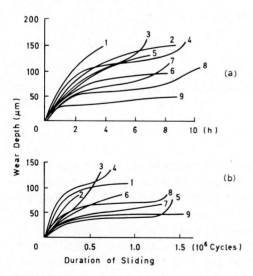

Fig. 6. Comparison of experimental bearing liners in (a) reciprocating line contact, 450 N load (≈ 80 MPa) and (b) journal bearing tests, 12.5×12.5 mm, 234 MPa, ±25°, 40 c min^{-1}. (From ref. 10.)

4. Influence of operating parameters on performance

In most flight-control bearing applications, the usual criterion of failure is unacceptable backlash and the relevant wear parameter is thus the depth of wear. What constitutes "unacceptable" depends on the particular application. In non-critical areas, it may be possible to tolerate wear depths up to the effective thickness of the liner, say 0.2 mm. In others, however, such as manual control-linkage bearings in helicopters, backlash becomes excessive above about 0.02 mm. Idealized representations of the relationships between wear depth and time are shown in Fig. 7(a) and illustrate the point that in the approach to failure there are two aspects of wear which must be taken into account; the initial wear depth and the steady-state rate of wear. Two extreme types of behaviour are given by the two curves in Fig. 7(a) and, depending on the particular depth chosen as the failure criterion, either liner could be the preferred choice.

Extensive wear testing has demonstrated [10] that the depth of initial wear depends primarily on the structure of a composite bearing liner and, in particular, on that of the reinforcing fabric. In contrast, the magnitude of the steady-state rate of wear depends more on the composition of the liner and, in particular, on the concentration of PTFE. The latter arises because during sliding the surfaces of both the liner and its counterface become modified by the generation of "third-body films" [11] which differ in composition and structure from both of the original materials. It is the properties of these third-body films which determine the magni-

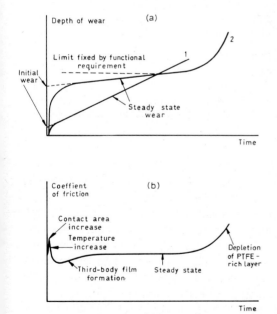

Fig. 7. Idealized variations of (a) wear depth and (b) coefficient of friction with time for thin, composite bearing liners.

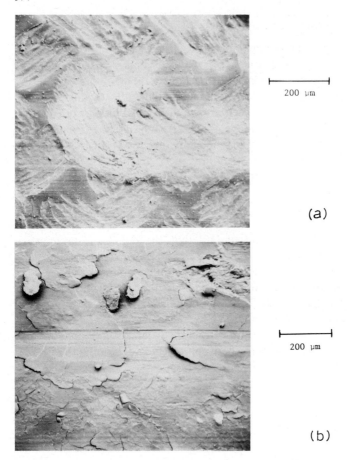

Fig. 8. Third-body surface films on a PTFE/glass fibre/resin, dry bearing liner after (a) continuous rotation and (b) oscillatory motion, amplitude 2.35 mm. Counterface AISI 440C stainless steel and load = 4.82 kN (mean stress ~ 15 MPa).

tude of the steady-state wear rate. These films also influence the level of the coefficient of friction and a typical variation with time is shown in Fig. 7(b). The initial transient in friction is complex and influenced by several factors. The early rise is associated with an increase in the real area of contact as the originally rough liner surface deforms and wears to a smoother topography. The subsequent decrease results from a rise in surface temperature which reduces the shear strength of the developing, PTFE-rich, third-body films and the final rise to constant level follows the development of these third-body films to a steady-state condition. Some typical examples of the third-body films on a composite liner containing PTFE fibre/glass fibre and resin are shown in Fig. 8 and the way in which their formation depends on the many parameters influencing wear will become evident during the following discussions.

4.1. Stress

Most simplified theories of wear lead to the conclusion that the volume of wear per unit distance of sliding is proportional to the absolute load; thus $V/s \propto L$. Stress, per se, is important only via the applied load and the magnitude of the apparent area of contact is irrelevant. In practice, and particularly with thin, composite liners, this idealized situation is not always valid. With porous bronze/PTFE/Pb bearings, for example, the specific wear rate (or wear coefficient), V/sL, remains constant only at low stresses and begins to increase rapidly above about 35 MPa [8]. This is partly a

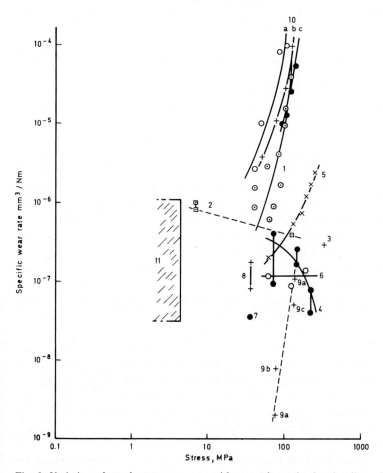

Fig. 9. Variation of steady-state wear rate with stress for a dry bearing liner similar to Fig. 2(a). 1, Reciprocating line contact; 2, 3 pin-on-disc; 3, journal bearings [13]. Oscillatory motion $\pm 25°$; 4, pad on reciprocating flat [14]. Stroke length 0.15 m; 5, pad on reciprocating flat [13]. Stroke length 0.24 m; 6, spherical bearings [15]. Oscillatory motion $\pm 25°$; 7, journal bearings [16]. Continuous rotation; 8, journal bearings [16]. Oscillatory motion, $\pm 25°$; 9, thrust washers [17] (a) unidirectional, (b) and (c) oscillatory motion; 10, journal bearings [18]. Oscillatory motion; 11, miscellaneous bearing tests of filled PTFE formulations [19]. (From ref. 8.)

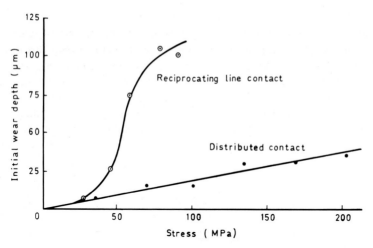

Fig. 10. Variations of initial wear depth with stress for a dry bearing liner similar to Fig. 2(a). (From ref. 20.)

consequence of the composite structure, which is stress-limited by the sinter bond between the bronze particles, and partly due to wear debris inducing additional surface damage and wear during its escape from the contact zone at high stress. The last contribution can be reduced by the provision of slots or holes within the bearing surface [12]. Similar trends between wear rate and stress are also sometimes observed with PTFE fabric bearing liners and Fig. 9 shows a compilation of data for one particular type of liner from a variety of bearing and accelerated wear tests. The very wide variability in both the magnitude of the wear rate and its trend with increasing stress is attributable to the fact that different types of motion were involved in the various test series, together with bearings of different types and sizes. As will be seen later, these factors all influence the wear rate through their effect on third body film formation.

In addition to a general increase in the specific wear rate with stress, the volume or depth of initial wear also tends to increase [20]. Examples are shown in Fig. 10 for one particular type of liner in two types of accelerated test. It may be noted that the magnitude of the stress effect is much more marked during testing in reciprocating line contact than in a distributed contact (three pin/disc). In the former, there is appreciable elastic penetration of the rotating counterface ring into the liner surface which facilitates disruption [10].

Variations in the coefficient of friction with stress, corresponding to those in wear, are shown by the insert to Fig. 11. For any one test series, friction decreases with increasing stress, but the overall level of friction depends on the bearing size and type. When, however, the coefficient of friction is plotted against the absolute load, a "master" curve can be obtained showing an almost monotonic decrease in friction with load. This trend is generally similar to that observed with many engineering plastics in bulk form [21], and with PTFE in particular [22], and results

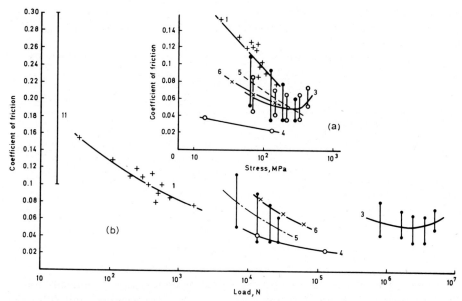

Fig. 11. Variation of friction coefficient with load and stress for tests itemized in the legend to Fig. 9. (From ref. 8.)

from the fact that the mode of asperity deformation is usually mixed elastic/plastic. However, the same trend could also arise, with PTFE-containing liners, from the formation of a low shear-strength, third-body film over the harder substrate; the classical process of thin-film, solid lubrication [23].

Different types of composite bearing liners will obviously exhibit different responses to stress and it is by no means safe to assume that their relative performance rankings will remain the same at all stresses. This has recently been confirmed [24] in some experiments with cylindrical bearings where an experimental liner, specifically developed to give an improvement in wear life over the original product from which it was derived of ×3 at 100 MPa, showed a reduction in life at 300 MPa of ×5.

4.2. Temperature

It is now well-established that the wear rates of all PTFE composites tend to increase with increasing temperature [25]. The extent of this increase is often much greater than can be accounted for on the basis of reductions in composite strength or stiffness and the main reason appears to be that it becomes more difficult to form coherent third-body films of PTFE at high temperatures [26]. Possible explanations for the latter could be that reductions in composite stiffness reduce the localized contact stresses to a level insufficient to consolidate the PTFE wear debris. Alternatively, as the elastic modulus of the PTFE decreases, the load could become increasingly supported by filler particles which might then abrade away any transfer film of PTFE on the counterface.

TABLE 4

Influence of high (150°C) and low (−50°C) temperature on wear of 9 types of bearing liners from different manufacturers

Spherical bearings, ball diam. = 31.75 mm, oscillatory motion ±25°, 10 cycles min⁻¹, load = 80 kN

Liner type		Ratio of wear rates	
		150°C/20°C	−50°C/20°C
Interwoven reinforcing fibre and	i	1.0	4.1
PTFE fibre with resin	ii	3.7	27
	iii	0.8	14.5
	iv	0.2	2.2
PTFE particles or flock dispersed	v	12.7	34
in fabric-reinforced resin	vi	12.9	3.5
	vii	4.8	8.0
	viii	1.9	3.8
	ix	1.3	1.3

The same trend of wear rate increasing with temperature is also generally found with thin, composite liners containing PTFE in both accelerated and full-scale bearing tests [8]. In addition, the depth (or volume) of initial wear also increases with temperature [27]. The proportionate change in the steady-state wear rate with temperature depends, not surprisingly, on the type of bearing liner. The first part of Table 4 shows some results obtained during spherical bearing tests [16] with nine different types of thin composite liners. The ratio of the wear rate at 150°C to that at 20°C shows considerable variability and two liner types do, in fact, show a reduction in wear rate at 150°C. Whilst the exact behaviour must clearly be influenced by the composition and structure of the liner, it has not, so far, proved possible to establish any general relationships.

A similar state of affairs applies to low temperature performance. The second half of Table 4 shows that, compared with room temperature, the wear rates of spherical bearings increase at −50°C, the amount again depending on the type of liner. A similar increase in wear at low temperatures (−40°C) has been reported [13] for a PTFE fibre/glass fibre liner during "pad-on-track" tests. The latter results have also demonstrated the additional point that the proportionate increase in wear did not vary appreciably with stress over the range 60–250 MPa. It is reasonable to envisage that one reason for the increased wear of all thin, composite liners at low temperatures is the greater brittleness of the synthetic resin matrix. However, another possibility is that the bearings may become contaminated by condensation of water vapour. The part played by fluid contamination in the wear of composite liners is discussed later.

4.3. Counterface effects

The role of the counterface in the friction and wear behaviour of plastics composites has probably received more research attention than any other parameter.

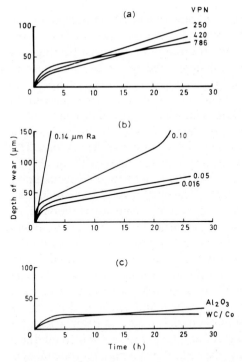

Fig. 12. Counterface effects on the wear of a dry bearing liner similar to Fig. 2(a) but with 10% of the PTFE warp fibres replaced by glass. [See Fig. 18(c).] (a) 440C Steel of three hardness levels. $R_a = 0.05$ μm. (b) 440C Steel of four roughnesses. Hardness = 750 VPN. (c) Plasma-sprayed WC/Co and Al_2O_3 coatings. $R_a \simeq 0.02$ μm. Three pin-on-disc tests. Load = 8 kN (stress = 133 MPa), sliding speed = 0.02 m s^{-1}.

Three aspects are involved; counterface type, hardness, and surface roughness. The usual counterface for spherical, airframe control bearings is AISI 440C stainless steel ($\sim 1\%$ C, $\sim 17\%$ Cr), hardened to between 600 and 740 VPN and finished to a surface roughness of less than 0.2 μm Ra. Hard, chromium-plated ball surfaces are also sometimes used.

The effects of counterface hardness on liner wear are illustrated in Fig. 12(a) for a PTFE fibre/glass fibre liner sliding against 440C steel tempered to different hardness levels. The steady-state wear rate increases as the counterface hardness decreases. The particular liner used in these tests contained some glass fibre in the rubbing surface and examination of the wear tracks on the counterfaces showed that the softer ones suffered some abrasion damage and an increase in roughness above the initial value. It is this increased roughness which is primarily responsible for the increased rate of wear. Although a high counterface hardness, for minimum liner wear, could well be less critical when using liners which do not contain abrasive components like glass, there is always the possibility of abrasive contamination occurring from the environment. A high counterface hardness is thus always a preferred choice.

The influence of changes in the initial counterface roughness on liner wear are shown in Fig. 12(b). These particular results, like those in Fig. 12(a), are from three pin-on-disc tests, but similar trends have been observed during journal bearing and pad-on-track tests [13,18]. Both the initial wear depth and the steady-state rate of wear appear to increase monotonically with increasing roughness; there is no suggestion of a minimum in the wear rate–roughness relationship, as is sometimes observed with other polymer–metal combinations [28–30]. A general explanation for these effects of counterface roughness can be provided as follows. In the early stages of sliding transfer occurs to the counterface and fills the surface depressions. The amount and rate of wear in these early stages thus increases with increasing roughness, as observed. In steady-state conditions, the rate of wear depends upon the topography generated during the early stages. This may involve some reduction in roughness due to transfer filling the surface depressions, as already mentioned, or there may be either an increase or a decrease in roughness due to abrasion by fillers or fibres within the composite liner. The extent of such abrasion will depend on the relative hardness of the fillers and the counterface and on the shape, size and orientation of the fillers or fibres. On the hardest counterfaces, significant damage and an increase in roughness is only likely to be detectable on surfaces which are initially very smooth; rougher ones will be polished and show a decrease in roughness. There will thus tend to be a minimum initial roughness below which no further reductions in wear are likely to occur. As Fig. 12(b) shows, for a counterface hardness of 780 VPN and a liner containing glass fibre, this minimum roughness is of the order of 0.05 μm Ra. The minimum initial roughness will tend to increase when sliding against softer counterfaces or when sliding against liners containing more-abrasive constituents. Conversely, the minimum roughness will decrease for non-abrasive liners and extremely hard counterfaces.

The very slight roughening of the smoothest surfaces of hardened 440C steel by a liner containing glass fibre suggests that further reductions in wear might be possible by using even harder counterfaces, providing that their surfaces could be finished to roughnesses less than about 0.05 μm Ra. Figure 12(c) shows a selection of results from some recent experiments on counterfaces coated with ceramics and cermets. By comparison with Fig. 12(b), it can be seen that the steady-state wear rate is appreciably lower against both coatings than against the smoothest surface of hardened 440C steel. The roughness values quoted for the coatings contain some contribution from the coating porosity and the "effective" values, relevant to liner wear, will be rather lower. A detailed topographical analysis [31] for one of the coatings (tungsten carbide) suggests that the "effective" Ra is about 25% less than the "apparent" Ra.

Although it is usual for dry-bearing liners to be mated against metal (or ceramic) counterfaces, there are some situations in which it might be advantageous for them to slide against the same or other liner materials. These could arise, for example, in all-composite (carbon fibre) structures where the presence of metals might initiate galvanic corrosion or where light alloy bearings are necessary for weight-saving and hard-facing or electroplating of the counterface is precluded. Very little information

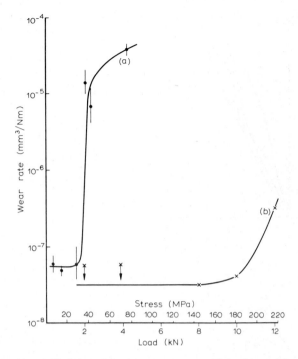

Fig. 13. Variation of steady-state wear rate with load (stress) for the dry bearing liner of Fig. 12 sliding (a) against itself and (b) against 440C steel. Three pin-on-disc tests. (From ref. 33.)

is available for composites sliding against themselves, although it is known that thermal effects due to frictional heating become much more severe than when a metallic counterface is used [32]. Some recent results for a PTFE fibre/glass fibre liner sliding against itself in a three pin-on-disc arrangement are shown in Fig. 13. At low stresses, the wear rates of the liner–liner combinations are not very much greater than those for the liner against steel, but at high stresses, the difference increases enormously. Thermal effects are insufficient to account for this marked increase and it appears instead that it is associated with penetration of the more rigid parts of the liner (glass fibre) on one surface into the softer parts (resin, PTFE) of the other [33].

4.4. Contamination by fluids

Most flight control bearings are potentially liable to contamination by a variety of fluids commonly encountered in service environments. This is recognized in bearing specification requirements, which include tests for compatibility with fluids; mineral oil and phosphate ester hydraulic fluids, a diester gas turbine lubricant, a glycol-based anti-icing fluid, and a standard test fluid which contains iso-octane, toluene and cyclohexane as its major constituents. The main concern of the specification requirement is with the effect of static immersion in fluids on resin softening and the

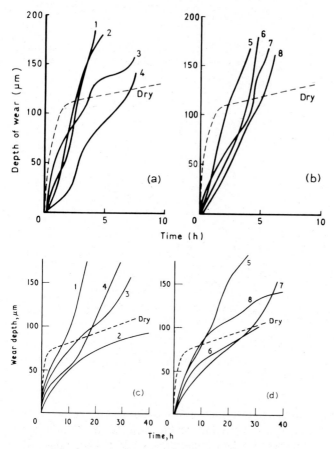

Fig. 14. Wear depth–time relationships for the bearing liner of Fig. 12 contaminated continuously by various fluids. Reciprocating line contact tests (a), (b) 90 MPa; (c), (d) 60 MPa. 1, Silicate ester fluid; 2, diester oil; 3, aviation kerosene; 4, de-icing fluid; 5, water; 6, mineral oil hydraulic fluid; 7, sea water; 8, phosphate ester hydraulic fluid. (From ref. 35.)

adhesive bond between the liner and its substrate. There have, however, been indications from service that the introduction of fluids during bearing operation can be much more serious than that resulting from static immersion. Attention has therefore been paid in recent years to assessing the dynamic performance of bearings in the presence of fluids [34,35]. Figure 14 shows some results obtained during accelerated tests (reciprocating line contact) on one particular bearing liner in the presence of a continuous supply of fluid. At 90 MPa, all the fluids are detrimental to performance and appear to prevent the establishment of the wear "plateau" characteristic of dry sliding. At a lower stress of 60 MPa, greater differences begin to emerge in the effects of the different fluids. Water and the silicate ester hydraulic fluid are still extremely deleterious, but the mineral and diester oils become rather less so. Broadly similar trends have been observed with other types of liner and in other conditions of sliding [35].

The mechanism by which fluid contamination leads to an increase in liner wear involves two main elements. Fluids inhibit the formation of third-body films on both the liner and its counterface and the effects of the initial counterface roughness on wear thus tend to persist for a considerably longer period. In addition, fluid penetrates into discontinuities in the liner surface leading to localized hydrostatic stresses during contact which facilitate surface disruption. Evidence for the latter is largely circumstantial and is based on the observation that a high viscosity fluid increases wear more than one of low viscosity. The latter can escape more easily from surface discontinuities by side leakage during the contact time and thus relieve the hydrostatic stresses.

The most obvious and immediate solution to overcome problems of fluid contamination is to incorporate some type of seal into the bearing assembly. Several manufacturers have developed sealed bearings and one of the most sophisticated designs for spherical bearings involves an elastomeric ring pressed against the ball surface by a flexible, annular shim which is electron beam welded to the outer race of the bearing. Performance data on sealed bearings are still very limited and there is considerable uncertainty as to how effective seals can be. It may also be argued that if a seal is sufficiently effective to prevent fluid from entering a bearing assembly, it will also prevent debris from escaping. In these circumstances, backlash measurements, which are commonly used in service to assess bearing wear, may underestimate the real extent of wear. In the extreme case, the possibility could arise that the whole of a bearing liner might be disrupted into debris without any detectable increase in backlash at all.

An alternative, longer-term solution to the fluid contamination problem is to develop composite bearing liners whose performance is relatively insensitive to fluids. One approach to this is to incorporate finely divided, polishing additives into the composite to generate an extremely smooth counterface surface and so avoid reliance on transfer films to produce low wear [36]. This method, however, cannot influence the disruptive effects of the hydrostatic stresses produced in surface cracks and the only real possibility here is to develop composites with fewer actual or potential, surface discontinuities. In this context, it has been noted [37] that composite liners containing PTFE flock dispersed throughout a fabric-reinforced resin [Fig. 2(b)] appear to be somewhat less susceptible to fluid contamination than those containing PTFE fibre interwoven within the reinforcing fabric.

4.5. Reversing loads

Most dry-bearing tests reported in the literature, together with all those needed for qualification to specifications, are associated with unidirectional loads. In many flight-control applications, however, bearings are subjected to fluctuating and even reversing loads; examples include actuator bearings for flaps, spoilers or rudders. Undercarriage linkages are a particularly severe example of reversing loads and the wear of some bearing liners in service has proved to be appreciably greater than that

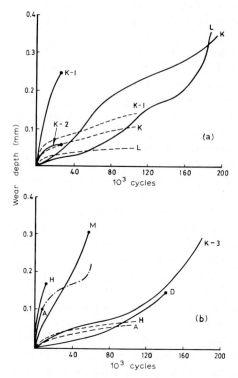

Fig. 15. Comparison of the performance of various liners under square-wave reverse loading (———) and unidirectional loading (– – – –). ● Signifies delamination failure. K, As Fig. 2(a) plus an all-glass backing fabric; K-1, as K but different manufacturer; K-2 as K but without backing fabric; K-3 as K but PTFE in monofilament form; L as K but 10% of PTFE warp fibres replaced by glass [see Fig. 18(c)]; A, as Fig. 2(b), polyester fabric, PTFE flock; D, as Fig. 2(b), aromatic polyamide fabric, PTFE flock; H, as Fig. 2(b), polyester fabric, granulated PTFE; M, injection-moulded liner, composition unknown. Spherical bearings, ± 172 MPa, 5.5 c min^{-1}, $\pm 25°$. (From ref. 37.)

which would have been predicted from wear data obtained under unidirectional loading [16]. A recent laboratory investigation [37] has confirmed this service experience. Figure 15 shows a comparison of wear depth–time relationships for a number of spherical bearing liners under unidirectional and reversing loads. The latter not only lead to appreciably greater wear, but an additional feature also emerges; some bearing types begin to exhibit delamination failures at, or near, the liner–substrate interface. An example is shown in Fig. 16. Delamination generally begins near the edges of the liner and gradually extends towards the centre. For the experiments in Fig. 15, testing was usually terminated following the first observation of delamination. It is interesting to note, however, that the bearing could nevertheless still remain usable for a limited period, as shown by the extension to curve A. Only when delamination reaches the centre of the bearing is failure rapid and catastrophic.

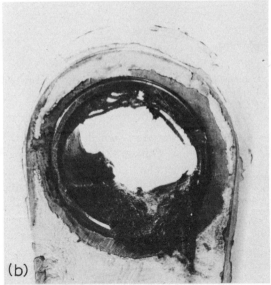

Fig. 16. (a) Edge delamination of liner A after 10^3 cycles of square-wave loading. (b) Total delamination and disruption of liner A from a service component susceptible to reverse loading. (From ref. 37.)

The critical element in the induction of delamination failure is a high rate of application of load following reversal. Bearing liners subjected to sinusoidal or triangular, reversing-load waveforms are much less prone to delamination than those which undergo square-wave loading. Following an analysis by Matthewson [38] of the stress distribution in a spherical contact against an elastic layer on a rigid

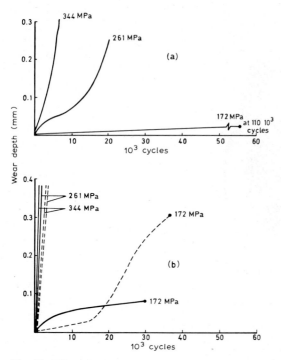

Fig. 17. Effect of stress on wear and delamination of liner D (see legend to Fig. 15. (a) Dry; (b) contaminated by water (-----) and mineral oil hydraulic fluid (———). (From ref. 37.)

substrate, it can be shown that the maximum shear stress, which occurs at the layer–substrate interface, increases very rapidly as Poisson's ratio of the layer approaches 0.5. Values of Poisson's ratio for polymers and composites tend to be rather uncertain, but it has been suggested [9] that 0.4–0.45 is not an unreasonable level in quasi-static conditions at high stresses. High rates of loading will increase the effective value of Poisson's ratio beyond this because polymers tend to become increasingly incompressible as the time of loading approaches the molecular relaxation time. A further conclusion from Matthewson's analysis is that the shear stress at the layer–substrate interface also increases with decreasing ratio of layer thickness to contact radius. As wear of a spherical bearing occurs, not only does the liner thickness decrease but the contact radius will also tend to increase as the liner achieves greater conformity with the ball. The susceptibility to delamination failure thus increases with time as well as with the rate of loading.

One final comment is in order about the combined effects on wear of square-wave reverse loading together with fluid contamination. This combination is extremely damaging to bearing performance, as shown in Fig. 17. At the lowest stress, 172 MPa, where the introduction of fluid does not greatly increase the rate of wear, delamination is initiated at an earlier stage. At high stresses, the main effect of the fluid is to increase the wear rate; insufficient time is available to initiate delamination before the liner is completely worn away. It is reasonable to envisage that this increased rate of wear is largely due to surface disruption by hydrostatic stresses within fluid-filled cracks. At high rates of loading, insufficient time may be available for side-leakage to occur and relieve these stresses.

4.6. Kinematic factors

For convenience, a number of parameters can be grouped together under the above, general heading; speed, type of motion, amplitude of oscillation and the degrees of freedom available within the bearing support system.

The effect of speed on the friction and wear of thin, composite, bearing liners has not been systematically explored, but several investigations have indicated a decrease in wear life with increasing speed [13,16]. It is, however, difficult to disentangle the intrinsic effects of speed on wear (or friction) from those caused by frictional heating and an increase in temperature. Whilst the bulk temperature of a component can easily be controlled during testing, the localised contact ("flash") temperature increases with speed and is uncontrollable. The coefficient of friction also tends to increase with speed; this trend is common to most plastics and elastomers [39] and is attributable to the increased rate of energy losses via viscoelastic fdeformation. With PTFE and its composites, speed also affects the nature of the third-body films produced during sliding. At low speeds, the transfer films on a counterface are thin and uniform, whereas at high speeds, when the time of contact becomes commensurate with the molecular relaxation time, the films are more irregular and contain larger PTFE fragments [40].

It is commonly believed that third-body films develop primarily through the movement and aggregation of wear debris within the contact zone. Since the "quality" of these films greatly influences the wear rate, it follows that all parameters which affect debris movement will also influence wear. Considerable work has been undertaken by Godet et al. [41] in recent years to examine this aspect. The escape of wear debris from within a cylindrical bearing contact will obviously be more difficult during oscillatory motion than during unidirectional sliding and, as Fig. 8 has already shown, the third-body films are thicker and more uniform during oscillation. Friction and wear also thus tend to be lower during oscillatory motion. The effects of the amplitude of oscillation on third-body film formation and wear have received particular attention [42]. When the amplitude is reduced, transfer film formation on the counterface becomes increasingly non-uniform, leading to a greater surface roughness and an increase in wear. The greatest effect occurs when the amplitude falls below the pitch of the structural elements in the composite liner. When, for example, the bearing liner contains a regular array of glass fibres in the surface layer, a characteristic pattern of markings can be produced on the counterface, as shown in Fig. 18. The spacing of these regions is the same as that of the glass fibre in the liner and surface examination shows that they contain little or no transferred PTFE or resin. Although the increases in liner wear associated with these uneven transfer films at low amplitudes appear to be generally small, greater increases might occur for bearings operating under "active control" where low

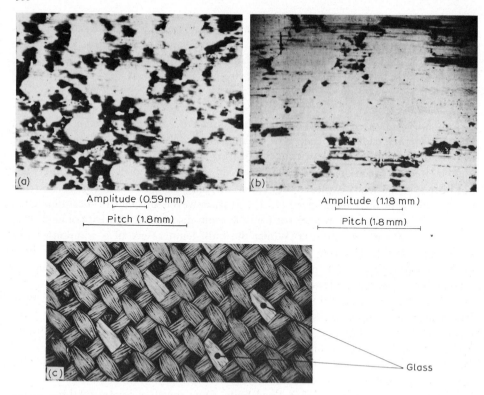

Amplitude (0.59mm)

Pitch (1.8mm)

Amplitude (1.18 mm)

Pitch (1.8 mm)

Glass

Fig. 18. (a), (b) Markings on 440C counterfaces after wear against liner L, shown in (c), at two amplitudes smaller than the pitch spacing of the glass fibre. (From ref. 42.)

amplitude oscillation is followed by rapid movement to a greater amplitude. This possibility has not yet been systematically examined.

Debris motion and third-body film formation may also depend on the bearing geometry (cylindrical or spherical) and it has been shown that the escape of debris is more difficult from spherical than from cylindrical bearings [43]. Debris escape from spherical bearings is also influenced by the degree of conformity between the curvature of the liner and that of the ball; the greater the conformity, the more difficult it becomes for particles to escape. At light loads and low amplitudes, which together are insufficient to permit aggregation of the debris into a coherent, third-body film, the escape of the debris leads to increased surface damage. Wear rates thus tend to increase with increasing bearing conformity and be greater for spherical than for cylindrical bearings. At large amplitudes and high stresses, however, these differences tend to disappear.

A great deal of detailed information on the movement of debris within bearing contacts has been obtained from model experiments using a chalk bearing sector oscillating against a glass ring through which the sliding interface can be examined

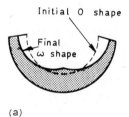

(a)

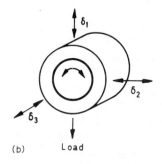

(b)

Fig. 19. (a) Shape of bearing section frequently developed during wear in oscillatory motion. (b) Degrees of freedom in a bearing support system.

microscopically [44]. The circumferential movement of particles due to the ring motion is changed at the centre of the sector where escape is only possible via transverse flow. The shape of the sector thus tends to be modified from an O shape to an ω shape, as shown in Fig. 19(a). Debris motion is also influenced by the degrees of freedom inherent in the bearing support system [Fig. 19(b)] and this, in turn, affects the distribution of stresses over the apparent contact area [45]. In unidirectional motion, and with movement available in direction δ_2, space can be opened at the rear of the contact, permitting the lateral escape of debris. The proportion of load supported by aggregated debris is thus reduced and the wear rate tends to increase. In oscillatory motion, movements in direction δ_2 again lead to an ω-shaped profile of wear damage on the bearing sector, which poses a number of problems in connection with the accurate measurement of wear and/or backlash [42]. In both unidirectional and oscillatory motion, reducing movement in direction δ_1 restricts debris movement and encourages aggregation, increases the proportion of load supported by the aggregated debris and so, in turn, reduces wear. Movements in direction δ_3 are largely irrelevant unless they become very large and contribute to the total distance of sliding. At the present time, it would appear that the influence of most of the above effects on the wear of thin, composite, bearing liners remains relatively small. However, they could well become of significance in the detailed design of high-precision bearing assemblies.

5. Materials development

For many years, there has been a long-term research objective to relate the structure and composition of plastic composites to their friction and wear behaviour. Some success towards this has been achieved with filled PTFE formulations [46], but for the thin-layer composites of the type under consideration here, progress has remained rather slow. A few conclusions have nevertheless emerged from recent accelerated wear tests and some of these are described below.

5.1. Effect of weave structure

One of the most fundamental questions to be asked about the composition of all dry-bearing composites is: what is the optimum content of solid lubricant, e.g. PTFE? It is intuitively obvious that the friction, and possibly the wear rate also, is likely to be high if insufficient PTFE is available to form a coherent third-body film. At the other extreme, too high a PTFE concentration will produce mechanically weak composites with limited load-carrying capacity and possibly high wear also. One way to change the PTFE content of composites incorporating PTFE fibre is to vary the weave structure of the fabric. Figure 20(a) shows the results of reciprocating line-contact tests on several liners incorporating PTFE and glass fibres but with different weaves; all were impregnated with the same phenolic resin. If the conventional double-weave material [as in Fig. 2(a)] is taken as a standard for comparison (curve 1), it can be seen that reducing the PTFE content by reducing fibre denier (curve 2) and by replacing the top PTFE weft fibres by glass (curve 3) both lead to increased wear. Greater wear is also obtained with a four-shaft satin weave (curve 4), but with PTFE warp fibres in a plain weave (curve 5), initial wear is reduced at the expense of a slight increase in the steady-state rate of wear. It is not possible to estimate the PTFE concentrations present in these composites with any great degree of certainty and, in any case, they vary with the depth of wear. However, on the basis of simple models of the weave structures [20], it would appear that the lowest steady-state wear rate is associated with a surface PTFE content of around 20% and a glass fibre concentration of around 50% [Fig. 20(b), (c)]. The presence of the glass fibre not only strengthens and stiffens the wearing surface layer [Fig. 20(d)], but broken-down glass debris will produce the same result in the third-body surface films. Since these films influence the wear rate so critically, the optimum composition of the bearing composite must also depend on all those parameters which affect film formation, e.g. type of motion, amplitude, bearing size, shape, etc. No single composite formulation, therefore, can ever be expected to give optimum performance in all applications, a conclusion which has been obvious for some time to most airframe bearing users.

5.2. Type of second fibre

In the earliest versions of thin-layer, dry-bearing composites, glass fibre was incorporated, partly as a reinforcement for the resin matrix and partly to facilitate adhesive bonding to a metal substrate. Since then, there have been many develop-

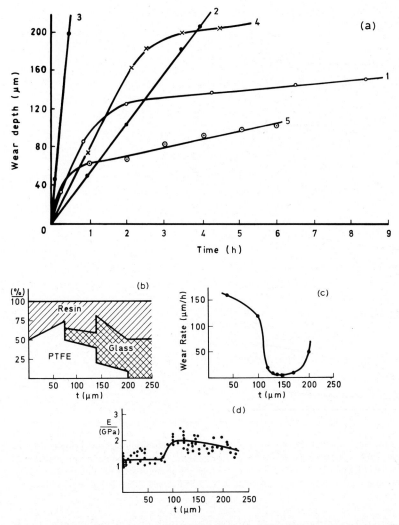

Fig. 20. (a) Wear behaviour of various PTFE fibre/glass fibre liners in reciprocating line contact, 450 N. 1, Double weave, PTFE warp, PTFE top weft, glass bottom weft [as Fig. 2(a)]; 2, as 1 but PTFE denier halved; 3, as 1 but both wefts glass; 4, four-shaft satin weave with PTFE warp and glass weft; 5, plain weave with PTFE warp and glass weft. (b) Variation in composition with depth for liner corresponding to curve 1, derived from an idealized model of the weave structure. (c) Variation of wear rate with depth (from curve 1). (d) Variation of elastic modulus with depth. (From ref. 10.)

ments in synthetic fibre technology and alternatives to glass fibre are now being incorporated into liners. Accelerated wear tests have shown that at least two types of synthetic fibre lead to better performance than glass fibre in several different weave structures. An example is given in Fig. 21(a) for a double-weave fabric, but generally similar results have also been found with four-shaft satin and plain-weave fabrics

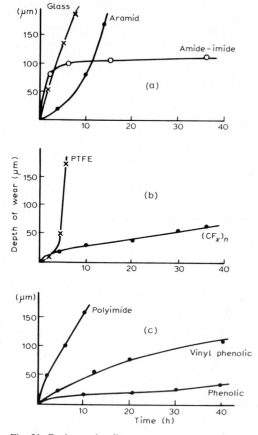

Fig. 21. Reciprocating line contact tests on various experimental composite liners. (a) Influence of the type of weft fibres in a double-weave fabric with PTFE warp fibres. Phenolic resin matrix. (b) Comparison between PTFE and $(CF_x)_n$ additions to a phenolic resin reinforced with a plain-weave, glass fabric. (c) Effect of resin type in a 2×2 matt fabric of twisted PTFE and aromatic polyamide fibres.

[10,20]. The particular combination of properties which makes polyamide–imide fibres so effective in reducing wear has not yet been established.

5.3. Other solid lubricants

PTFE is the solid lubricant most commonly incorporated into plastic bearing composites, but there are, of course, many other solids available (see Table 1). MoS_2 is an obvious possibility and experiments have been made to examine the effects of MoS_2 concentration in a phenolic resin reinforced with an all-glass, plain-weave fabric [10]. Even with an MoS_2-to-resin ratio as high as $2:1$ by volume, low wear in reciprocating line-contact conditions could only be obtained for limited periods. It appeared that the MoS_2-rich third-body films generated by sliding were prone to the formation of "blisters" which were then easily disrupted, exposing lubricant-depleted areas. In contrast to this behaviour of MoS_2, graphite fluoride appears much more promising. Figure 21(b) shows some results obtained with an aramid-reinforced, phenolic resin composite where the $(CF_x)_n$ gives markedly lower wear than PTFE particles at the same concentration.

5.4. Resin type

In principle, there are many potentially attractive resins which might lead to improved dry-bearing composites as alternatives to the phenolics which are still so widely used. However, extensive testing in reciprocating line-contact conditions at room temperature has, so far, failed to isolate any which consistently give lower wear than phenolics. An example illustrating this is given in Fig. 21(c) for a matt-weave fabric composite, but similar trends have been observed with other fabrics [20]. Whilst high-temperature resins could well become superior at temperatures where the stability of phenolics is limited, by far the most important resin-related factor yet emerging is the "quality" of impregnation of the fabric. The presence of voids has been shown to have a markedly adverse effect on wear [27]. This suggests that the exact method by which composite bearing liners are fabricated could be just as important to wear performance as their composition and structure. Optimization of the composite fabrication procedure will therefore always need to be undertaken when any change in structure or composition is made.

6. Life prediction

The performance criterion most widely used for dry bearings is the PV factor, the product of pressure and velocity in whatever units are appropriate, e.g. MPa $\times$ m s^{-1}. This criterion is used in two ways; as a limiting PV above which friction, temperature and/or wear increase rapidly, or as a PV factor corresponding to an arbitrarily specified wear rate [47]. The uncertainties associated with the use of PV factors for predictions of bearing life are now well-known, the most important arising from the fact that, for any given PV product, the bearing surface temperatures will not be the same for high P and low V as for low P and high V. In general, limiting PVs decrease at high speeds due to temperature increases and decrease at high loads because of limitations to material strength. When, however, the imposed conditions of sliding are remote from these extremes, the PV concept can provide a useful first level of design guidance. For porous bronze–PTFE–Pb composites, it has been shown [48] that, after the initial period of "bedding-in", the wear rate up to a depth of about 40–50 μm is approximately proportional to PV. Since the initial wear rate for this type of composite is very rapid, the useful life of bearings is thus approximately inversely proportional to PV. The range of applicability of this inverse relationship can be extended [49] by applying correction terms to the PV factor to take into account other parameters which affect wear, such as temperature, heat dissipation through the bearing housing, type of motion, presence of fluids, counterface type, and roughness, etc.

An essentially similar approach to the above is also valid for other types of dry bearing [19], including the thin-layer composites of interest here, provided that the volume (or depth) of initial wear remains low in comparison to the total permissible wear, e.g. curve 1 in Fig. 7(a). Problems arise, however, when this proviso no longer holds, as in curve 2 of Fig. 7(a). In the latter instance, one approach adopted [50] has been to establish an experimental "master curve" relating wear depth to the logarithm of the sliding distance for a particular type of bearing liner at a "standard" level of stress. The loads, amplitudes, and time intervals to be experienced by a bearing in service are then converted to stress and sliding distance values and the latter "normalised" to the standard stress and summed, thus enabling wear to be predicted from the master curve.

All the above, and other methods so far devised for predicting dry-bearing lives, remain approximations because it is seldom, if ever, possible to quantify the effects of every one of the parameters which might affect wear. Interpolation and extrapolation from user experiences thus continues to play an extremely important role in bearing selection procedures. Inevitably, this means that the introduction of new bearing composites into service is a very slow process with a time scale measured in years rather than months. In addition, this time scale tends to be extended even further by the rigorous specification and standardization requirements which are mandatory for airframe control bearings in critical areas. It is for this reason that, more than a decade after the introduction of improved, second-generation, composite liners, many bearing applications still continue to demand first-generation products.

7. Future trends

Within the general area of airframe bearings, the main incentive for further performance improvements is likely to remain with military aircraft where space and weight considerations combine to raise both stresses and temperatures. Some increases in temperature could also result from the introduction of active control technology and the consequent increases in the speed of bearing movements. In so far as bearing strengths are concerned, some improvements are possible by the introduction of carbon fibres [51]. The scope, however, is not indefinite because of limitations within other parts of the bearing assembly, e.g. shaft distortions and fatigue. The main restriction to the development of higher-temperature composite liners is the lack of an effective solid lubricant to replace PTFE above about 280°C. MoS_2 has so far proved disappointing as a solid lubricant additive to reinforced resin composites, even at ordinary room temperatures [10]. On the other hand, WSe_2–Ga–In mixtures have shown promise in carbon fibre-reinforced polyimides above 300°C [52] and graphite fluoride, $(CF_x)_n$, retains its lubricating ability up to 400°C [53] or possibly even 500°C [54]. For prolonged bearing operation in excess of around 400°C, composites based on even the most thermally stable organic resins are largely precluded and the two most important material groups remaining are inorganic coatings and carbon composites. Plasma-sprayed mixtures of Ni–Cr, Ag, CaF_2 and glass have been developed [55] to give reasonably low friction and wear over the entire temperature range from -100 to $+870°C$, but friction coefficients at ordinary temperatures are appreciably higher than those characteristic of PTFE-containing composites and load-carrying capacities tend to be lower. Carbon fibre-reinforced carbon composites exhibit high compressive strengths and stiffnesses and, although developed primarily for brake applications [56], their friction and wear properties are readily modified by the introduction of additives [57]. There is, in fact, a great deal of technological information available on all kinds of brake materials which has so far remained largely untapped as a source for the further development of high temperature bearings.

In addition to the above demands for higher performance limits, a more general requirement exists in both civil and military aircraft for increasing bearing life and reliability still further. The emphasis here is on extending maintenance periods and so reducing the total life costs of an aircraft. The main limitation to developing composites with lower wear rates is the lack of basic understanding of relationships between composition, structure and performance. This topic still offers major opportunities for research, particularly in elucidating many of the details associated with third-body, debris films. Until considerably more information becomes available on how these films are formed, maintained, and destroyed, progress in developing improved composite bearing liners is likely to remain slow and empirical.

References

1 J.K. Lancaster, Tribology, 6 (1973) 219.
2 N.G. Crum, A Review of the Science of Fibre-Reinforced Plastics, HMSO, London, 1971.
3 C. White, Br. Pat. 845 547, 1960.
4 J.V. Shepherd, Br. Pat. 1 233 103, 1971.
5 H.E. Sliney and T.P. Jacobson, Lubr. Eng., 31 (1975) 609.
6 D.C. Mitchell and G.C. Pratt, Proc., Lubr. Wear Conf., London, 1957, Institution of Mechanical Engineers, London, 1957, pp. 416–423.
7 G.C. Pratt, Plast. Inst. Trans. J., 32 (1964) 255.
8 J.K. Lancaster, Tribol. Int., 12 (1979) 65.
9 J.K. Lancaster, J. Phys. D, 15 (1982) 1125.
10 J.K. Lancaster, in R.G. Bayer (Ed.), Selection and Use of Wear Tests for Coatings, ASTM STP 769, 1982, pp. 92–117.
11 M. Godet and D. Play, Colloq. Int. CNRS, 23 (1974) 361.
12 J.C. Anderson and E.J. Robbins, in Wear of Materials 1981, American Society of Mechanical Engineers, New York, 1981, pp. 539–544.
13 K.A. Rowlands and S.A. Wyles, British Aircraft Corporation Rep. PRO 251, 1973.
14 K.J. Cheeseman, Private communication, 1974.
15 M.B. Harrison, Private communication, 1977.
16 D.J. Wade, Private communication, 1978.
17 J.C. Anderson, Private communication, 1977.
18 W.D. Craig, Lubr. Eng., 18 (1962) 174.
19 A Guide on the Design and Selection of Dry Rubbing Bearings, Engineering Science Data Unit, London, Item 76029, 1976.

20 J.K. Lancaster, Proceedings of the 8th Leeds–Lyon Symposium on the Running-in Process in Tribology, Lyon, 1981, Butterworths, London, 1982, pp. 33–43.
21 M.W. Pascoe and D. Tabor, Proc. R. Soc. London Ser. A, 235 (1956) 210.
22 A.J.G. Allen, Lubr. Eng., 14 (1958) 211.
23 F.P. Bowden and D. Tabor, The Friction and Lubrication of Solids, Oxford University Press, Oxford, 1950, pp. 111 et seq.
24 J. Davies, Private communication, 1981.
25 D.C. Evans and J.K. Lancaster, in D. Scott (Ed.), Wear. Treatise on Materials, Science and Technology, Academic Press, London, Vol. 13, 1979, pp. 85–139.
26 D.C. Evans, Proc. Inst. Mech. Engr. Tribol. Group Conv., Swansea, April 1978, Paper C26/78.
27 R.B. King, Wear, 56 (1979) 37.
28 R.B. Lewis, in J.V. Schmitz and W.E. Brown (Eds.), Testing of Polymers, Vol. 3, Wiley, New York, 1967, pp. 203–219.
29 D.H. Buckley, in L.-H. Lee (Ed.), Advances in Polymer Science and Technology Symposium Series, Vol. 5B, Plenum Press, New York, 1974, pp. 601–603.
30 D. Dowson, J.M. Challen, K. Holmes and J.R. Atkinson, Proceedings of the 3rd Leeds–Lyon Symposium on Wear of Non-metallic Materials, Leeds, 1976, Mechanical Engineering Publishers, London, 1978, pp. 99–102.
31 A.B. Birkett and J.K. Lancaster, Proceedings of the JSLE International Tribology Conference, Tokyo, 1985, pp. 465–470.
32 D.C. Evans and J.K. Lancaster, Proceedings of the 3rd Leeds–Lyon Symposium on Wear of Non-metallic Materials, Leeds, 1976, Mechanical Engineering Publishers, London, 1978, pp. 288–290.
33 A.B. Brentnall and J.K. Lancaster, in Wear of Materials, 1983, American Society of Mechanical Engineers, New York, 1983, pp. 596–603.
34 F.J. Williams, Rockwell Int. Rep. TFD-75-1362, 1975.
35 R.W. Bramham, R.B. King and J.K. Lancaster, ASLE Trans., 24 (1981) 479.
36 D.C. Evans, Proc. 2nd Int. Conf. Solid Lubr., Denver, 1978, ASLE, SP6, 1978, pp. 202–211.
37 J.K. Lancaster and D.J. Wade, Proc. 3rd Int. Conf. Solid Lubr., Denver, 1984, ASLE, SP-14, pp. 296–307.
38 M.J. Matthewson, J. Mech. Phys. Solids, 29 (1981) 89.
39 J.K. Lancaster, in A.D. Jenkins (Ed.), Polymer Science, Vol. 2, North-Holland, Amsterdam, 1972, Chap. 14, pp. 959–1046.
40 C.M. Pooley and D. Tabor, Proc. Roy Soc. London Ser. A., 329 (1972) 251.
41 M. Godet, D. Play and D. Berthe, J. Lubr. Technol. 102 (1980) 153.
42 J.K. Lancaster, R.W. Bramham, D. Play and R. Waghorne, J. Lubr. Technol., 104 (1982) 559.
43 D. Play and B. Pruvost, J. Tribol., 106 (1984) 185.
44 Y. Berthier and D. Play, Wear, 75 (1982) 369.
45 I. Kohen, D. Play and M. Godet, Wear, 61 (1980) 381.
46 J.K. Lancaster, Proc. Int. Conf. Tribology in the 80s, Cleveland, 1983, NASA, Conf. Publ. 2300, 1984, pp. 333–365.
47 Dry Rubbing Bearings – a Guide to Design and Material Selection, Engineering Science Data Unit, London, Item 68018, 1968.
48 G.C. Pratt, in E.R. Braithewaite (Ed.), Lubrication and Lubricants, Elsevier, Amsterdam, 1967, Chap. 8, pp. 376–426.
49 DU Dry Bearings, Designers' Handbook No. 2, Glacier Metal Co. Ltd., Wembley, Gt. Britain, 5th edn., 1973.
50 Aerospace Bearing Design Manual, Ampep Ltd., Clevedon, Gt. Britain.
51 H.E. Sliney, T.P. Jacobson and H.E. Munson, NASA Tech. Note TN-D-7880, 1975.
52 M.N. Gardos and B.D. McConnell, ASLE Prepr., 81-3A-3 to 81-3A-6, 1981.
53 R.L. Fusaro and H.E. Sliney, ASLE Trans., 16 (1973) 189.
54 D. Play and M. Godet, Colloq. Int. CNRS, 233 (1975) 441.
55 H.E. Sliney, Tribol. Int., 15 (1982) 303.
56 J.V. Weaver, Aeronaut. J., 76 (1972) 695.
57 J.K. Lancaster, Proceedings of the 3rd Leeds–Lyon Symposium on Wear of Non-metallic Materials, Mechanical Engineering Publishers, London, 1978, pp. 187–195.

Chapter 12

Self-Lubricating Composites for Extreme Environmental Conditions

MICHAEL N. GARDOS

Hughes Aircraft Company, P.O. Box 902, El Segundo, CA 90245 (U.S.A.)

Contents

Abstract

At very low or very high temperatures, in cryogenic environments, in vacuum, or in air, only solid lubricants can be used with confidence. The environmental range of conventional or specially designed solid lubricants can be extended by the formulation of self-lubricating composites consisting of matrix binders and fillers. Both the binder(s) and filler(s) can assume multiple physical-chemical roles, usually as synergistic combinations of brittle and ductile phases. In particular, the proper chemistry, crystal structure, volume percent and distribution of the binder and the filler enable tailoring of the strength, modulus, fracture toughness, load-carrying capacity, thermal conductivity, oxidation resistance and surface shear behavior of each composition. Smart compounding results in properties unattainable where the individual constituents are used by themselves.

In this chapter, the friction and wear mechanisms of a variety of self-lubricating composites are examined as functions of composition, temperature, atmosphere, and counterface type. It is shown that a composite and its mating surface comprise an inseparable tribological system under any and all conditions. The most important and decisive factor in composite-versus-counterface performance is the effectiveness of a preferentially accumulated, low shear strength surface layer on the worn composite. This layer must also be willing and capable to form low shear strength, low surface energy transfer films on the counterface. Such thin layers, especially if they are even and homogeneous, can control (mostly lower, but sometimes raise) the pressure–velocity-induced flash temperatures of the moving mechanical assembly couple. As long as the environment is such that it does not degrade the structural and chemical integrity of these films as well as that of the yet-unworn composite substrate and its load-support capacity, and the mating counterface remains un-damaged by any of the composite constituents, a tribologically advantageous system can be created.

1. Introduction

For decades, the lubrication of moving mechanical assemblies (MMA) at the opposite ends of the environmental spectrum, i.e. in or near cryogenic fluids or the cold vacuum of space to high temperature air, has traditionally fallen within the realm of aerospace tribology. The work performed during the Mercury, Gemini and Apollo spacecraft projects established the general boundaries of oils, greases and solid lubricants for limited space and re-entry use. The ensuing, explosive develop-ment of satellite communications and energy conservation through higher power density engines is, however, rapidly outpacing our ability to handle satisfactorily friction and wear problems of long-life MMA operating in extreme environments. There is an alarming rate of increase in the failure of these advanced mechanisms. Driven by pressures to meet deadlines, the resolution of the problems is attempted mostly by quick fixes rather than through systematic research on wider temperature range and better environmental stability lubricants. The penalty paid is the inability to predict accurately the wear life of expensive, performance-limiting and sometimes impossible-to-repair hardware with sufficient reliability.

In view of the real need, the shortcomings in extreme environmental tribology can no longer be blamed on the lack of a ready market and profit potential usually associated with limited military uses, one-of-a-kind planetary probes, or unmanned orbital observatories. These vehicles may not lend themselves to mass production, but there is an increasing number of communication satellites which are now sold off-the-shelf with 10–15 year guarantees. The recent successes of various American and Russian manned orbital laboratories, the U.S. Space Shuttle, the Western European Ariane satellite launcher and serious thoughts on materials processing in space are harbingers of the growing commercialization of space and its various technologies. Advanced military applications lie superimposed on the commercial incentives.

The high temperature MMA which operate key re-entry vehicle aerodynamic control surfaces have themselves been redesigned and adapted to air-breathing propulsion systems planned for terrestrial or airborne applications. The current trend is toward high power density engines. High power density is synonymous with higher temperatures and higher mainshaft speeds. Reduced engine size combined with maximized combustion temperatures lead to thermal and load-bearing capacity requirements beyond the limits of bearing steels and superalloys, as well as conventional liquid and solid lubricants. It is not uncommon to predict or measure prototype turbine and adiabatic diesel engine MMA temperatures in the 500–900°C range.

At very low or very high temperatures in cryogens, in vacuum, or in air, oils and greases exhibit inherent physical and chemical limitations which cannot be overcome by clever molecular engineering. In such environments, only solid lubricants can be used. Unfortunately, the utilization of the currently most often employed solid lubricants has also been stretched to the point where even the best MMA design iterations fail to circumvent the limits of the available materials technology.

One feasible method of extending the triboenvironmental range of solid lubricants is the formulation of self-lubricating composites. These materials usually consist of two or more components. Some act as binders, others as fillers. Both the binders and the fillers can assume multiple physical-chemical roles, usually as synergistic combinations of brittle and ductile phases. In particular, the proper chemistry, crystal structure, volume percent and distribution of the binder and the filler enable tailoring of the strength, modulus, fracture toughness, load-carrying capacity, oxidation resistance and surface shear behavior of each composition. Smart compounding results in properties unattainable by the use of the individual constituents by themselves.

The development of extreme environment composites may be driven by progressively more stringent MMA requirements, but the interaction between components is governed by the principles of science. Satisfying the growing industrial need can be successful only if the costly and time-consuming loop of empirical blending, followed by testing, and further followed by more trial-and-error compounding is avoided. The object is to gain as much fundamental understanding as possible, at least on the phenomenological level, of the interaction between the various composite phases and that of the composite against its counterface. The complex wear mechanisms which occur in a cryogenic environment, in vacuum at various temperatures, and in high temperature air must be properly dissected, followed by the reassembly of the fundamental building blocks into more desirable combinations. The results are reduced friction and wear in any triboenvironment and better predictability of MMA performance.

This chapter is not intended to be a comprehensive overview of all extreme environment, lubricative composites, or the research related thereto. The study of past accomplishments is useful only if systematic patterns are identified. Therefore, my main endeavor is an attempt to elucidate the rudiments of synergisms between the composite phases and between the composite and its counterface, in the various environments of interest. Based on the information presented here, promising possibilities for designing improved, severe environment MMA may emerge.

2. Composites for extreme triboenvironments

2.1. Preferred lubrication and design methods

The use of thin, sacrifical solid lubricant layers deposited by powder burnishing, by spraying/baking, or by sputtering is relegated to short wear life MMA only. For extended service, solid lubricant replenishment techniques must be employed. One extensively used replenishment technique is the film transfer process from a self-lubricating composite to a mating metallic or a ceramic-type bearing surface. This technique itself can be further divided into two parts: films formed by (a) a single-transfer mechanism where only a composite/counterface pair (e.g. a metal pin/composite bushing or a composite piston ring/metallic cylinder liner) is involved in forming a friction- and wear-reducing film on the unlubricated member; or (b) a double-transfer mechanism, where the single-transferred, low shear strength composite film developed on a bearing component is further transferred to the surfaces of another, yet-unlubricated bearing part integral within a MMA. One example of this double transfer is the bearing ball film. This layer is formed by rubbing (rolling) a ball against the pockets of a composite separator for further film transfer to the races. Another is the film transferred from a composite idler gear to the rest of the yet-unlubricated gear train members.

The composite and its mating surface comprise an inseparable tribological system. The choice of the counterface cannot be considered separately from the type of component blended into a self-lubricating composite. One of the greatest difficulties today lies in selecting the most likely composite for a given application from scores of proprietary products. The second most difficult task is to match the selected composite against a number of possible counterface materials. The number of possible mating bearing surfaces is usually restricted by the specific MMA design, the triboenvironment of interest and, to a very large extent, the cost of the part. To further complicate the matter, the loads and speeds in dynamically stressed components are seldom, if ever, constant. In particular, the loads are often frequency dependent. It is important to know the magnitude of load and load application frequencies (which are, of course, related to speed), because friction coefficients, traction coefficients, and stress failure of the composite and its counterface are functions of both the load magnitude and the frequency of application.

Although this chapter leaves the design of composite MMA parts to those better versed in the art, it suffices to say that meaningful dynamic analysis of a tribosystem requires modeling of the mass properties, as well as the elastic and thermal characteristics of the materials system. The MMA elements must be modeled analytically in sufficient detail to characterize stiffness, mass properties, and damping losses of the most prevalent contact geometries in a given environment. The analytical models must include the appropriate number of degrees of freedom in solid body dynamics to obtain meaningful results. From the dynamic analysis, the materials system requirements are defined, mainly in terms of adhesive, abrasive or fatigue wear of both the composite and the countersurface. Then, and only then, can the trade-offs of design configurations be made in view of the advantages and limitations of the available and probable material candidates. Materials science and the design process for MMA are inseparable.

2.2. Cryogenic applications (unreinforced polymers)

The first major impetus given to the development of cryogenic self-lubricating composites was provided by NASA Lewis Research Center tribologists during the early 1960s. Their object was to solid lubricate steel bearings which operated liquid fuel (hydrogen) and oxidizer (oxygen) pumps in rocket engines [1–6]. The design of the pumps is such that the turbopump bearings are fully submerged in the cryogens streaming through the bearings, while operating at high loads and speeds. Using the double-transfer solid lubrication system, various types of reinforced PTFE fabricated into self-lubricating ball bearing separators were investigated. The most successful versions appeared to be those filled with chopped or woven glass fibers. A two-dimensional glass weave-reinforced PTFE still serves today in composite separators lubricating the space shuttle LOX turbopump bearings. In view of the more stringent operating conditions of the current application, the performance of this simple composite material may be described, at best, as marginal [7].

Other drivers for cryogenic friction and wear research have been the need for free-turning rocket fuel and oxidizer valves [8] and, more recently, for stick–slip-free metal/insulator parts in superconducting magnet windings [9–11].

At first glance, the review of the referenced literature yields little fundamental information on the materials science-related aspects of cryogenic tribosystems. Mostly, one finds only empirical data generated by improperly instrumented bearing testers or friction and wear testers of limited test specimen configuration(s). The interpretation of the data is often qualified due to the experimental problems inherent with the test apparatus and techniques. In general, there is (a) marginal ability to approximate realistic MMA contact conditions [most apparatuses are pin(s)-on-disc testers]; (b) difficulty in excluding contaminants (i.e. air and moisture) from the specimen chamber; (c) inability to separate loading and friction events from spurious signals introduced by the characteristic (often unchecked and uncontrolled) frequencies of the long test machine shafts immersed in the cryogenic Dewars, or the consequent "ringing" of the force transducers themselves; and (d) inability to measure the true sliding interface temperature, which is one of the controlling factors of cryogenic composite performance.

To better understand the behavior of single phase or composite solid lubricants at very low temperatures, there is no other choice but to supplement the available (meager) tribodata with mechanical property measurements conducted at cryogenic temperatures [12–16]. Any correlation between the two sets of information can only help to unravel at least the rudimentary rules-of-thumb elucidating cryogenic tribocomposite performance.

As shown by Reed et al. [12], the reduced elongation and thermal expansion of PTFE with decreasing temperature is commensurate with a very large reduction of strain as a function of increasing stress. There is also a substantial increase of strength, elastic modulus and hardness. The shear modulus and thermal conductivity of progressively more crystalline PTFE are proportionally higher at lower temperatures.

Yet, even when filled with glass, PTFE exhibits the lowest tensile modulus and compressive strength among a large variety of reinforced polymeric composites in a cryogenic environment [13]. The performance of neat or reinforced PTFE is examined best, therefore, by first assessing the molecular behavior of the polymeric binder, as the function of decreasing temperature and increasing tribo-stresses. The objects of interest are (a) conformation effects on the relaxation of the long-chain molecules; (b) the physico-chemical interaction between the filler and the matrix; and (c) the compatibility of the composite with its countersurface, mainly in terms of the presence or absence of a transfer film and composite-induced surface damage on the counterface.

First, let us take the deformation rate dependence of the mechanical properties of epoxy resins at cryogenic temperatures as a reasonable model [15]. At a strain rate higher than an intrinsic limit, a polymer fractures without showing a yield point and it fractures at a much reduced breaking strain and stress. A progressive reduction of temperature to approximately 4 K ($T_{LHe} = 4.2$ K) induces the same behavior [see Fig. 1(a) and (b)]. The elastic modulus also increases with the strain rate, just as it does at reduced temperatures [Fig. 1(c)]. This behavior can be explained if the molecular relaxation time is assumed to correspond with the reciprocal of the strain rate and a direct function of increasing temperature. At a lower rate or higher temperature, the stored energy can be relaxed and the sample deforms plastically to exhibit a large breaking stress and strain. There is a severe reduction in relaxation at high rates and low temperatures; the breaking stress becomes much smaller, i.e. the sample is weaker and exhibits a lower fracture toughness [Fig. 1(d)].

There is considerable evidence that toughness, ductility and high impact strength have a great deal to do with the unusually good dissipation capabilities linked to very specific molecular chain motions. These motions are helped or hindered by the available free volume in a particular polymer. As discussed by Roe and Baer [17], many tough polymers have a free volume in the glassy state which is larger than normal [18]. It was proposed that the impact strength of a polymer varied in proportion to the free volume. The large free volume of tough polymers was also noted independently by Mercier et al. [19]. Litt and Tobolsky [20] defined a new type of free volume, $f = (V_a - V_c)/V_a$, called "excess volume", where V_a and V_c are the specific volumes of the amorphous and crystalline phases, respectively. On examining a large number of amorphous polymers, it was found that polymers were ductile at room temperature if $f = 0.07$. These observations resulted in the speculation that toughness was due to chain mobility, which was enhanced by free volume.

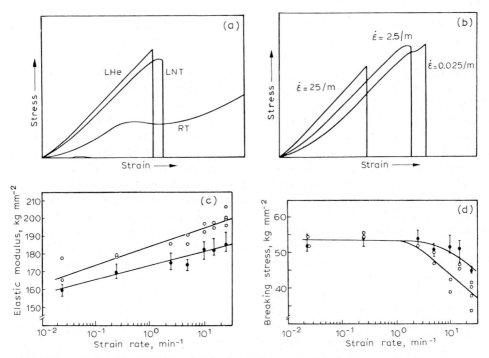

Fig. 1. Cryogenic behavior of a commercial bisphenol-A epoxy resin hardened by an amine/borate complex [15]. (a) Compressive stresses–strain curves with a strain rate of 0.25 min^{-1} measured at LHeT, LNT, and RT. (b) Compressive stress–strain curves with various strain rates measured at LNT. (c) Strain rate dependence of the elastic modulus measured at LNT (●) and LHeT (○). (d) Strain rate dependence of the breaking stress measured at LNT (●) and LHeT (○). LHeT = liquid helium temperature; LNT = liquid nitrogen temperature; RT = room temperature.

Other authors have shown that many polymer systems which are tough below the glass transition temperature also have large secondary relaxations at even lower temperatures [15,21–23]. This has lead some to conclude that ductility is the result of molecular mobility, which is demonstrated by low temperature relaxations. Data by Turley [24] have shown that transitions are effective in causing ductility only if the relaxation involves main chain motion.

Let us now examine PTFE in light of the above discussion. The –CF$_2$–CF$_2$– chains ($M = 10^5$–10^6) have little or no crosslinking or branching. As indicated by steric studies, the chains are not straight. They are helically twisted and at low temperatures there may be a decrease in the left–right defects in the helix [25]. At higher temperatures, the large fluorine atoms apparently prefer the *trans* configuration not only because of the van der Waals repulsion, but also due to the fact that the negative (fluorine) ends of the strongly polar C–F dipoles want to get as far apart as possible.

When the temperature is lowered, molecular motions are likely to be frozen into configurations progressively poorer in the *trans* state. At very low temperatures, only extremely localized motions are possible. Progressively lower free volumes will prevent chain relaxation by twisting due to crowding by the nearest neighbors.

At cryogenic temperatures down to 4 K, localized motions can also be hindered by influences other than intrinsic chain packing or backbone stiffness. During 1 Hz internal friction experiments on some poly-α-olefins at cryogenic temperatures, Pineri [26] found unexpected internal friction damping in a 50–60 K helium gas environment due to frozen-in nitrogen gas. Purification of the helium eliminated any such effect. This phenomenon may well be the reason why the behavior of glass-reinforced polymeric composites can become erratic and unpredictable below 77 K, especially around 20 K [13]. It is significant, however, that in the absence of such internal blockers, the barrier at cryogenic temperature between right- and left-handed helices of PTFE [25] and the energy of activation for the rotation of alkyl groups in polyethylene [26] was estimated around 2 kcal mol^{-1}. Such a low value is unusually small for the energies found for this type of motion and can be easily overcome by localized frictional heat generation. This possibility assumes special importance when the effects of hard fillers on adiabatic heating and friction are discussed in the next section.

Clearly, the molecular relaxation of PTFE or other lubricative polymers is inhibited at low temperatures and at high strain rates. Tanaka et al.'s band structure wear model for PTFE [27] can now be extended by suitably adapting it to a cryogenic environment. According to that model, the mechanism of PTFE wear can be explained by applying the model of a banded structure. Generally, the wear proceeds through a destruction of the banded structure in discrete, roughly 30 nm thick flakes. The width of the bands is influenced most by the cooling rate of the rubbery, melted gel of the polymer. The wear rate of PTFE quenched into ice water from a melt (i.e. a more amorphous form) was considerably lower than that of the slow-cooled or commercially available counterparts. The band widths of the rapidly quenched version were as small as 100–200 nm. Friction, on the other hand, appeared to be affected little by the band width or crystallinity.

It is by now widely known that the thin PTFE films transferred to glass or metallic surfaces cause PTFE to, in effect, slide against itself with the resultant low friction coefficient and low bulk wear. Replenishment in the form of larger rather than smaller fragments will increase the overall wear rate, especially if the polymer is more brittle than ductile [28]. Furthermore, as expected, the energetics of crack formation and propagation to form larger PTFE fragments are favored at higher sliding speeds [29]. The same phenomena are magnified at cryogenic temperatures. The formation of ribbon or flake-like debris and gross plastic deformation of the PTFE surface becomes closer to an abrasive wear mechanism, producing fine, powdery debris at 77 K [8]. That is the reason why this polymer is pulverized most efficiently by grinding and milling under liquid nitrogen.

If Brainard and Buckley's hypothesis that the adhesion of PTFE to metallic surfaces occurs by the formation of a metal–carbon bond [30] is accepted, larger-than-40 nm fragments produced at cryogenic temperatures should not show reluctance to adhere to a metallic counterface. PTFE transfer films can indeed be generated at cryogenic temperatures. Apparently, the few hundred dynes cm^{-1} reduction in the surface energy of metals from their melting point to absolute zero does not appreciably reduce this transfer; neither do the low temperature conformational changes in the polymer chain affect it adversely.

However, any strongly adhering, large fragment will act abrasively against the mating bulk polymer at any temperature. Recent research by Czichos [31] has shown that, at room temperature, a PTFE fragment transferred onto glass can plough a wear groove into the interface of its mother polymer matrix. The transferred fragments can be more abrasive than the bulk only if they are denser and harder. This seems to be reasonable at any temperature, because crazing and cracking have already occurred at the weakest matrix sites prior to removal.

The role of chain orientation on surface wear is yet to be elucidated. However, the chains on any PTFE surface must be oriented uniaxially by repeated, unidirectional sliding to obtain low friction [32,33], even at 4 K [11]. Although the room temperature friction of the most often quoted $f_k = 0.04$ cannot be reattained, f_k can still be reduced from approximately 0.2 to as low as 0.15 at liquid helium temperatures [11]. The high unit loads on the transferred, asperity-like fragments enhance this alignment on the particle surface; the torn, more disordered bulk acts as a burnishing tool. At the same time, the bulk surface is being burnished and aligned, while suffering wear damage. Further light was shed on the toughness of the transferred particle by Dimnet and Georges [34]. It appears that small agglomerates of relatively soft materials, such as polymers, can plastically deform a steel substrate onto which they were transferred. The fragments may even damage that hard surface under load and shear. The degree of damage depends on the thickness of the "lump" and the localized coefficient of friction. In the case of PTFE at cryogenic temperatures, the size of the "lumps" and the friction are relatively high.

In summary, it is not difficult to envision increased friction and wear of PTFE at cryogenic temperatures. To a significant extent, this increase is attributed to the generation of excessive amounts of large, abrasive but still counterface-adherent "transfer film" particles. More wear and higher friction are especially pronounced at high sliding or rolling speeds, under high frequency oscillation, and high frequency/high magnitude impact loads.

If low temperature ductility is associated with chain motion-induced relaxation processes (i.e. low internal friction) and the alignment of chains is a prerequisite for low friction but not necessarily low wear, a combination of circumstances must be found where low friction and low wear will manifest themselves simultaneously. Furthermore, if the formation and replenishment of a smooth, polymeric film transferred onto the counterface is necessary for continued low friction and wear of the mother matrix, the polymer and its reinforcement (i.e. the filler type and volume content) must be tailored to yield the best overall performance. The key step appears to be the formation of a thin, even and adherent transfer film free of large lumps or fragments.

2.3. Cryogenic applications (reinforced polymers)

At first glance, cryogenic tribobehavior predicted on the basis of even the most thorough overviews of polymer and polymeric composite tribology [35–37] appears risky because they omit discussions of low temperature friction and wear mechanisms. After a more thorough scrutiny of the available low temperature data (including current and comprehensive cryogenic tribotest results [10,11], the familiarity of patterns vindicates the rules of polymer science and tribology valid in any environment. Involving special hypotheses formulated for low temperatures only is not warranted.

For example, Kensley et al. [10,11] summarized that, at cryogenic temperatures, metals rubbing against soft, non-crosslinked polymers exhibit "semi-stable" (i.e. some stick–slip-type) sliding at $f_k = 0.1$. Metals sliding on hard, glass-fiber-laminated epoxy composites showed stable (i.e. stick–slip-free) behavior at high ($f_k = 0.5$) friction levels. Solid lubricants, e.g. PTFE or MoS_2 bonded to hard substrates slide stick–slip-free at low levels of friction ($f_k = 0.1$). The machined surfaces of epoxy laminates reinforced with two-dimensionally woven (2D) glass fibers displayed abrasive checkerboard patterns where the machined glass fibers were cross-cut perpendicular to the mating, relatively soft copper surface. The neighboring patterns of fibers parallel with the direction of sliding were non-abrasive to the counterface. In fact, they comprised the load-bearing portion of the entire surface. Abrasive damage to the counterface was circumvented by applying a thin epoxy topcoat, obtaining completely stable sliding behavior.

Although Kensley et al. addressed only the friction and not the wear life of particular sliding combinations, Vorobev et al. [39] examined both aspects of PTFE-based composites and PTFE-coated metal tape sliding against steel and hard cermet surfaces in liquid nitrogen. The latter combination exhibited the lowest friction ($f_k = 0.07$), one-tenth of the wear of pure PTFE and roughly one-seventh of the wear of coke, graphite and boron nitride-reinforced PTFE. At the moderate unit loads and speed ($P = 50–100$ kgf cm^{-2}; $V = 1.25$ m s^{-1}), the wear resistance of the composites varied only 10–30%.

It appears that the appropriate lubricant filler and the geometry or composition of the reinforcing agents are important in influencing the viscoelastic behavior of the polymer matrix caught between the reinforcement and the counterface.

When an increasing number of hard inclusions are embedded in the surface of a softer matrix and each progressively larger set of equal inclusion sizes carries the same normal load, the peak octahedral shear stress is brought toward the surface. Photoelastic analysis (Fig. 2), confirmed by computer diagnostics, showed this to be true [40]. If any hard, abrasive filler can be brought to the state where the surface-exposed portion can be flattened by an adroit run-in procedure without damaging the interface, the flatly worn hard phase provides an ideal underlay for the polymer oversmear. The high unit loads and the adiabatic heating of the localized contact areas (i.e. all thermal energy is expended in heating the surfaces only) provide the activation energy necessary for chain alignment.

As shown by Tanaka et al. [41,42] and Gardos [38,43], the chopped fibers in

Fig. 2. Distribution of stress in surface layer and depth of the region of maximum subsurface stress with a progressively large number of hard surface inclusions under the same total load from (a) to (c). (From ref. 40.)

PTFE composites preferentially accumulate on the surface and support the applied load effectively, but only if this aspect ratio is sufficiently high. At cryogenic temperatures, as well as in any other environments, 'the friction depends on the viscoelastic behavior of the polymer (and lubricant additive) oversmear and the friction behavior of the occasionally exposed, bare, partially or fully oversmear-coated filler particle.

If the oversmear is very thin, the contact area is determined primarily by the yield pressure of the pseudo-hard coat of flattened hard phase (e.g. glass) embedded in the worn composites surface, while the force required to shear the junctions is determined primarily by the polymeric oversmear. The sliding contact may be modeled by the equation

$$f_k = \frac{S_f}{P_s} = \frac{f_{k(f)} P_f}{P_s} \tag{1}$$

where f_k is the coefficient of friction of the tribosystem (i.e. composite and counterface), S_f the shear strength of the oversmear film, P_s the yield pressure of the substrates, P_f the yield pressure of the oversmear film, and $f_{k(f)}$ the coefficient of friction of the oversmear film.

In the limiting case, as the film thickness approaches zero, S becomes the bulk shear strength of the polymer film and P is the yield pressure of the substrate (assuming S to be independent of pressure, which does not always hold). It follows that the coefficient of friction of a tribosystem depends on a thin oversmear film formed on the exposed filler and transferred onto a sufficiently strong and hard counterface. It is equal to the product of the film material in the bulk form and the ratio of the hardness of the film material to the essentially glass substrate on one hand and to the counterface on the other. Unfortunately, the hardness and wear resistance of favored cryogenic alloys, such as austenitic steels (e.g. 321 CRES) and nickel-based alloys (e.g. Inconel 718) are too low to withstand the abrasiveness of the

TABLE 1

Estimated hardness of E-glass composite reinforcement compared with that of various counterfaces [a]

Material	Hardness		
	Moh's	Knoop	Rockwell
E-Glass	7.0 +	900	$R_c 60$
321 Stainless steel	4.5 +	400	$R_c 30$
Inconel 718	6.0 +	500	$R_c 42$
440C Stainless steel	7.0 +	900	$R_c 63$
TiC	9.0 +	2500	$R_c 83$

[a] Fully hardened.

glass filler, as shown in Table 1. Gardos [43] found that the (hard) aluminosilicate E-glass filled PTFE composite can damage even fully hardened 440C steel if the steel counterface is not protected by a solid lubricant film tough enough to withstand the abrasion until flattened alignment of the sharp glass tips by wear occurs. Rough, damaged counterfaces will, in turn, increase the polymeric composites' wear rate to propagate a vicious cycle of a self-aggravated wear process. The successful use of hard cermets as cryogenic composite counterfaces [39] suggests the applicability of surface hardening methods and processes, especially in the case of the softer cryogenic alloys.

The simultaneous accumulation of the load-carrying hard phase and the low shear strength film in some ideal combination and the reduction of counterface damage is the tough challenge facing the tribologist. Usually, a compromise is struck in providing an acceptable balance specifically designed for an MMA configuration. The successful use of epoxy composites at cryogenic temperatures indicates that PTFE may not be the only viable polymer candidate.

It seems that any matrix, polymeric or otherwise, must exhibit the right micro-fragmentation characteristics combined with acceptable surface energetics. Specifically, the fragments should be of very small particle size with sufficient cohesion to form an even, low shear strength film either on the worn composite's surface or on the counterface. The strength of the bulk composite substrate must be greater than the powdered composite film composed of the tightly packed but displaceable microfragments. With epoxies, controlling the molecular weight and the degree of crosslinking, ergo the rigidity and fragmentation characteristics, is possible [44]. This degree of control is much more limited in the case of PTFE. The resulting network structure and film-forming characteristics can be further modified by other lubricative pigments, such as graphite, MoS_2, boron nitride, or PTFE. The true role of these pigments with the single possible exception of PTFE, is, however, obscure. It cannot be said with certainty that these pigments take an active part in the lubrication process by reducing wear and friction through preferential accumulation in the oversmear and in the transfer film, or that they are there as crystal structure, strength, modulus, differential thermal expansion, thermal conductivity or isothermal shrinkage controllers and modifiers. Furthermore, the thickness and the shear strength of the oversmear and the transferred film depends greatly on the reinforcement phase meant strictly for enhancing the hardness and the load-carrying capacity of the surface. The hardness, abrasiveness, and inherent wear rate of this phase (e.g. glass versus bronze or stiff-backboned polymers such as polyimide) always present the tribologist with the dilemma of finding an ideal trade-off between counterface damage reduction and gainful accumulation on the sliding surface to (a) support both the load and the oversmear in accordance with eqn. (1), (b) yield low friction but not necessarily low wear, or, (c) give low wear but not necessarily low friction.

At cryogenic temperatures, surface wear can be enhanced by surface fatigue manifested by the untimely removal of the accumulated reinforcement. Lavengood and Anderson [45] concluded that composite fatigue life at very low temperatures is strongly affected by the composite interfacial stress which arises due to differential thermal contraction of the reinforcement and other fillers and the matrix. Cooling the composite increases the compressive forces at the interface and improves the fatigue life and reduces surface wear. Where fatigue strains serve to decrease the interfacial stresses, especially when the filler contracts more than the matrix on cool-down, the reverse is true. The hard phase is subsequently plucked from the surface, severely increasing the wear rate.

Certain types of glasses, which have anomalously negative pressure derivatives of bulk and shear moduli (the higher the pressure the easier the glass becomes to compress and shear), also show negative thermal expansions at low temperatures [46,47]. Therefore, similar amorphous materials (e.g. some silica-based glasses) or even crystalline materials (e.g. BeF_2) with low coordination numbers can be selected so that the bending vibrations lead to this desirable property. The abrasiveness versus the counterface can then be balanced with the inherent wear rate and chemical compatibility of the special filler with the polymer matrix to select the best reinforcement for cryogenic tribocomposites. It is noteworthy that Brown [48] found the friction of borosilicate glass drops from 0.73 at $-55°C$ to 0.30 at $-180°C$, commensurate with an improvement in stick–slip characteristics.

Little has been discussed in the literature about the effect of the medium on the friction and wear of composites (i.e. running in cold gas or in liquid cryogen). Yakovenko et al. [49], in describing the fatigue life of brittle superconducting alloys in normal and superconducting states, mentioned that changes in fatigue life during temperature decrease may not have been attributed solely to the microplasticity of the materials, but also to the differing influence of liquid or gaseous helium environment. Kannel et al. [50] have recently shown that, in spite of the low viscosity of cryogens, some elasto-hydrodynamic film can be formed with, for example, LN_2. The film thicknesses measured were found to be unrealistically large (e.g. 20 m s^{-1} speed, 0.24 GPa, film thickness = 1.07 μm). In any case, one would expect that the cavitation caused by vaporization of the cryogen at the "flash" temperatures (developed on the surface of low thermal conductivity composites) would increase the wear of the composite and reduce the effectiveness of transfer film formation.

Friction and wear in a gaseous cryogenic environment would be expected, therefore, to be lower than in cryogenic fluids.

The fundamental themes discussed above are not relegated to cryogenic temperatures only. As further dealt with in the following sections, the interactions between the brittle and ductile phases in the composite are characteristically similar in any triboenvironment.

2.4. Vacuum applications

In view of the previous discussion, it is not surprising that since the beginning of unmanned and manned spaceflight, PTFE-based self-lubricating composites have been used the most. They comprise the largest and most reliable data base. Yet, in spite of such popularity and the large number of commercial products available, there have been large batch-to-batch variations in performance [51,52].

The critical nature of space applications demand predictability of composite friction and wear more so than in the case of more easily replaceable (even if not less expensive) parts used in terrestrial or airborne applications. Therefore, this section will deal mainly with the formulation-dependent aspects of wear quantization of PTFE-based and other composites in inert gases and in vacuum, at or somewhat higher than room temperature and at loads/speeds commensurate with most moving mechanical assemblies used in spacecraft.

Composite applications in space do involve cryogenic temperatures, either because the MMA is located at the outer periphery of the spacecraft in the shadow of the sun or because the machine element is operated in a cryogenically cooled environment. However, the most common uses include relatively low load (seldom exceeding 27.6 MPa $\cong$ 4000 p.s.i.) and low speed single transfer-film-type mechanisms operating near or not much above room temperature (54°C $\cong$ 130°F). In double transfer applications, the Hertz stresses are kept below 0.83 GPa $\cong$ 120 k.s.i., to compensate for the relatively low load-carrying capacity of most transfer films under concentrated contact conditions.

The reduced structural strength and the increasing creep and wear rate of neat polymers at the higher than cryogenic temperatures forces the tribologist to heavily depend on a wide variety of small metallic and non-metallic particles, whiskers or chopped fibers dispersed in the polymer matrix to alter its properties. Continuous fibers, in the form of random mats, woven cloths and three-dimensional (3D) woven structures, can further enhance certain attributes such as strength, modulus, and, consequently, stiffness.

It must be kept in mind, however, that the rules applicable to structural composites do not necessarily apply to the formulation and behavior of the self-lubricating kinds. Although the composite must be strong enough to withstand the thermomechanical stresses to which the given MMA component is exposed, wear and friction really depend on what is going on at or immediately below the surface. In the case of neat polymers, just as with metal alloys, sliding will induce a surface smear layer different from the parent matrix. Tribomechanical action and spot melting will introduce such secondary structures that do not necessarily form when fillers are present [30,53,54]. The layers, if formed, heavily influence tribological behavior, especially transfer film formation. Briscoe [55] tried to extend some of the arguments used here in the previous section (in terms of chain alignment as a function of molecular shape, free volume and activation energies for configurational changes and shear) to general film transfer, but with little success of finding more than often-conflicting rules of thumb.

Basically, all that can be said with certainty is that it depends on the load/speed/surface temperature and atmospheric triboenvironment, whether the viscoelastic and environmental stability of the polymer matrix or the effects of the reinforcing agents dominate the overall friction and wear behavior. Ideally, the polymer alone should exhibit inherently low friction, wear and willingness to form a low surface energy, well-adhering, thin transfer film on the counterface. There are only a few classes that qualify in this respect. For low-to-moderate temperature use, PTFE is at the head of the class, followed by the polyacetals, nylons, high density polyethylenes, and polyesters. Within some limitations, their tribological properties may be further enhanced by certain lubricant additives and strength/modulus increasing fillers.

The neat polymers usable at higher temperatures, on the other hand, act more like brake materials than lubricants. The stiff-backboned polyimide molecules, the polyphenylene sulphides, the polyamide–imides, and polysulfones must be compounded with various solid lubricant additives, if not to increase their inherently higher modulus and strength, certainly to decrease their friction and wear.

Regardless of the polymer type or temperature capability, the best matrices are capable of forming or promoting the formation of a transfer film. Such films, ranging in thickness up to as much as 1.5 μm, cannot be thought of as evenly thin, completely homogeneous, never changing entities. They are generated by the transfer and compaction of debris with a wide range of particle size, adherency to the substrate and cohesiveness, formed differently at various loads and temperatures. Single transfer films are formed under the usually low unit load contact between the metallic (or ceramic) member and the composite. Double transfer films caught between the high Hertzian stresses of ball/race or gear versus gear contacts are generated and behave differently.

In the initial formation stages of single- and double-transfer films, the latter reach a more complete, less patchy state sooner. This is due to the crushing, homogenizing, and compacting action of the traction (i.e. rolling/skidding) loads. The concentrated contacts demand that either the polymer film alone exhibits sufficient load-carrying capacity not to extrude from under the contacts, or the worn filler provides sufficient reinforcement. Pure PTFE films or those containing insufficient reinforcement behave much like a soft, asphalt road on a hot summer day grooved by car tires. Only macadam roads sufficiently reinforced with rocks can withstand the loads without excessive deformation and extrusion.

The same concentrated contacts, however, tend to fatigue even the reinforced films, along with some of the underlying substrate. The highly compacted and now partially fatigued double-transfer film contains a lot of glass, metal, or other filler

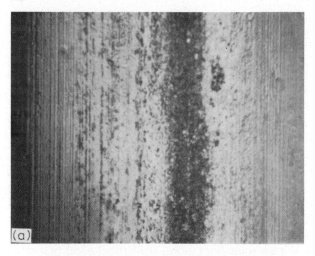

Fig. 3. PTFE/glass/MoS$_2$ double transfer film removed by tape from a gimbal bearing raceway after vacuum testing. (From ref. 56.)

debris. As shown by photomicrographs of debris removed from the ball path of gimbal bearings operated in vacuum (see Fig. 3, taken from ref. 56, the companion paper of ref. 43), the film appears to be hard and brittle. The sharp edges of the peeling film that tend to develop on the ball will wear the mating ball pocket more than a young and smooth double-transfer film, or any lower density, less abrasive single-transfer film [38]. The fatigued debris flakes circulate within the bearing and become transported back to the ball pockets (Fig. 4), only to return to the ball path again. These fragments can temporarily act as hard pseudo-asperities, actually promoting the micropitting wear of the ball and the races through the development of the localized micro-Hertzian stress fields.

As discussed in the previous section, the retention of evenness and homogeneity

BALL POCKET

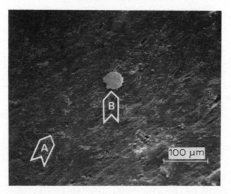

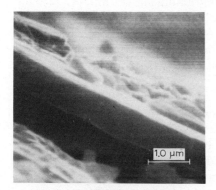

AREA A

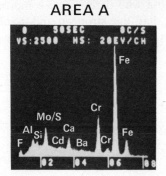

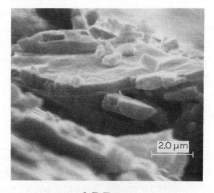

AREA B

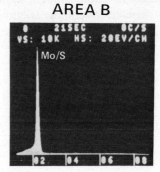

Fig. 4. Recirculated double transfer film wear debris in PTFE/glass/MoS₂ composite separator ball pocket.

of the transfer film is just as important in vacuum applications as it is in a cryogenic environment.

The wear of the composite and the resulting formation of a transfer film are heavily affected by the atmospheric environment. Fortunately, more data exist on

TABLE 2

Average wear volumes ($\times 10^{-5}$ cm^3) of various reinforced PTFE compositions (from ref. 57)

Material	Air, dew point (°F)		Nitrogen, dew point (°F)			
	+ 50	− 50	+ 50	− 20	− 50	− 90
Carbon graphite specimen	86	160	17	11	1550	1755
25% Glass fiber TFE specimen	88	23	46	157	167	116
60% Bronze TFE specimen	38	89	420	48	40	73
32% Carbon TFE specimen	110	60	20	15	35	10

this subject than are found in the cryogenic literature. As shown by Fuchsluger and Taber [57], for glass, bronze, and carbon graphite-filled PTFE, the mode of wear and transfer film formation are different in room temperature air and nitrogen containing various amounts of moisture. The summary of their findings in Tables 2–4 shows the see-saw effects of filler–matrix dominance of tribobehavior, as one environmental regime changes into another. The key point raised was the overall importance of forming a transfer film on the counterface, which keeps the normally detrimental, abrasive effects of the filler to a minimum.

Gardos [43] further demonstrated that the wear and transfer film formation tendencies of glass fiber-reinforced, MoS$_2$-containing PTFE composites appear to be the same in dry argon and in high vacuum. The abrasive nature of certain types of reinforcement was also reconfirmed. The case in point is the comparison of two commercially available composites, contemplated for use as self-lubricating retainer candidates for space bearing use [43,56]. Examination of the composites' physical properties in Table 5 might indicate (at least to those who dare screen composite types without obtaining friction and wear data) that the stronger and harder Composite B with more MoS$_2$ would wear less. This does not happen: the wear of

TABLE 3

Transfer film formation characteristics of various reinforced PTFE compositions (from ref. 57)

Material	Air, dew point (°F)		Nitrogen, dew point (°F)			
	+ 50	− 50	+ 50	− 20	− 50	− 90
Carbon graphite specimen	Yes	Yes	Yes	Yes	No	No
Grey iron drum	Yes	Yes	Yes	Yes	Yes	Yes
25% Glass fiber TFE specimen	Yes	Yes	No?	No?	No?	No?
Grey iron drum	Yes	Yes	Yes	Yes	Yes	Yes
60% Bronze TFE specimen	Yes	Yes	No	No	No	No
Grey iron drum	Yes	Yes	Yes	Yes	Yes	Yes
32% Carbon TFE specimen	No	No	No	Yes	Yes	Yes
Grey iron drum	Yes	Yes	Yes	No?	No?	No?

TABLE 4

Description of wear tracks on the counterface with various reinforced PTFE compositions (from ref. 57)

Material	Air, dew point (°F)		Nitrogen, dew point (°F)			
	+50	−50	+50	−20	−50	−90
Carbon graphite specimen	High polish, fine scoring	Polished track, fine scoring	High glaze, few score marks	High glaze, few score marks	Dull abraded	Dull abraded
Grey iron drum	Slight glaze	Slight glaze	Polished tan/blue film, black deposit	Polished tan/blue film, black deposit	Smooth polished	Smooth polished
25% Glass fiber TFE specimen	Smooth polished tan debris	Smooth polished tan debris	Smooth polished black debris	Smooth polished black debris	Smooth polished black debris	Smooth polished black debris
Grey iron drum	Polished tan/blue film	Polished tan/blue film black deposit	Polished blue/grey film, black deposit	Polished blue/grey film, black deposit	Polished dark brown to black deposit	Polished dark brown to black deposit
60% Bronze TFE specimen	Glazed bronze to black film	Glazed bronze to black film	Smooth bronze particles abraded	Polished bronze streaked	Polished bronze streaked	Polished bronze streaked
Grey iron drum	Smooth dark deposit	Smooth dark deposit	Smooth bronze black color	Smooth bronze transfer	Smooth bronze transfer	Smooth bronze transfer
32% Carbon TFE specimen	Polished carbon partly visible	Polished carbon partly visible	Polished carbon partly visible	Glazed surface	Glazed surface	Glazed surface
Grey iron drum	Polished tan film, black deposit	Polished tan film, black deposit	Polished blue/grey film	Polished	Polished	Polished

TABLE 5

Selected physical properties of two commercially available PTFE/glass/MoS$_2$/self-lubricating composites (from refs. 38 and 43)

Material	Approximate composition (% by wt.)	Density (kg m^{-3})	Tensile strength (MPa)	Hardness (durometer D)
Composite A [a]	65–75 PTFE 18–28 chopped glass fibers 5–7 MoS$_2$	2.30	13–16	55–63
Composite B [b]	60–70 PTFE 20–30 random fiber (fiberglass mat) 10–15 MoS$_2$	2.22	28–48	73 +

[a] Rulon A + 5% MoS$_2$, Dixon Corp., Bristol, RI, U.S.A.
[b] Duroid 5813, Rogers Corp., Rogers, CT, U.S.A.

Composite B is greater than that of Composite A by about a factor of two. Friction and wear tests with both materials using a standard area contact test configuration in the oscillatory single transfer mode (Fig. 5), sliding in an argon test atmosphere against bare and MoS$_2$-burnished, hard stainless steel (440C; R_c 63) revealed that the geometry of the glass fiber filler had an overpowering influence on both the load, speed, and time-dependent rate and magnitude of wear. In particular, the thin glass fiber strands in Composite B are less than 1 μm in diameter, while the chopped glass fibers in the other are over an order of magnitude thicker. The well-worn surfaces of both composites that rubbed against MoS$_2$-burnished steel surfaces (Fig. 6) reveal

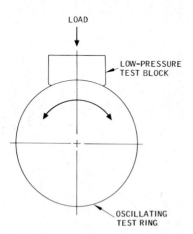

LOAD

LOW-PRESSURE TEST BLOCK

OSCILLATING TEST RING

Fig. 5. Area contact configuration of the LFW-1/Faville-1 friction and wear tester.

417

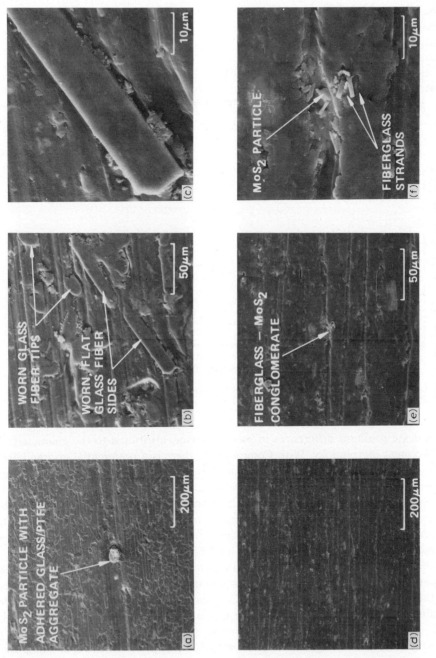

Fig. 6. Well-worn composite surfaces against MoS₂-burnished 440C stainless steel (R_c 63). (a), (b), (c) Composite A. (d), (e), (f) Composite B. (From ref. 43.)

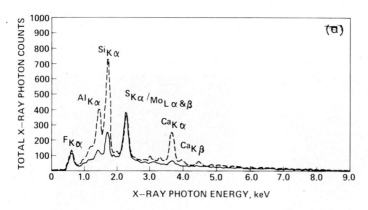

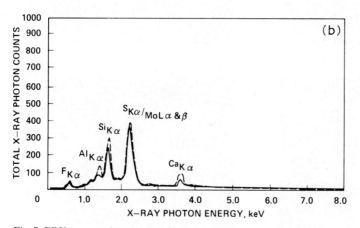

Fig. 7. EDX spectra of (a) the Composite A and (b) the Composite B surfaces. ———, New block; – – –, run-in block. (From ref. 43.) Run-in block surfaces are shown in Fig. 6.

this and the resultant differences in wear behavior, attributed to the characteristic reinforcement morphology. With Composite A, there is a preferential accumulation of the wear-flattened, relatively large glass fibers reinforcing the worn composite's surface which is confirmed by the energy dispersive X-ray (EDX) spectra in Fig. 7. The thin and brittle fiber ends on the Composite B surface also fragment, but the broken tips are incapable of the same preferential accumulation and are unable to provide the same load-carrying support. On the contrary, these small fragments turn into mobile abrasive particles, ploughing into the soft PTFE–MoS$_2$ matrix.

The abrasive influence of these small, thin fragments is, however, negligible compared with the sharp edges of the larger chopped fibers' ends. It was observed that the appearance of the as-machined Composite A and B blocks under high SEM magnification shows little difference, due to a thin PTFE oversmear that apparently forms during block grinding. This thin "camouflage" quickly disappears on sliding, exposing the ends and edges of the relatively large chopped glass fibers in Composite

A. Undoubtedly, the degree of damage is load-dependent. In contrast, the freshly machined Composite A surface sliding against the MoS_2-burnished steel undergoes a more orderly reinforcement wear process. The residual MoS_2 film, however thin or uneven it is (as burnished films normally are), appears to protect the metal from abrasive damage until a more or less satisfactory wear and flattened alignment of the glass particles occur. Instead of metal stock removal and the formation of a roughened surface, the high load capacity MoS_2 film facilitates plastic deformation of the underlying asperities. The smoothened and work-hardened steel substrate becomes a more ideal mating surface to the now run-in composite.

The additional beneficial effects of burnishing, especially in the case of Composite A, show themselves indirectly in Fig. 8 where load and speed are correlated with block temperature and friction versus the steel countersurfaces. In line with the above findings, no Composite A transfer film could be detected on the unburnished steel (continuous abrasive removal). The Composite A film on burnished steel, as well as those of Composite B in all cases, was patchy but macroscopically complete under moderate (approximately 20 ×) magnification. It could be surmised that thin, sputtered layers of MoS_2 on the steel counterface would be even more effective, provided the coated bearings were kept in a dry, inert environment. The qualitative statements above become quantitative by examining the wear equations in Table 6. These equations were formulated for both materials in a simple as well as in an advanced form. The former is expressed as one version of the widely used Archard equation

$$W = KPVt \tag{2}$$

where W is the weight loss, K the wear coefficient, P the unit load, V the sliding velocity, and t the test duration.

The advanced relationships were previously suggested by Rhee [58]

$$W = K'P^a V^b t^c \tag{3}$$

where W, K', P, V and t are as described previously and the exponents, a, b and c, are constants, experimentally determined for each material.

The lower wear and lower sensitivity to load of Composite A ($P^{1.29}$ versus $P^{1.44}$, see Table 2) may be attributed to the preferential accumulation of the wear-flattened, relatively large glass fibers reinforcing the worn composite's surface.

The larger-diameter-glass-fiber-reinforced PTFE appears to be more vulnerable to repeated cyclic shear stress under load ($V^{2.42}$ versus $V^{0.39}$, see Table 6).

It is also apparent that treating the steel surfaces with MoS_2 not only reduced the wear of Composite A by over a factor of two, but significantly improved the statistical validity of the wear data.

On the other hand, MoS_2 burnishing had little effect on the wear of Composite B; no protection of the steel surface is needed against the thin, fragile, glass fiber strands. One may suppose, therefore, that the wear of Composite B approaches that of neat PTFE more closely than Composite A.

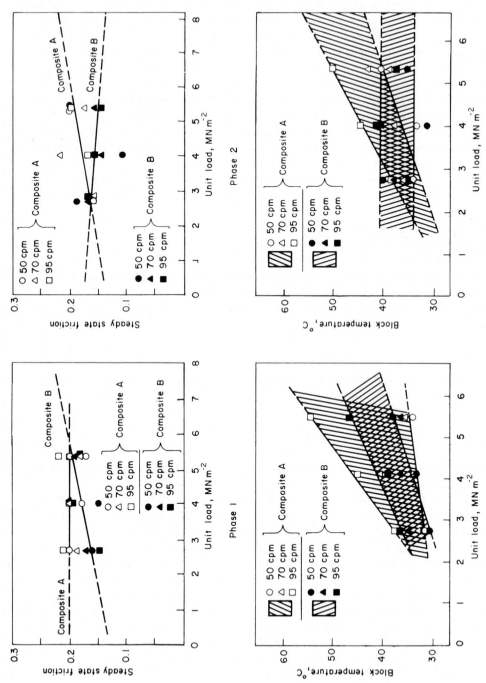

Fig. 8. Effects of MoS_2 burnishing on the load/speed/temperature interaction of Composite A and B surfaces sliding against 440C steel (R_c 63) in an oscillatory mode. Left, unburnished steel counterface; right, burnished steel counterface. cpm = cycles min^{-1}. (From ref. 43.)

TABLE 6

Wear equations for the single-transfer/oscillatory mode of composite A and B (from refs. [38] and [43])

Composite	MoS$_2$ burnish	Equation type [a]	W (g)	Correlation coefficient
A	No	Archard	$5.87 \times 10^{-13}\ PVt$	0.66
		Rhee	Not obtainable [b]	
	Yes	Archard	$2.24 \times 10^{-13}\ PVt$	0.87
		Rhee	$4.85 \times 10^{-17}\ P^{1.29}\ V^{2.42}\ t^{1.27}$	0.93
B	No	Archard	$4.82 \times 10^{-13}\ PVt$	0.98
		Rhee	$4.52 \times 10^{-13}\ P^{1.41}\ V^{0.93}\ t^{0.80}$	0.98
	Yes	Archard	$5.44 \times 10^{-13}\ PVt$	0.92
		Rhee	$1.40 \times 10^{-12}\ P^{1.44}\ V^{0.39}\ t^{0.74}$	0.93

[a] Speed range = 45.9–87.3 mm s^{-1}; load range = 2.76–5.52 MPa; for wear prediction use P(gf cm^{-2}); V(cm s^{-1}); t(s).

[b] Six out of the nine weighings after 2 h showed negative weight loss (i.e. weight gain), preventing regression analytical work.

It may be more than a coincidence that in the wear laws formulated by Uchiyama and Tanaka [59] for pure PTFE, the pressure exponent is close to what was measured for Composite B: the linear wear rate of α_T was expressed as $\alpha_{29^\circ C} = kp^{1.26}$ and $\alpha_{80^\circ C} = kp^{1.63}$, where k varies with sliding speed.

At any given load, the speed–wear curves are camelback functions, rising to a hump around 10–100 mm s^{-1} sliding speed, decreasing at around 300–500 mm s^{-1} and increasing again beyond that. The wear of neat PTFE, therefore, is not linear either with pressure or with speed. Neither is it linear with time, at any load or speed. Furthermore, the wear curves shift towards high speed and high wear rate with increasing temperature within the range 50–150°C, but remain quite similar in form. It so happens that this range is above the glass transition temperature, T_g, of PTFE (-13°C) and coincidences with the secondary (β) phase of transitions (45–107°C) and the primary (α) transition of around 127°C. The wear curves beyond 200°C to the upper test limit of 350°C are much different, indicating a drastically changing wear mechanism, probably due to the proximity of the melting point, T_m, of PTFE at 327°C.

As previously discussed, the abrasiveness of certain neat PTFE particles themselves [31] and their increased size as a function of stress rate could at least partially account for the greater-than-1 load exponent of neat PTFE and for some of the speed effects.

It is of further interest to examine quantitatively the influence of less abrasive reinforcements in PTFE. Additional wear equation work in our laboratories, conducted similar to the previous effort but now using a nichrome alloy (Inconel 718; R_c 40) as a softer counterface, again revealed the abrasive nature of glass. As shown in Table 7, the speed exponent is still predominant but less than in the case of Composite A rubbing against the harder stainless steel counterface (see Table 6).

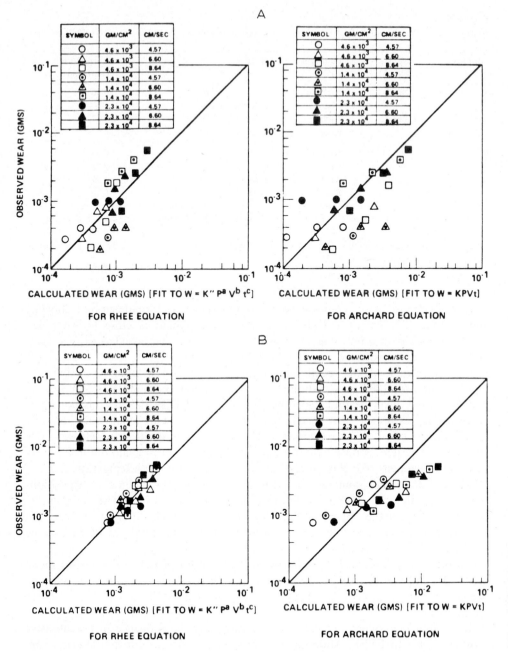

Fig. 9. Scatter about the line at equal values for (a) 75% PTFE–25% chopped glass fiber, (b) 80% polyacetal–20% PTFE, (c) 80% PTFE–20% polyimide self-lubricating composite wear equations. (Oscillatory mode, see Table 7.)

TABLE 7

Oscillatory mode, Rhee-type equation wear coefficients and load/speed/time exponents of fiber-reinforced and all-polymeric, self-lubricating composites

Composition	K''	P exponent	V exponent	t exponent
75% PTFE–15% glass fiber–10% MoS_2 [a]	1.8×10^{-9}	1.44	0.39	0.74
80% PTFE–20% glass fiber–5% MOS_2 [b]	4.1×10^{-13}	1.29	2.42	1.27
75% PTFE–25% chopped glass fiber [c]	3.40×10^{-10}	0.60	1.68	0.49
80% PTFE–20% polyimide [d]	2.50×10^{-8}	0.78	0.26	0.26
80% Polyacetal–20% PTFE::[e]	1.05×10^{-7}	0.37	0.09	0.51

[a] Composite A from Table 5.
[b] Composite B from Table 5.
[c] Fluorogold by the Fluorocarbon Co., Los Alamitos, CA, U.S.A.
[d] Rulon J by the Dixon Corp., Bristol, RI, U.S.A.
[e] Delrin AF by E.I. DuPont de Nemours & Co., Inc., Wilmington, DE, U.S.A.

The load and time exponents are much less than before, but the K factor is considerably larger. It is suspected that, as the effects of load, speed, or time become reduced (i.e. progressively less than 1) within a PV range where no drastic wear mechanism changes take place, the regression analysis automatically increases the

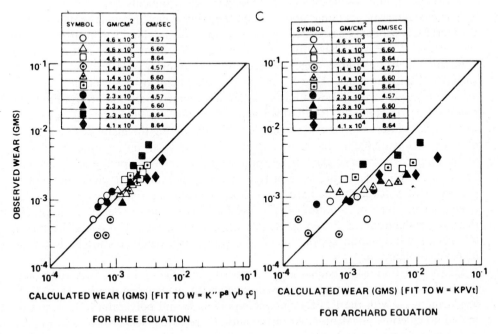

Fig. 9 (continued).

magnitude of K. If the opposite occurs, the wear coefficient may be small, but any substitution of higher loads and speeds rapidly increases the magnitude of the predicted wear.

The data in Fig. 9 again indicate that the Rhee equation predicts wear behavior better than the linear, Archard relationship. There is also some indication in Tables 6 and 7 that acceleration of wear at increased loads and speeds can be achieved only if the respective exponents associated with the composite are sufficiently large. For the all-polymeric materials, accelerating wear by the application of higher loads and speeds appears to do little to increase the wear rates above what might be found in real-life testing. With abrasive-filled materials, it may be possible to test for wear in an accelerated manner, but reduced accuracy of wear prediction should be anticipated due to the larger scatter of the data base.

These findings are all in line with one major wear mechanism consistently emerging in composite friction and wear: (a) the preferential and fast accumulation of any abrasive and/or lubricative phase on the sliding surface, (b) serious and rapid damage of an unprotected, mating counterface (if it is sufficiently softer than the reinforcement phase and is not covered with a tough enough transfer film) at and beyond some PV limit characteristic to a given sliding couple, and (c) the continued, irregular wear behavior and film-forming tendency of the composite in continuous sliding against the now-damaged counterface. The characteristically rougher wear surfaces of the glass-filled composite, compared with the far smoother ones of the all-polymeric materials (see Fig. 10) again attest to the validity of this well-established rule-of-thumb applicable in a wide range of environments.

The characteristic PV limit above which the reinforcement begins to damage the counterface is probably associated with the removal of the thin oversmear film from the surface of the worn hard filler and the counterface, previously discussed here with eqn. (1). For example, there is indeed massive transfer of PTFE to clean glass surfaces [29,60], provided there is sufficient displacement of the friction couple to allow oversmear of the exposed low shear strength matrix onto the protruding hard phase [61]. Consequently, PTFE filled with relatively large diameter, but properly worn-in glass fibers should exhibit very low friction among the many different types of composite available in the open market. Composite wear, on the other hand, is a more complex function of glass phase/polymer phase removal, preferential accumulation of components on the wear surface, matrix/fiber adhesion which determines the degree of plucking of the hard phase from the resin matrix, glass wear behavior effects on the type of counterface and the flash temperature characteristic of the thermal conductivity of the conglomerate composite surface. The overall result of this complex interaction appears to be impossible to predict; it must be tested in realistic configuration.

Grove and Budinski's comparison of experimental data [62] provides additional proof of the insight to the superimposed phenomena, as depicted in Figs. 11 and 12. Replacing glass with the PTFE-compatible polyimide and polyacetal (also found in the all-polymeric equivalents used in the author's research, see Table 7 and Fig. 9) will reduce the wear rate, but not necessarily the friction.

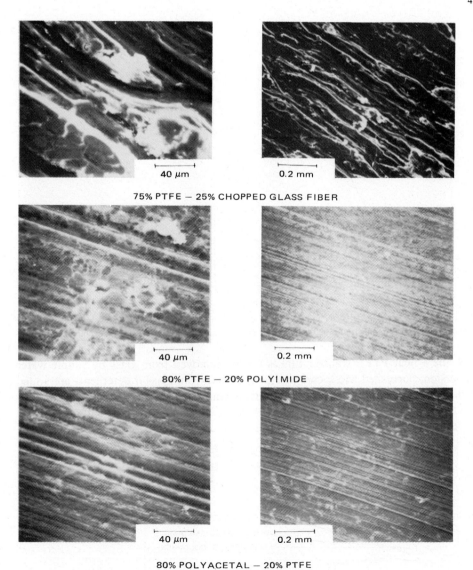

75% PTFE — 25% CHOPPED GLASS FIBER

80% PTFE — 20% POLYIMIDE

80% POLYACETAL — 20% PTFE

Fig. 10. SEM photomicrographs of worn composite rubshoe surfaces used in oscillatory wear equation development. (See Table 7.)

Despite all the complexities involved in PTFE-based composite wear, these materials have remained the mainstay of aerospace lubrication. From the qualitative beginnings of composite screening [64,65] to the most recent quantitative design information [43,56,66], the quest has continued for an understanding of the balance between the wear of the composite, the resulting transfer film, and its ability to prevent substrate wear.

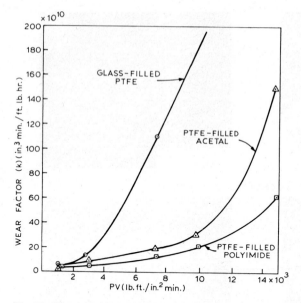

Fig. 11. Wear of reinforced PTFE composites as a function of PV. $V = 100$ ft. min^{-1}; data developed with 25 mm i.d. bushings with a 1020 steel shaft of 90 HB and 0.5 μm c.l.a. finish. (From ref. 62.)

While a neat PTFE transfer film is readily removed from glass by simple water immersion, burnished MoS$_2$ can only be removed from glass if detergents are added to the water and if scrubbed vigorously. One would then pre-suppose that, if the low

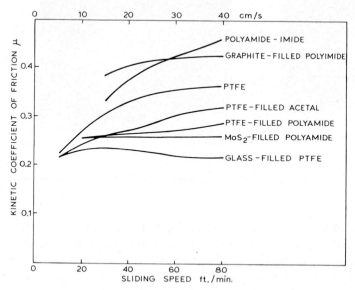

Fig. 12. Coefficient of friction of various polymers and composites sliding against 1020 steel. 0.5 N normal load; 0.4 μm c.l.a. finish; cylinder-on-cylinder contact. (From ref. 62.)

shear strength oversmear would consist of other, more adherent and higher load capacity solid lubricants such as the lamellar, layered transition metal dichalcogenides (MoS_2, $MoSe_2$, $NbSe_2$, etc.), both the wear and the friction could be reduced simultaneously. This is especially true in dry, inert gases or in vacuum, where the moisture desorbed from MoS_2 renders it one of the lowest shear strength lubricants found for these applications.

However, despite the fact that the MoS_2 content of Composites A and B was close to a minimum value which appears to be needed to induce a meaningful reduction in the friction and wear of PTFE and nylon [13], the role of MoS_2 here is nebulous. Neither the EDX spectra in Fig. 7 nor those of transfer films removed by adhesive tape from a variety of metallic bearing surfaces could reveal preferential accumulation of MoS_2 either on the worn composite's surfaces or in the transfer films. The MoS_2 content of Composite A, by itself, was clearly unable to protect the steel surface against the composite's own abrasive glass fibers. It is surprising that compounding houses continue adding MoS_2 to such blends. The inability of this solid lubricant to bestow protection to the counterface against wear, when mixed into certain glass-reinforced plastics, has already been recognized by Theberge et al. [63].

Science repeats itself, giving us the same warning over and over again. If composite wear were such that there had been preferential accumulation of a high load-carrying lubricant additive on the composite surface and this accumulation also enhanced the formation of a high load-carrying transfer film, any such composition would satisfy the tribologist. As is shown in the next section, certain high temperature composites not based on polymers are also excellent lubricants in vacuum, because not only do they satisfy the above criteria, but the desorption of air, gases, and moisture further lowers friction and wear.

2.5. High temperature applications (in air)

Self-lubricating composites, just like other forms of solid lubricants, do not fall neatly into temperature ranges definable with sharp upper or lower thermal limits. With respect to a search for that elusive upper limit, static exposure of lubricant candidates to air at elevated temperature serves only as a rough screening technique because their tribooxidative stability and wear life have a great deal to do with load–speed-induced flash temperatures and the dissipation of heat from the interface. These surface temperatures, in turn, are extremely dependent on the physico-chemical characteristics of the friction couple, the configurational design, and other materials and systems-oriented parameters.

Furthermore, those lubricants that are traditionally utilized at elevated temperatures more often than not fail to act lubricatively at room temperature. In a similar vein, those normally applied for low-to-moderate temperature use tend to soften and creep, or turn into volatile compounds or abrasive oxides at high temperatures.

In this portion of the chapter, the composites applicable for use in high temperature air will be labeled with somewhat arbitrary upper temperature limits, usually associated with the thermooxidative stability of one key component (in the case of polymers, for example, it is their thermal performance capability). This upper limit is often exceeded in real use because the designer presumes the existence of general environmental temperatures measured by thermocouples positioned some distance away from the sliding/rolling interface. In reality, thermooxidative degradation is a function of the PV-induced flash temperatures, the availability of oxygen to the interacting surfaces, the thermal conductivity of the machine elements, air cooling, or convective heat transfer, etc. No predictions as to the life of composite MMA parts can be made, unless these parameters are measured by appropriate instrumentation.

2.5.1. Polymeric composites (to approximately 316°C ≅ 600°F)

The appropriate lubricant filler and the geometry or composition of the reinforcing agents are important in influencing the tribological behavior of polymeric composites used in elevated temperature air, but the controlling parameters are the viscoelastic and environmental stability of the polymer matrix. It will be shown, however, that these factors are intimately interrelated. The perennial competition between the matrix and the fillers to control wear continues at elevated temperatures also.

The development of high-performance aircraft and high power density machinery requires self-lubricating, polymeric composite bearing materials with higher load-carrying capacity at elevated temperatures. These improved materials are needed for applications in journal, spherical, and ball bearings as well as in gears and dynamic seals. The previously described neat polymers are rarely suitable above 204°C (400°F), due to either "cold flow" (creep) under heavy loads or thermal degradative wear, or the combination of both.

Current needs now stretch to a dynamic load capacity goal of 172 MPa at temperatures up to 316°C. Polymers such as the polyphenylene sulphides and polyamide–imides serve acceptably, but only at considerably below this temperature and load limit. The polyimides are generally more thermally stable and are the most widely used neat resins for the more demanding applications. There is, however, a large difference between the condensation-type and the addition-type polyimides, as well as those within each of these categories with respect to ability as extended thermal stability matrices [67–69].

Self-lubricating composites prepared from condensation-type polyimides are well-known in the industry. These composites exhibit lower elevated temperature strength and wear life than addition-types, largely because of the presence of voids attributed to the liberation of gases during curing. Furthermore, composites prepared from thermoplastic polyimides which depend upon a high glass transition temperature (as high as 371°C) to provide the desired thermal properties exhibit the disadvantage of requiring unacceptably high fabrication temperatures, as well as the shortcoming of excessive deformation under load (creep) at high temperatures.

In those instances where conventional addition-type polyimides derived from bismaleimides have been used, the composites were found to be limited to use below 290°C because the functional groups through which they cure are aliphatic and remain aliphatic after cure. Acetylene-terminated polyimide oligomers have been unusually successful as lubricant matrices up to 316°C [69–71]. Work with these and other, more conventional polyimides showed, however, that those exhibiting good thermal stability and strength in high-temperature structural composites are not necessarily easily compounded. Even the processable ones may not produce viable self-lubricating composites. At present, only high-temperature friction and wear tests can reveal the true performance of such matrix polymers. Also, once a good polymer is selected, its continuing availability must be assured. Many times in the recent past, manufacturers discontinued products on which tribologists and formulation houses had learned to depend.

Regardless of the high temperature polymer type or temperature capability, the best matrixes are capable of forming or promoting the formation of a transfer film.

As previously mentioned with PTFE, the extremely small size (30 nm) of the flakes formed at relatively low speeds and the low activation energy of slippage between bands (21–29 kJ mol^{-1} [11]) coupled with PTFE's ready transfer to surfaces are the main reasons for its success as an outstanding film former. With polyimides, studies indicate that the flow properties on the frictional surfaces, as a function of temperature, determine the polymer's tribological property [27]. It was also apparent that large elongation at break and a relatively low modulus of a fully crosslinked polyimide are desirable traits for a good matrix. These findings agree with this writer's in that acetylene-terminated polyimides do exhibit reasonable elongation and a non-brittle nature. As with PTFE, polyimides are equally willing to transfer to oxidized or clean metallic surfaces. Higher contact loads lead to longer polymer chain fragments that do transfer [68,72–75].

Based on some of this knowledge, more scientific compounding of high temperature, polymer-based composites was started at the beginning of 1976. An engineering study was started to develop an improved self-lubricating composite with a dynamic load-carrying capacity goal of 172 MPa (25 000 lbf in.$^{-2}$) at temperatures up to 316°C (600°F). At that time, no such commercial or experimental material was available which exhibited proven usability anywhere near these requirements.

By use of literature data and a hypothesis based on the philosophy expounded here, a series of model screening compounds was prepared, leading eventually to the formulation of a working candidate. It consists of a simple orthogonal, 3D carbon (low modulus-low strength) fiber weave reinforced, acetylene-terminated polyimide self-lubricating composite, fortified with powdered versions of the Ga/In/WSe$_2$ lubricative compact and dibasic ammonium phosphate [(NH$_4$)$_2$HPO$_4$] as solid lubricant and carbon fiber adjuvant additives, respectively.

This composite repeatedly exhibited a minimum of 207 MPa (approximately 30 000 lbf in.$^{-2}$) ASTM compressive strength and a radial wear rate of 2.5×10^{-8} M s^{-1} (3.6×10^{-3} in. h^{-1}) at a load/unidirectional speed combination of 27.58 MPa/0.76 m s^{-1} (4000 lbf in.$^{-2}$/150 ft min^{-1} = 600 000 PV) with a steady-state coefficient of friction of 0.04 at 316°C (600°F) in air, against 0.13 μm (approximately 5 μin.) r.m.s. Rene 41 (a nickel–chromium, high temperature alloy) surfaces.

The same composite, encased in steel bushings as a thick (5×10^{-3} m $\cong$ 0.2 in.)

430

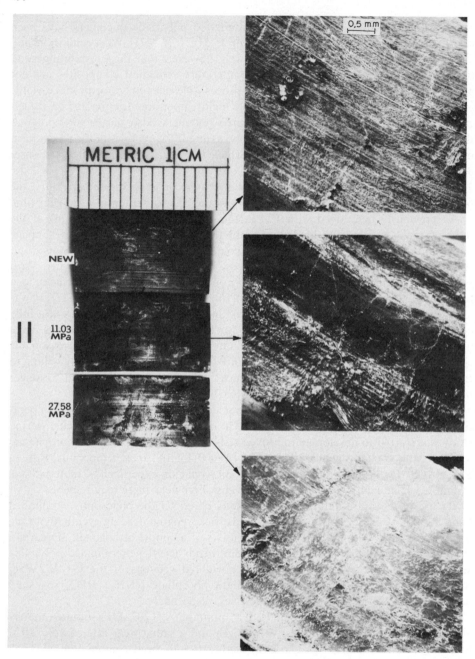

Fig. 13. Preferential accumulation of tungsten and phosphorus-containing lubricant constituents on used rubshoes (see also Fig. 14) as indicated by EDX photon count dot maps. (From ref. 69.)

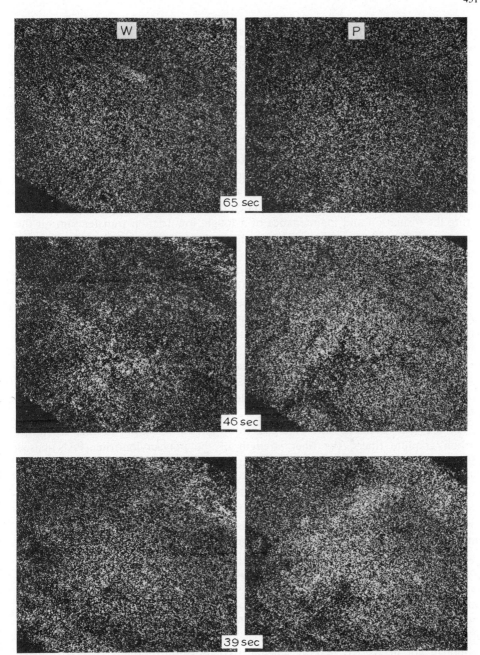

Fig. 13 (continued).

liner, repeatedly withstood dynamic loads in excess of 207 MPa (approximately 30 000 lbf. in. $^{-2}$) against 0.13 μm (approximately 5 μin.) r.m.s. chrome-plated, hard steel pins, sliding at an average oscillatory speed of 8.8 × 10^{-3} m s^{-1} (1.76 ft min^{-1}) with a steady-state coefficient of friction of less than 0.20. It also exhibited an acceptable wear rate in air at the same elevated temperature [$PV \cong 53\,000$ (in British units) [69].

One might think of this material as a high temperature analog of a glass-reinforced, MoS_2-containing composite. The glass and the MoS_2 are replaced by a multi-directional weave of high-strength-producing carbon fibers rendered non-abrasive by another oxidatively stable, low shear strength solid lubricant and a fiber adjuvant. These additives, unlike MoS_2 in the case of PTFE composites, did preferentially accumulate on the worn, polyimide-based composite's surface, facilitating the formation and maintenance of a tough, low friction transfer film on the counterface. This film was able to protect the mating metallic surfaces from abrasive damage, which certain carbon fibers tend to induce (Fig. 13).

Having determined the best (i.e. acetylene-terminated) polyimide matrix, the emphasis was placed on the correct identification of the fiber/additives synergism.

One major problem appeared to be the higher modulus–lower strength (more graphitic) fibers imparting the best high temperature tribological properties, yet the highest compressive (load-carrying) strength was provided by the rather abrasive, lower modulus–higher strength fibers.

Further work along these lines, which dealt with formulating even more advanced composites for service with steel and hot-pressed silicon nitride counterfaces [76–80], confirmed the perennially reoccurring problem of reinforcement abrasiveness and its often dire consequences.

As the reinforcement was changed from low to medium to high modulus carbon/graphite fibers, the fiber strength of the polyacrylonitrile (PAN)-based material went from a low-to-high-to-medium strength, describing a camelback curve. The medium modulus, high strength fiber yielded only incrementally greater composite strength than the other two types. However, this reinforcement was so abrasive that it consistently removed any transfer film which otherwise may have formed on the metal alloy counterfaces. This led to high friction, high wear and high interface temperatures. The high modulus–medium strength fiber seemed to be the best compromise, at least permitting the formation of transfer films on metallic counterfaces. Yet, the accumulation of additive components on the composite surface previously noted with the least abrasive, low strength–low modulus fiber-reinforced composite in Fig. 13 was not observed. Although the medium strength–high modulus fibers allowed some transfer film formation, they probably exceeded some abrasiveness threshold limit to permit such accumulation. It would be expected (although not measured) that the chemical composition of these varietal transfer films would be different.

The abrasiveness of even the best graphite fibers need to be mitigated by the proper designs of the reinfocring weave and by the use of adjuvants. Antioxidants, as well as high tribooxidative stability, solid lubricant pigment additives, are also needed, especially in the case of soft counterfaces.

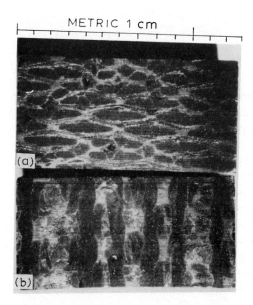

METRIC 1 cm

(a)

(b)

Fig. 14. 3D graphite fiber-reinforced polyimide composite rubshoes. Ground curved test surface (a) perpendicular and (b) parallel to the main trend of fiber lay. The molding force was applied from the top in (a) and into the page in (b). (From ref. 38.)

The tribologist and the polymer technologist both benefit when flat 3D weaves are processed into flat sliding pads or half-shell bearing liners. It was demonstrated that, on incorporation of the right kind(s) of solid lubricant additives, 3D carbon/graphite weave-reinforced versions can perform satisfactorily where the majority of the fiber lay is in the plane of sliding, normal to the applied load. This configuration also happens to develop during pressing and curing of the impregnated preforms. Between flat platens and hemi-cylindrical molds, the round fibers in the X–Y plane tend to flatten inherently when the molding load is applied in the Z direction (Fig. 14). We have already discussed the abrasiveness of glass fibers perpendicular to the plane of sliding (see Sects. 2.3 and 2.4). The phenomenon is quite similar here: the preponderance of fibers lying parallel with the counterface are the least abrasive. It is unfortunate that, in the case of bearing separators, the reinforcing cylindrical graphite fiber weaves are such that drilling the composite ball pocket holes will inherently expose a large percentage of abrasive fiber ends at the ball-to-pocket interface [76,80].

The chemical factors involved in reducing fiber abrasiveness and thus lowering interface temperatures consist of (a) artifically replacing the desorbed moisture from graphite, (b) employing antioxidants to incrementally reduce the tribooxidative reactions of the solid lubricant pigment and reduce similar degradation of the polymer matrix, (c) using lubricant pigments whose oxides can act as a protective shield against removal by rubbing and a barrier against oxygen diffusion to the yet unoxidized portion of the lubricant, and (d) understanding the effects of the counterface chemistry (i.e. when the mating surface is not a metal alloy but a ceramic).

Let us now examine these chemical factors employed to reduce composite friction and wear.

2.5.1.1. Graphite adjuvation. It has been known for some time that graphite loses its ability to lubricate in the absence of moisture. Water intercalates the graphitic structure, increasing the van der Waals gaps sufficiently to reduce packet-to-packet attraction. In vacuum or in high temperature air, where the desorbed moisture causes graphite to become abrasive and to "dust" with an increased wear rate, it needs to be "rejuvenated" with "adjuvant" compounds.

At first glance, one would expect that the conventional halide, carbonate, phosphate, or metallic adjuvants simply intercalate the layers. Conte [81] has recently shown the value of several metal chlorides as effective intercalants, increasing the wear life of bonded solid lubricants formulated with the intercalated graphites. This increase in wear life was explained as a function of increased lattice expansion and thus lower shear strength. Graphite motor brushes and other parts operating at reduced pressures or elevated temperatures [82,83] contain a variety of such adjuvant compounds.

In composites, however, the graphite fiber reinforcement depends on adjuvants blended into the resin matrix. Even if it were possible to obtain intercalated graphite fibers commercially, they might be insufficiently strong to serve as a reinforcement. As shown in refs. 69, 77 and 78, physically mixing the insoluble $(NH_4)_2HPO_4$ into the resin matrix by way of a solvent slurry and treating it to the molding/.curing temperature (approx. $370°C$) will decompose the adjuvant. Gases (e.g. NH_3) are formed and a liquid (probably a mixture of phosphoric acids and pyrophosphates) is generated. The decomposed adjuvant tends to solidify and preferentially accumulate first around the reinforcing fibers then on the surface of the rubbed composite (see Fig. 14). It appears that only the exposed surface of the graphite fibers can become adjuvated during tribological action, clearly an advantage where the bulk strength of the composite must be maintained.

The preferentially accumulated phosphate adjuvant on the composite surface and Conte's increased wear life bonded solid lubricants [81] all indicate that accumulation of a low shear strength layer on the composite surface during rubbing reduces both the interface temperatures (due to the lower friction) and the surface attrition of the composite. If the layer also shows some tenacity to the yet unworn composite substrate and willingness to transfer to the mating counterface, the friction and wear of the entire system is reduced.

Certain pyrophosphate networks can also scavange metallic impurities in graphite. Some metallic impurities react with graphite at elevated temperature, reducing its structural strength and thus its wear. So adjuvants can undertake another role, in addition to any possible intercalation; on the other hand, certain metallic intercalants may become harmful at high temperatures.

The presence of any lubricant pigment added to the composite can enhance the beneficial mechanisms, but only if these mechanisms are not prevented from acting by the abrasiveness of the graphite fibers and the oxidation of the pigment(s).

2.5.1.2. *Solid lubricant pigment and counterface chemistry.* Ideally, the oxide layers formed under quiescent conditions on a solid lubricant pigment particle should be in accordance with parabolic or even higher-order reaction kinetics. This means that the oxide layers should be effective diffusion barriers against further oxygen intrusion, aimed to destroy the yet unoxidized, underlying pigment portion. As shown by the author [84], any protective oxide layer plays an important part on reducing the static oxidation of selenides, which normally follow linear kinetics. Fischer and Sexton [85] have recently hinted that removal of these oxide layers by tribological action will turn higher-order kinetics into pseudo-linear ones, due to the continuous removal of the protective scale and exposure of the yet unoxidized substrate.

It follows that the tenacity of the oxide layer to its pigment substrate and the inherent friction of its oxide are key factors in contributing to the localized friction conditions and thus the magnitude of the flash temperatures. The surface energy and thermal conductivity of the counterface are, in turn, the main influencing factors that contribute to the inherent tribooxidation kinetics of the pigment, as shown in refs. 77, 78 and 80. Clearly, any increased flash temperature-induced oxidation of any pigment and the subsequently increased friction further aggravate the oxidation of the residual pigment, leading to run-away tribooxidative destruction.

The presence of the adjuvant $(NH_4)_2HPO_4$ notwithstanding, the lamellar selenides of the $Ga/In/WSe_2$ self-lubricating compact rubbing against a low thermal conductivity, low surface energy counterface such as hot-pressed silicon nitride have indeed become rapidly oxidized to the relatively high friction oxides of gallium and indium. The high surface temperatures developed on the ceramic counterface were thus sufficiently elevated to form a SiO_2-rich oxidation layer of the silicon nitride on the ceramic counterface. This layer, in turn, was removed by the abrasive graphite fibers. As the ceramic debris was transferred to the rubbing surface of the composite, the interface was transformed into a ceramic versus ceramic contact situation. The resulting thermal runaway was immediately translated into rapidly and simultaneously increasing friction and wear of the composite and, to some extent, the ceramic counterface also.

On the other hand, a mixture of layered transition metal dichalcogenides based mainly on MoS_2 was able to provide low friction and thus low surface temperatures, retarding the oxidation of MoS_2. The apparently greater tenacity of the MoO_3 protective layer forming on MoS_2 and its lower friction, in contrast to the gallium–indium oxides, produced the overall result of some formation of a transfer film on the ceramic. This led to the formation of a thinner and far less complete ceramic debris layer on the composite surface and lower overall composite and counterface wear and friction.

What is most interesting is that both types of the lubricant pigments performed identically well against the high surface energy and high thermal conductivity steel substrate, except at high speeds. The high PV-generated flash temperatures there have already indicated the selenide lubricant to be more marginal in high temperature performance than the MoS_2-containing lubricant mixture.

Certain adjuvants may be able to undertake the role of a pigment antioxidant and vice versa. Once case in point is a Belgian patent [86] filed by Russian chemists. There, a method is described whereby the MoS_2 pigment of a poly(phenylquinoxaline)/MoS_2 self-lubricating composite was modified by treatment with di-heptylphosphinic acid so that the composite contained 0.01% phosphorus. This small amount was capable of significantly reducing friction and wear. Another is the often-used combination of MoS_2/Sb_2O_3. The good performance of antimony tri-oxide has been confirmed by several researchers [87–89] where the mechanism was identified as the actions of an antioxidant. Sb_2O_3 apparently has an oxygen scavenging role. It is currently believed that the synergistic interaction between MoS_2 and Sb_2O_3 lies in the preferential oxidation of the latter to Sb_3O_4 in mixtures of the two additives. This reduces the oxidation rate of MoS_2 to MoO_3 and to molybdenum sulfate compounds, thereby keeping the friction low and the wear life extended.

$(NH_4)_2HPO_4$ and Sb_2O_3 are routinely used as fire retardants in conventional plastics [90]. The effectiveness of the latter may, at least partly, arise from its low volatility. As stated by Thomas [91], the sublimated Sb_2O_3 vapor barrier remaining close to the surface prevents the oxygen from attacking the polymer surface at ignition temperatures. Oxygen scavenging should be very effective in the vapor phase. It is, therefore, possible that there is an additional beneficial effect between the phosphates and antimony trioxide and the polymer base and the pigment through reduced oxidative degradation of both the pigment and the polymer.

It is of interest that Sliney and his co-workers at the NASA Lewis Research Center, Cleveland chose to concentrate on chopped graphite fiber-reinforced polyimides containing no additives or adjuvants [92,93]. They achieved high load-carrying capacity by bonding thin liners of these composites against hard metal backings. The reported friction coefficients were typically 0.15 ± 0.05 from room temperature to 360°C during service against metallic counterfaces. Sliney [94] stated that, in his experience, the addition of a solid lubricant to a composite structure did not reduce the coefficient of friction, it only lowered the mechanical strength of the composite. In this writer's opinion, this may be true with excessively filled composites operating at low PV, especially in the case of low speed applications, where preferential accumulation of low shear strength surface layers is very slow. It is puzzling, however, that, during molding and curing of composite liners and spherical bearing elements, the well-known "skin effect" (i.e. lube pigment or, in the present case, chopped fiber starvation at the mold interface [95]) did not cause excessive friction due to the accumulation of a resin skin on the composite's frictional surfaces during the molding process.

2.5.2. Self-lubricating compacts (to approximately 540°C ≅ 1000°F)

The explosive aerospace boom of the early 1960s prompted a few serious efforts to develop self-lubricating materials with a wider temperature range of application than polymeric composites could withstand. The work was nearly all empirical, with very little fundamental studies, because the development of the new materials was indeed driven by a need for immediate use in moving mechanical assemblies.

The most notable type of these new composites was prepared by powder metallurgical techniques. Metal powders were initially mixed with lubricant pigments, compacted by pressing, then "cured" at elevated temperatures in various gaseous environments. As shown here, only now has some research been done to demonstrate that the resulting solids were far from physical mixtures of the original compositions. The solids consolidated from powders at elevated temperatures undergo complex chemical reactions and phase transformations, which decide the overall usefulness of the particular compact. The complexity and the lack of control of these reactions and transformations also render batch-to-batch quality control of a compact difficult.

Twenty years ago, Campbell and van Wyk [96] used a mortar grinder to mix various refractory metal powders with MoS_2 and other layered transition metal dichalcogenides in a solvent slurry, which was then dried in a vacuum oven and screened through a fine mesh. The powdery aggregate was first cold-pressed (coined), ground into a powder again, than hot-pressed in induction-heated graphite dies. The pressing pressures and temperatures varied from 100 to 12 000 p.s.i. and from 816 to 1871°C (1500 to 3400°F) and pressing lasted from 3 to 4 h. This final pressing and curing is done under an argon atmosphere in graphite dies to prevent any appreciable oxidation of the constituents or the products.

The diffusion of carbon from the graphite dies into the powder mixture results in the formation of various refractory metal carbides. The breakdown or dissociation of some MoS_2 provides additional molybdenum for carbide formation and permits the formation of simple and complex refractory metal sulfides. The relative combination of the strength-reinforcing and load-carrying carbide hard phase in a residual refractory metal matrix, along with the low shear strength residual MoS_2, MoS_x (where x is less than 2) and the other simple and complex refractory metal sulfides leads to a more-or-less simultaneous manifestation of low wear and low friction.

Certain formulations also contain a small amount of copper and/or silver to further reduce friction and wear at elevated temperatures. At the higher end of the useful temperature limit, Ag and Cu tend to diffuse to the compact surface. The enrichment of the sliding surface with preferentially migrated, low shear strength metals such as Ag/Cu works synergistically with the preferentially accumulated, residual MoS_2 and other low shear strength sulfides.

Among the many combinations investigated, five of the materials are now commercially available under the Molalloy trade name (Pure Industries, St. Marys, PA). Their use ranges from ball bearing separators and high load plain/spherical bearing ($f_k \cong 0.04$) to motor brushes and clutch facings ($f_k \cong 0.5$). It is interesting that, depending on the composition, not only can the compact strength and friction be tailored, but also the electrical resistance can be varied. Jones [97] demonstrated, for example, that Boeing Compact 046-45 (Molalloy PM-105), an 80% MoS_2, 15% Mo, 5% Ta compact with an electrical resistance of 6.4×10^{-4} ohm-cm was an excellent low friction and wear potentiometer contact candidate for ultrahigh vacuum use. The same material was utilized successfully as the motor brush material of the Surveyor moon vehicle's surface sampler device [98,99]. More recently, the

438

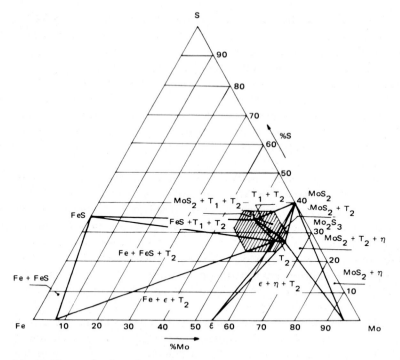

Fig. 15. The ternary phase diagram of iron, molybdenum, and sulfur. (From ref. 101.)

spherical bearings in the Space Shuttle door hinges employed Molalloy compact liners. These hinges continue to operate flawlessly in space [100].

Unfortunately, the successful and widely accepted use of such compacts has yet to lead to the elucidation of the friction and wear mechanisms as a function of stoichiometry and crystal structure. One can only estimate the possible chemical reactions by thought experiments and other related work.

Similar "alloying" of MoS_2 was accomplished by the research firm Hydromecanique et Frottement (HEF), St. Etienne, France, in cooperation with local universities. By the development of the Fe/Mo/S ternary phase diagram (see Fig. 15 [101]) they have discovered that the composition bounded by the $Fe_{0.66}Mo_3S_4$–$FeMo_5S_6$–$FeMo_3S_4$ triangle yielded a solid lubricant material that exhibited better friction and wear characteristics than MoS_2. This improvement was manifested in air, inert gas, and in vacuum [102]. A material within the designated compositional bounds was produced in compact form; it has also been magnetron sputtered in the HEF laboratories.

The French confirmed previous Japanese work [103], which showed that the MoS_x (where $x \cong 1.8$) that forms in situ in the Fe–Mo–S system imparts excellent wear resistance to the entire composition. More recently, this compact from Japan was shown to have low friction and wear in vacuum also [104]. Apparently, the substoichiometric moieties of MoS_2 themselves assume a layered structure, imparting good tribological properties to a compact. Similar effectiveness of substoichiometric MoS_2 was confirmed by reactive sputtering technologists [105].

The MoS_2/metal powder compacts can, therefore, yield a variety of properties, depending on the relative amount of the constituent and the extent of the curing reactions. The complexity of the reactions indicates that thermodynamic equilibrium is difficult to attain by thermal curing because the variety of starting compositions, intermediates and their gradual change in stoichiometry and crystal structure lead to continuous change in the compact chemistry. This alteration varies when the period of cure is extended and/or the curing temperature(s) is changed. Regardless of how far the reactions advance, the important steps in composite behavior again appear to be (a) the preferential accumulation of low shear strength constituents on the composite surface, (b) the formation of the transfer film on the counterface, and (c) sufficiently high load-carrying capacity of the refractory metal-based matrix. As discussed in ref. 99, the high MoS_2 content compact motor brushes of the Surveyor Surface Sampler did lay down a complete transfer film on the mating copper commutator counterface.

In high temperature air, however, the exothermic oxidation and other reactions of the MoS_2-containing compacts generated enough heat and gaseous products to decompose the polyimide matrix of self-lubricating composites discussed in ref. 69 (see also Sect. 2.5.1.) The one containing the highest quantity of MoS_2 caused the greatest degradation and the highest friction/wear due to the voids generated by the gaseous products and the higher shear strength of the refractory metal oxides (especially MoO_3). While in vacuum or in inert gases the manufacturer-recommended Molalloy upper temperature limit of 540°C could be safely extended further, high PV applications in air should be limited to no more than approximately 427°C (800°F) or preferrably even less.

Not until recently was some research done to reveal the complexities of reactions and crystalline phase transformations which occur during self-lubricating compact preparation.

As discussed by Gardos [84], the chemistry and the crystal structure of the well-known Ga/In/WSe_2 compact inherently limits the usefulness of this material either in vacuum or in air due to the thermodynamic instability exhibited at elevated temperatures.

In the late 1960s, Boes and Chamberlain intermixed WSe_2 ($-200 + 325$ mesh powder) with a liquid gallium–indium eutectic alloy [106,107]. The starting composition was 90 WSe_2/10 Ga–In, where the eutectic itself consisted of 80 Ga–20 In (all percentages by weight). The powdered lubricant/metal aggregate was compacted in steel dies at room temperature and high pressure followed by a stepwise curing sequence at progressively increased temperatures to 500°C (1050°F) in air. The mechanism of interaction between the constituents is such that 5–25 wt.% of the metal eutectic reacts with the solid lubricant. The resultant material was metallic gray, medium hard (scleroscope hardness = 37), and difficult to machine. It resisted, however, oxidizing environments at temperatures two and three times higher than the parent lubricant, WSe_2.

The compact was shown to be effective as a wide-temperature-range (i.e. 300°C to 540°C) bearing retainer insert in metallic cage shrouds [108–110]. It also exhibited good friction and film transfer characteristics in the author's laboratory work and in other tribological examination by the industry at room ambient temperatures [111–113]. As previously discussed here, it was also demonstrated [69] that the pulverized compact constituted the most thermally stable and tribologically effective additive in the 3D carbon weave-reinforced, high-temperature, polymeric self-lubricating composites used in 316°C applications sliding against metallic counterfaces.

Unusually good thermooxidative stability data on the compact were more recently presented by Kiparisov et al. [114] at test temperatures to 800°C (1472°F).

The major problem in trying to understand the advantages and shortcomings of this compact has been the dearth of data related to materials chemistry and crystal structure. The role of these fundamentals remained unclear, especially in terms of high temperature lubricative and wear tendencies. Apart from the well done, but less than complete, original analysis [107] and the more recent SEM/EDX/X-ray diffraction work by Russian researchers [114], all other data in the literature represented engineering-type friction, wear, and bearing tests.

As demonstrated by the author [84], the active ingredients within the compact are the lamellar gallium and indium selenides formed during the cure of this material. These solid lubricants work in conjunction with the residual WSe_2 not yet reduced by the Ga/In during the cure cycle. The binder encapsulant holding the lubricative entities together is the residual Ga/In eutectic partially reacted to substoichiometric selenides of mostly indium. The high-temperature, tribooxidative resistance of the compact is attributed to a protective mechanism, whereby the substoichiometric indium selenides preferentially oxidize during curing in high temperature air. The oxide network becomes a diffusion shield to the encapsulated lamellar ingredients. The practical use of the compact is limited, however, by inherent thermodynamic instability problems at elevated temperatures due to incompleted reactions and phase transformations, as well as the removal of the protective oxide layers by tribological action.

Essentially, the migration of Ga and In to the surface of the compact in high temperature air and the removal of the formed oxides not only promotes surface attrition, but may weaken the Ga/In-containing substoichiometric and more amorphous phase. This phase could be considered a protective, encapsulating layer network and a matrix binder for the desirable gallium– indium selenide and WSe_2 layered compounds formed during compact preparation. During relatively short-term static exposure, the protective oxides of mostly indium (especially In_2O_3, the densest, most stable oxide of indium) prevent any further, heavy oxidation of the lamellar bulk. If rubbed away, the reformation and repeated removal of this oxide layer will negate the role this film normally exhibits in the static heating mode. Sublimation of some of the other surface reaction products at high sliding surface temperatures further increases the surface attrition (i.e. apparent wear). The level of friction provided by the oxides can safely be assumed to be higher than those provided by the interlayer shear of the lamellar plates within the yet-unoxidized portions of the selenides. The combined effect can only be determined by high temperature tribotesting, in realistic test machines.

Limitations of the compact are due to reduction and oxidation reactions which take place between pigment and binder components. This five-component (i.e. gallium selenides/indium selenides/unreacted WSe_2/unreacted Ga–In binder/metallic tungsten) system and its oxides are thermodynamically unstable at the elevated temperatures required by most engineering applications. Instability is exacerbated by improper preparation technique and variations in the diffusion cross-section of the compact slug. Proper curing of the compact can mitigate, but not alleviate, the fundamental shortcomings.

2.5.3. CaF$_2$ / BaF$_2$-impregnated porous nichrome (to approx. 900°C ≅ 1650°F)

During the past two decades, space vehicle reentry science dictated an especially intensive search for oxidation-resistant solid lubricants with higher temperature capabilities. This search uncovered several of the alkaline earth fluorides as chemically stable candidates for use up to temperatures as high as 816°C (1500°F). Several fluoride lubricant compositions show promise to equal and eventually surpass the performance of the currently employed oxidation-resistant high temperature solid lubricants, such as cermets, zirconia and alumina formulations, or impregnated graphites.

Fluoride lubricant research pursued two approaches: (a) bonding of fused fluoride coatings to certain high temperature alloys [115–125] and (b) impregnating suitable porous metal matrices with specific fluoride compositions [126–131]. Fundamental considerations appear to favor the use of composites (by recognizing, for example, the inherently sacrificial nature of bonded films), provided that lubricant resupply from within the composite, material stability and strength are sufficient to insure long-term performance.

Preliminary test machine examination and bearing tests of a porous nichrome alloy vacuum-impregnated with a calcium fluoride–barium fluoride eutectic (38 wt.% CaF_2–62 wt.% BaF_2) were especially successful [126,127], enough to warrant commercial licensing of the composite [132,133].

As described by Sliney in his recent overview of high temperature lubricants [134], the fluoride eutectic melts at 1022°C but undergoes a brittle-to-ductile transition in the region of 500–600°C. The increased plastic flow during the microfragmentation process provides a relatively low shear strength, thermally and oxidatively stable fluoride layer on a sliding surface, whether it is plasma sprayed, sputtered, or originates from a porous metal (or perhaps ceramic) matrix.

The author demonstrated [135] the importance of this accumulated layer on a surface of CaF_2/BaF_2-impregnated, porous nichrome composite (AmCerMet $^{®}$, the trade name of Astro Met Associates, Cincinnati, OH). It was shown that the as-machined composite contains mostly the porous metal on the surface, leading to rapid seizure and failure when such a composite slides against a super-alloy counterface. The composite operated successfully only when a post-machining treatment provided a thin, fluoride eutectic layer on the frictional surfaces prior to the very first movement in service.

During continued operation, the high flash temperatures caused by the inherently higher friction of the fragmenting eutectic and the occasional metal-to-metal contact affect the differences in the thermal expansion of the nichrome sponge and the eutectic. This difference becomes pronounced, especially if the environmental temperatures were already high. The eutectic exudes preferentially to the surface and, with possible occasional spot melting, provides an aggregate, low shear strength surface layer richer in the eutectic than a cross-sectional surface of the composite itself. At the top temperature limit of use, however, the nichrome softens excessively. This leaves the composite with little structural strength and load-carrying capacity.

It is not difficult to conceive other porous metals or ceramics impregnated with CaF_2/BaF_2 eutectic, provided there were no adverse chemical reactions(s) between the harder, porous matrix and the more ductile lubricant phase.

2.5.4. Graphite composites (to approximately 650°C ≅ 1200°F)

The previous discussion on graphite adjuvation and the data in Paxton's book [82] can only be extended by repeating that little fundamental information exists on the relative effectiveness of the various antioxidants and friction/wear modifiers in graphite. Bhushan and Sibley [136] did indeed demonstrate differences in graphite composite ball separator behavior among five different compositions when tested up to 540°C. It is the author's opinion that no amount of adjuvation can prevent the inherently high oxidation rate of graphite and the segregation and degradation of the adjuvants at around 650°C. These mechanisms reach catastrophic proportions at that temperature.

Ideas of exploiting the high specific strength of carbon/carbon composites and have them double as tribological surfaces have not been successful. As shown in ref. 137, this type of composite wears more than graphite. At least 60% of the total wear loss is caused by oxidation to CO and CO_2. The oxidatively weakened surface and subsurface facilitate stock removal by the rubbing action at interfacial temperatures as high as 1000°C (1832°F) [138]. The loss of creep rupture strength and load-carrying capacity of unprotected carbon/carbon composites is most affected by elevated temperature air in the range 500–550°C (932–1022°F) at applied stresses lasting for at least 10 h [139]. While protective coatings such as CVD SiC sealed with tetraethylorthosilicate increase the materials' resistance to oxidation, they work well only in static applications. It appears that these composites also need systemic adjuvation if they were to be used in moving mechanical assemblies in high temperature air.

3. Conclusions

The science of tribology has never been more interdisciplinary than in the case of self-lubricating composites used in extreme environments. Chemists, physicists, surface scientists, compounding and physical property test technologists, as well as mechanism designers and computer diagnosticians of moving mechanical assemblies must all interact with the tribologist to bring any viable design to fruition.

A wide variety of data and technical information on composites operating at cryogenic temperatures, in inert gases, air, and vacuum at moderate temperatures, and in high temperature air have been digested here to show the few generalized rules applicable to any composite in any environment. The most important factor in composite behavior is the effectiveness of preferentially accumulated, low shear strength layers on the composite. As long as the operating environment is such that (a) the yet unworn composite body has sufficient load-carrying capacity to support this surface layer, (b) the composite reinforcement does not damage the mating counterface, and (c) a low surface energy, low friction transfer film forms on that counterface, a tribologically advantageous system can be created.

Naturally, these rudimentary rules-of-thumb should be further elucidated by using better testing techniques and appropriate surface/subsurface analyses. It will not be long before computer scientists will enter the picture, trying to simulate the materials-related interactions among the composite constituents. A priori predictions by software of the strength/surface shear/wear/thermal expansion and conductivity behavior of future composite bearing materials, followed by advanced testing will lead to a holistic composite design concept.

There is little doubt that the simultaneous performance of advanced materials development, computer simulation of moving mechanical assembly dynamics, and systems analytical test techniques will comprise the blueprint of any further studies of extreme environment (or, for that matter, any other kind of), self-lubricating composites.

List of symbols

f	free volume
f_k	coefficient of kinetic friction
k, K, K'	wear coefficients
M	molecular weight
p, P	unit load or yield pressure
P_s	bulk yield pressure of a substrate
P_f	yield pressure of a thin film on a substrate
S	bulk shear strength
S_f	shear strength of a thin film
t	test duration
T	temperature
T_g	glass transition temperature
T_m	melting point
V	velocity
V_a	specific volume of amorphous phase
V_c	specific volume of crystalline phase
W	weight loss
α_T	linear wear rate at a given temperature

Acknowledgements

The author would like to express his appreciation to Ms. Ronnie Flynn for word processing the manuscript and thank Hughes Aircraft Company for the permission to publish this chapter.

References

1 R.E. Cunningham and W.J. Anderson, NASA Tech. Note D-2637, 1965.
2 D.E. Brewe, H.W. Scibbe and W.J. Anderson, NASA TEch. Note D-3730, 1966.
3 D.E. Brewe, H.H. Coe and H.W. Scibbe, NASA Tech. Note D-4616, 1968.
4 H.W. Scibbe, D.E. Brewe and H.H. Coe, Proc. Bearing Conf., Dartmouth College, Hanover, NH, September 4–6, 1968.
5 D.W. Wisander and R.L. Johnson, NASA Tech. Note D-5073, 1969.
6 H.H. Coe, D.E. Brewe and H.W. Scibbe, NASA Tech. Note D-5607, 1970.
7 B.N. Bhat and F.J. Dolan, NASA Tech. Memo. TM-82470, 1982.
8 W.A. Glaeser, J.W. Kissel and D.K. Snediker, Proc. Conf. Polym. Sci. Technol., April 1–4, 1974, Los Angeles, CA, pp. 651–662.
9 E.L. Stone and W.C. Young, paper presented at the CEC–ICMC, August 21–24, 1979, Paper No. HB-6.
10 R.S. Kensley and Y. Iwasa, Cryogenics, 20 (1980) 25.
11 R.S. Kensley, H. Maeda and Y. Iwasa, Cryogenics, 21 (1981) 470.
12 R.P. Reed, R.E. Schramm and A.F. Clark, Electrical Properties of Selected Polymers, Cryogenics, 13 (1973) 67.
13 M.B. Kasen, Cryogenics, 15 (1975) 327.
14 M.B. Kasen, R.E. Schramm and D.T. Read, in K.L. Reifsnider and K.N. Lauraitis (Eds.), Fatigue of Filamentary Composite Materials, Society for Testing and Materials, Philadelphia, PA, 1977, pp. 141–151.
15 S. Nishijima and T. Okada, Cryogenics, 20 (1980) 86.
16 P.J. Klich and C.E. Cockrell, AIAA J., 21 (1983) 1722.
17 J.M. Roe and E. Baer, Int. J. Polym. Mater., 1 (1972) 133.
18 R.F. Boyer, Rubber Chem. Technol., 36 (1963) 1393.
19 J.P. Mercier, J.J. Aklonis, M. Litt and A.V. Tobolsky, J. Appl. Polym. Sci., 9 (1965) 447.
20 M. Litt and A.V. Tobolsky, J. Macromol. Sci. Phys., 1 (1967) 433.
21 L.E. Nielsen, Mechanical Properties of Polymers and Composites, Marcel Dekker, New York, 1975.
22 Heijboer, J., J. Polym. Sci. Part C, 16 (1968) 3755.
23 C.D. Armeniades, I. Kuriyama, J.M. Roe and E. Baer, J. Macromol. Sci. Phys., 1 (1967) 777.
24 S.G. Turley, as quoted by R.F. Boyer, 1968 International Award in Plastics and Science Engineering Lecture, New York, May 8, 1968.
25 A. Barrejea and P.L. Taylor, J. Appl. Phys., 58 (1982) 6532.
26 M. Pineri, Polymer, 16 (1975) 595.
27 K. Tanaka, Y. Uchiyama and S. Toyooka, Wear, 23 (1973) 23.
28 B.J. Briscoe, in K.L. Mittal (Ed.), Physiochemical Aspects of Polymer Surfaces, Vol. 1, Plenum Press, New York, 1983, pp. 387–412.
29 K.R. Makinson and D. Tabor, Proc. R. Soc. London Ser. A., 281 (1964) 49.
30 W.A. Brainard and D.H. Buckley, Wear, 26 (1973) 75.
31 H. Czichos, Proc. ASLE Third Int. Conf. Solid Lubrication, August 5–10, 1984, Denver, CO, ASLE SP-14, 1984.
32 B.J. Briscoe and T.A. Stolarski, Nature (London), 281 (1979) 206.
33 B. Briscoe, Tribol. Int., 14 (1981) 231.

34 J. Dimnet and J.M. Georges, ASLE Trans., 25 (1982) 456.
35 B.J. Briscoe, in K.L. Mittal (Ed.), Physicochemical Aspects of Polymer Surfaces, Vol. 1, Plenum Press, New York, 1981, pp. 387–412.
36 J.P. Giltrow, Composites, 4 (1973) 55.
37 J.K. Lancaster, Tribology, 5 (1972) 249.
38 M.N. Gardos, Tribol. Int., 15 (1982) 273.
39 E.L. Vorobev, Yu. F. Malyshkin, V.M. Grushevskii and M.V. Belenkii, Friction and Wear of Steels, Hard Alloy, and Polytetrafluoroethylene Materials at Low Temperatures, Khim. Neft. Mashinostr. (Sov. Chem. Pet. Eng.), (12) (1968) 20 (Engl. Transl. UDC 620.178.16-974).
40 I.M. Fedorchenko, Poroshk. Metall. (Sov. Powder Metall. Met. Ceram.), 18 (1979) 256.
41 K. Tanaka, Y. Uchiyama, S. Ueda and T. Shimizu, in T. Sakurai (Ed.), Proc. JSLE–ASLE Int. Lubr. Conf., Elsevier, Amsterdam, 1975.
42 K. Tanaka and S. Kawakami, Wear, 79 (1982) 221.
43 M.N. Gardos, Lubr. Eng., 37 (1981) 641.
44 H.S. Chu and J.C. Seferis, in J.C. Seferis and L. Nicolais (Eds.), The Role of the Polymeric Matrix in the Processing and Structural Properties of Composite Materials, Plenum Press, New York, 1983, pp. 53–125.
45 R.E. Lavengood and R.M. Anderson, Technical paper presented at the 24th Annu. Conf. Soc. Plast. Ind., Sect. 11-E, 1969.
46 C.R. Kurkjian, J.T. Krause, H.J. McSkimin, P. Andreatch and T.B. Bateman, in R.W. Douglas and B. Ellis (Eds.), Amorphous Materials, Wiley, New York, 1972, p. 463.
47 M.P. Brassington, A.J. Miller, J. Pelzl and G.A. Saunders, J. Non-Cryst. Solids, 44 (1981) 157.
48 E.D. Brown, ASLE Trans., 12 (1969) 227.
49 L.F. Yakovenko, V.V. Pustovalov and N.M. Grinberg, Mater. Sci. Eng., 60 (1983) 109.
50 J.W. Kannel, T.L. Merriman, R.D. Stockwell and K.F. Dufrane, NASA Contract. Rep., CR-170865, Battelle Columbus Laboratories, 1983.
51 R.D. Taber and J.H. Fuchsluger, Lubr. Eng., 31 (1975) 75.
52 J.H. Fuchsluger and R.D. Taber, Paper presented at the Fall Conf. Am. Soc. Qual, Control., Chem. Div., Charlotte, NC, October 1976.
53 M. Clerico, Wear, 64 (1980) 259.
54 R. Rameshand Kishore and R.M.V.G.K. Rao, Wear, 89 (1983) 131.
55 B.J. Briscoe, Adhesion 5: Proc. 8th Ann. conf. on Adhesion and Adhesives, 1981, pp. 49–80.
56 C.R. Meeks, Lubr. Eng., 37 (1981) 657.
57 J.H. Fuchsluger and R.D. Taber, J. Lubr. Technol., 93 (1971) 423.
58 S.K. Rhee, Wear, 16 (1970) 431.
59 Y. Uchiyama and K. Tanaka, Wear, 58 (1980) 223.
60 C.M. Pooley and D. Taylor, Proc. R. Soc. (London) Ser. A, 329 (1972) 251.
61 J.K. Lancaster, R.W. Bramham, D. Play and R. Waghorne, Paper presented at the ASLE–ASME Lubr. Conf., 5–7 October 1981, New Orleans, LA, ASME Preprint No. 81-Lub. 6.
62 T.H. Grove and K.G. Budinski, in R.G. Bayer (Ed.), Wear Tests for Plastics: Selection and Use, ASTM STP 701, January, 1980.
63 J.E. Theberge, P.J. Cloud and B. Arkles, Plast. World, 34(10) (1976) 46.
64 G.R. Smith and C.E. Vest, Lubr. Eng., 27 (1971) 12.
65 C.E. Vest, Lubr. Eng., 30 (1974) 246.
66 K.T. Stevens and M.J. Todd, Tribol. Int., 15 (1982) 293.
67 N. Bilow, A.L. Landis and T.J. Aponyi, Proc. 20th Natl. SAMPE Symp., San Diego, CA, April 29–May 1, 1975.
68 R.L. Fusaro, ASLE Trans., 25 (1982) 465.
69 M.N. Gardos and B.D. McConnell, ASLE SP-9, 1982.
70 N. Bilow and M.N. Gardos, US Pat. No. 4075111, 1978.
71 M.N. Gardos and A.A. Castillo, US Pat. No. 4376710, 1983.
72 R.L. Fusaro, ASLE Trans., 21 (1978) 125.

73 D.H. Buckley and W.A. Brainard, in Lieng-Huang Lee (Ed.), Advances in Polymer Friction and Wear, Vol. 5A, Plenum Publishing Corp., New York, 1976, pp. 315–328.

74 D.H. Buckley, NASA Tech. Memo TM-82645, 1981.

75 J.M. Burkstrand, J. Appl. Phys., 52 (1981) 4795.

76 M.N. Gardos, A.A. Castillo, J.W. Herrick and R.A. Soderlund, Proc. ASLE 3rd Int. Conf. Solid Lubr., August 5–10, 1984, Denver, CO, ASLE SP-14, 1984, p. 248.

77 P. Sutor and M.N. Gardos, Proc. ASLE 3rd Int. Conf. Solid Lubr., August 5–10, 1984, Denver, CO, ASLE SP-14, 1984, p. 258.

78 M.N. Gardos and P. Sutor, Proc. ASLE 3rd Int. Conf. Solid. Lubr., August 5–10, 1984, Denver, CO, ASLE SP-14, 1984, p. 266.

79 S.A. Barber and J.W. Kannel, Proc. ASLE 3rd Int. Conf. Solid Lubr., August 5–10, 1984, Denver, CO, ASLE SP-14, 1984, p. 275.

80 C.R. Meeks and M.W. Eusepi, Proc. ASLE 3rd Int. Conf. Solid Lubr., August 5–10, 1984, Denver, CO, ASLE SP-14, 1984, p. 285.

81 A.A. Conte, ASLE Trans., 26 (1983) 26.

82 R.R. Paxton, Manufactured Carbon: A Self-Lubricating Material for Mechanical Devices, CRC Press, Boca Raton, FL, 1979.

83 R.I. Christy, Proc. ASLE 3rd Int. Conf. Solid Lubr., August 5–10, 1984, Denver, CO, ASLE SP-14, 1984.

84 M.N. Gardos, Paper presented at the 39th ASLE Annual Meeting, May–10, 1984, Chicago, IL, ASLE Prepr. No. 84-AM-6C-1.

85 T.E. Fischer and M.D. Sexton, Proc. Conf. Phys. Chem. Solids, Paris, France, September 1983.

86 V.V. Korshak, I.A. Gribova, M.I. Kabochnik, A.N. Krasnov, A.N. Chumaevskaya, O.V. Vinogradova, S.V. Vinogradova, E.S. Kronganz and A.M. Berlin, (Institute of Heteroorganic Compounds, Academy of Sciences, USSR), Belg. Pat. 834, 735 (Cl. C08L), 22 Apr. 1976, App. 22 Oct. 1975, 44 p.; see also Chem. Abstr., 86 (1977) 44520.

87 M.I. Nosov, Khim. Tekhnolog. Topl. Masel, 7 (1978) 43.

88 G.H. Kinner, J.S. Pippett and I.B. Atkinson, Royal Aircraft Establishment Tech. Rep. 76-26, February 1976.

89 R.S. Harmer and C.G. Pantano, AFML-TR-77-227, Final Tech. Rep., USAF Contract No. F33615-76-C-5160, Univ. of Dayton Res. Inst., December 1977.

90 L.I. Naturman, SPE J., 17 (1961) 965.

91 H.L. Thomas, Ind. Res. Dev., 23 (1981) 141.

92 H.E. Sliney and T.P. Jacobson, Lubr. Eng., 31 (1975) 609.

93 H.E. Sliney, Lubr. Eng., 35 (1979) 497.

94 H.E. Sliney, Proc. Int. Conf. Tribology in the 80s, NASA Lewis Res. Center, April 18–21, 1983, NASA CP-2300, 1984.

95 G.J.L. Griffin, ASLE Trans., 15 (1972) 171.

96 M.E. Campbell and J.W. van Wyk, Lubr. Eng., 20 (1964) 463.

97 J.R. Jones, Wear, 11 (1968) 355.

98 E.R. Rouze, M.C. Clary, D.H. LeCroisette, C.D. Porter and J.W. Fortenberry, NASA Contract. Rep. CR-93205, Jet Propulsion Laboratories, Pasadena, CA, 1968.

99 Anon., Hughes Aircraft Co., Rep. P70-54 (SSD 00628), Culver City, CA, 1971.

100 Anon., Molysulfide ® Newslett., 23(3), November 1981, Climax Molybdenum Co., Greenwich, CT, U.S.A.

101 Anon., Journal du Frottement Industrial, No. 1 (1978) 2. (Centre Stephanois de Recherches Mecaniques, HEF, St. Etienne, France.)

102 Anon., Frottement et Lubrication dans l'Industrie IV – Traitments de Surface et Economies d'Energie, 69 Ecully, par les Chercheurs du Centre Stephanois de Recherches Mecaniques, HEF, St. Etienne, France.

103 Y. Mizutani, Y. Imada and K. Nakajima, Proc. JSLE–ASLE Int. Lubr. Conf., 9–11 June, 1975, Tokyo, Japan.

104 Y. Mizutani, Y. Shimura and Y. Nagasawa, Proc. ASLE 3rd Int. Conf. Solid Lubr., August 5–10, 1984, Denver, CO, ASLE SP-14, 1984.

105 K. Reichelt and G. Mair, J. Appl. Phys., 49 (1978) 1245.

106 D.J. Boes, ASLE Trans., 10 (1967) 19.

107 D.J. Boes and B. Chamberlain, ASLE Trans., 11 (1968) 131.

108 D.J. Boes, J.S. Cunningham and M.R. Chasman, Lubr. Eng., 27 (1971) 150.

109 A.E. King, Proc. Natl. Aerosp. Electron. Conf., Dayton, OH, May 15–17, 1972, pp. 136–142.

110 J.F. Dill, D. Brandes, P. Kamstra and R. Solomon, Paper presented at the AIAA/ARE/ADME 19th Joint Propulsion Conf., June 27–29, 1983, Seattle, Washington, AIAA Prepr. No. 83-1130.

111 J.R. Jones and M.N. Gardos, Lubr. Eng., 2 (1971) 47.

112 J.R. Jones and M.N. Gardos, Proc. Int. Conf. Solid Lubr., 24–27 August, 1971, Denver, CO, ASLE SP-3, pp. 185–197.

113 P. Martin, Jr. and G.P. Murphy, Lubr. Eng., 29 (1973) 484.

114 S.S. Kiparisov, G.A. Shvetsova, T.A. Lobova, L.M. Sergeeva, A.Z. Pimenova and G.A. Volodina, Poroshk. Metall., 185(5) (1978) 88.

115 H.E. Sliney, NASA Tech. Note TN-D-478, 1960.

116 H.E. Sliney, NASA Tech. Note TN-D-1190, 1962.

117 H.E. Sliney, T.N. Strom and G.P. Allen, NASA Tech. Note TN-D-2348, 1964.

118 H.E. Sliney, T.N. Strom and G.P. Allen, ASLE Trans., 8 (1965) 307.

119 K.M. Olson and H.E. Sliney, NASA Tech. Note TN-D-3793, 1967.

120 J.E. Krysiak, AFFDL-TR-68-147, January, 1969.

121 H.E. Sliney, NASA Tech. Memo. TMX-2033, 1970.

122 M.T. Lavik, B.D. McConnell and G.D. Moore, J. Lubr. Technol., 94 (1972) 12.

123 J. Amato and P.C. Martinengo, ASLE Trans., 16 (1973) 42.

124 H.H. Nakamura, W.R. Logan, Y. Harada, T.P. Jacobson and H.E. Sliney, Proc. 6th Annu. Conf. Composites Adv. Ceram. Mater., Cocoa Beach, FL, January, 1982.

125 T.P. Jacobson and S.G. Young, NASA Tech. Publ. TP-1990, 1982.

126 H.E. Sliney, ASLE Trans., 9 (1966) 336.

127 H.E. Sliney and R.L. Johnson, ASLE Trans., 11 (1968) 330.

128 H.E. Sliney and R.L. Fusaro, NASA Tech. Note TN-D-636, 1971.

129 H.E. Sliney, ASLE Trans., 15 (1972) 177.

130 J. Amato and P.C. Martinengo, Paper presented at the 1973 ASLE/ASME Lubr. Conf., Oct. 16–18, Atlanta, GA, ASLE Prepr. No. 73LC-4B-1.

131 H.E. Sliney and J.W. Graham, Paper presented at the ASLE/ASME Lubr. Conf., Oct. 7–9, 1974, Montreal, Canada.

132 Tech. Concentrates, Chem. Eng. News, 51 (Aug. 27) (1973) 10.

133 Anon., Lubr. Eng., 29 (1973) 423.

134 H.E. Sliney, Tribol. Int., 15 (1982) 303.

135 M.N. Gardos, ASLE Trans., 18 (1975) 175.

136 B. Bhushan and L.B. Sibley, ASLE Trans., 25 (1982) 417.

137 C.C. Li and J.E. Sheehan, Proc. Int. Conf. Wear Mater., March 30–Apr. 1, 1981, San Francisco, CA, American Society of Mechanical Engineers, New York.

138 H.W. Chang and R.M. Rusnak, Carbon, 16 (1978) 309.

139 Yu.N. Rabotnov, Ye.I. Stepanychev, V.S. Kilin, S.A. Kolesnikov, V.S. Matytsin, I.M. Makhmutov and V.I. Rezanov, Mekh. Polim., 1 (1978) 45.

AUTHOR INDEX

The number directly after the initial(s) of the author is the page number on which the author (or his work) is mentioned in the text. Numbers in parentheses are reference numbers and indicate that an author's work is referred to, although his name may not be cited in the text. Numbers in italics give the page on which the complete reference is listed.

454

SUBJECT INDEX